1st International Conference on Innovative Informational and Engineering Technologies (IIET-2020)

AA007020

IOP Conference Series: Materials Science and Engineering
Volume 873

Stavropol, Russia
28 - 29 May 2020

ISBN: 978-1-7138-2808-2
ISSN: 1757-8981

Printed from e-media with permission by:

Curran Associates, Inc.
57 Morehouse Lane
Red Hook, NY 12571

CURRAN ASSOCIATES INC. proceedings.com

Some format issues inherent in the e-media version may also appear in this print version.

Institute of Physics
Dirac House, Temple Back
Bristol BS1 6BE UK

Phone: 44 1 17 929 7481
Fax: 44 1 17 920 0979

techtracking@iop.org

Additional copies of this publication are available from:

Curran Associates, Inc.
57 Morehouse Lane
Red Hook, NY 12571 USA
Phone: 845-758-0400
Fax: 845-758-2633
Email: curran@proceedings.com
Web: www.proceedings.com

TABLE OF CONTENTS

COMPUTER AND INFORMATION TECHNOLOGIES

MODELLING AND MANAGEMENT

IOP Conf. Series: Materials Science and Engineering **873** (2020) 011001 doi:10.1088/1757-899X/873/1/011001

Stavropol State Agrarian University

Saint Petersburg State University

INNOVATIVE INFORMATIONAL AND ENGINEERING TECHNOLOGIES IIET-2020

Stavropol, Russia, May 28-29, 2020

The aim of the 1st International Conference on *Innovative Informational and Engineering Technologies (IIET-2020)* is to share the results and achievements of academics, practitioners and researchers in application of advanced scientific innovative technologies in computer engineering, communication, manufacturing, process automation and other fields.
The topics of interest are presented in 3 sections, including *Computer and Information Technologies, Engineering, Modelling and Management.*
On the basis of submitted works at the conference, 41 selected papers have been included in this volume of IOP Conference Series: Materials Science and Engineering.

CONFERENCE COORDINATORS

OLEG MALAFEYEV, St. Petersburg State University, Russia
IRINA SAMOYLENKO, Stavropol State Agrarian University, Russia
IRINA ZAITSEVA, Stavropol State Agrarian University, Russia

IOP Conf. Series: Materials Science and Engineering **873** (2020) 011002 doi:10.1088/1757-899X/873/1/011002

Peer review statement

All papers published in this volume of *IOP Conference Series: Materials Science and Engineering* have been peer reviewed through processes administered by the proceedings Editors. Reviews were conducted by expert referees to the professional and scientific standards expected of a proceedings journal published by IOP Publishing.

IOP Conf. Series: Materials Science and Engineering **873** (2020) 012001 doi:10.1088/1757-899X/873/1/012001

Internet of Things: Examination of efficiency and safety

Y V Alyshev, O N Maslov, I S Shatalov

Povozhskiy State University of Telecommunications and Informatics, L. Tolstoy str., 23, Samara, 443010, Russia

E-mail: shatalovivv@gmail.com

Abstract. The paper studies the problem of complex safety in personal area network and local area network in the Internet of Things project's boundaries. The electromagnetic safety analysis of various wireless picocell technologies have realized by statistical simulation with applying the criterion of functional of expected utility. The results of the statistical model are presented. The model's stochastic variables are «projected economical effect» and «environmental and ergonomic risk»; the determinate variable is internet of things realization cost. Recommendations for the selection of personal area network and local area network elements based on the Bluetooth protocol are given. The model of equivalent radiator for estimation of Internet of Things objects safety is offered. It is shown that the forced switching of all Internet of Things elements to radiation mode is an «energy threat» for sensors in the room.

Keywords: Internet of Things, Personal Area Network, Local Area Network, wireless picocell technologies, safety examination, statistical simulation method, functional of expected utility, environmental and ergonomic risk.

1. Introduction

Picocell (PCS) technologies are the basis for the wireless integration of electronic devices of various types and purposes - radiotelephones, computers, and digital video cameras, including the elements of the Smart House [1-5]. The integration of PCS-technologies with the Internet within the framework of Internet of Things (IoT) conception [6-9] implements a qualitatively new, synergetic stage of their joint development.

IoT infocommunication networks can be divided into four groups:

-Personal Area Networks (PAN) are located in close proximity to both "things" and ordinary "biological" Internet users;

-Local Area Networks (LAN) are located on the territory within one or a small group of buildings (residential building, office, enterprise, company);

-Metropolitan Area Networks (MAN) are combining LANs within the city;

-Wide Area Networks (WAN) are connecting network elements dispersed over distances of the order of $10^2...10^4$ and more kilometres from each other.

The range and versatility of the operation mode of PCS-modules (and also acceptable radiation intensity at low power consumption; resistance to interference; lack of influence on radio-electronic equipment (REE) and other electronics; low cost, etc.) make them indispensable at the PAN level networks. At the LAN level, the most winning option seems to be their use in highly efficient and economical multi-storey building management systems.

IOP Conf. Series: Materials Science and Engineering **873** (2020) 012001 doi:10.1088/1757-899X/873/1/012001

The global objectives and goals of the IoT project are discussed in [1-2]. Design methods for sensor networks are described in sufficient detail in [3-4] and [10], as well as in [5], which presents theoretical design principles. Key protocols are described in publications [6–7]; standardization problems are briefly discussed in [8]. A wide range of practical possibilities and prospects for the use of IoT are demonstrated in [9-10], as well as in [11-13].

In the literature issues of ensuring complex IoT security are considered in less detail, a typical example is [14]. Another important drawback is the lack of analysis of the alleged effectiveness of the IoT project; for example, it can be done from the standpoint of both classical theories of expected utility: the von Neumann – Morgenstern objective theory [15] and the Savage subjective theory [16], as well as the theory of prospects and risk [17]. This is largely due to the difficulty of choosing a quantitative criterion for assessing the expected utility of the IoT project by analogy with other applications of the theory of expected utility [18], as well as the corresponding software and experimental data.

Similar problems emerge when developing systems for protecting information from leakage into the external environment through “random antennas”. The effective solution is the computer method of statistical simulation modeling (SSM) [19-21], which combines analytical calculations with probabilistic combinatorics according to the Monte Carlo methods [22-23]. The initial data for simulation can be obtained by experimental and heuristic methods [24].

This work illustrates the application of the SSM method and the functional of expected utility (FEU) for solving the scientific and technological problems: examination of the effectiveness and safety of the IoT project. The proposed method combines the advantages of mathematical modeling and probabilistic combinatorics using data characterizing the economic, environmental-ergonomic, and information security of IoT systems and networks.

2. Composition and structure of FEU

We are using the multi-factor model of FEU. It is formed according to the scheme: “relative gain X_k plus relative saving of resources Y_k plus change in risk V_k” [9] as a criterion for evaluating the effectiveness of a collective project carried out under the conditions of Germeier’s game by N participants with opposing interests [10]. Let K be the total number of "scenarios" (options for implementing the IoT project) available for decision makers; k [1; K] is the scenario number. Let F_k denote the general positive effect (gain) predicted by the decision makers with a total cost of resources equal to G_k, which can be provided by them with probability P_G. The probability values P_F and P_R take into account, accordingly, the correctness of the prediction of the decision makers to get at the project exit according to the k-th scenario the gain F_k and the specified (environmental-ergonomic, economic, informational) risk R_k, and the probability value P_k is the chance of the decision maker to obtain permission to carry out the project under this scenario.

A component of the FEU showing savings of resources is represented as $Y_k = P_{Nk} G_0 / \sum_{n=1}^{N} \left(P_{Gn} G_n + P_{Un} U_n\right)$, where P_{Gn} is probability of obtaining the resource G_n; P_{Un} is probability of effective project promotion by allocating funds U_n for its advertising for the n-th project participant, n [1;N]; P_{Nk} is probability of agreement between all N participants to start a project (note that accounting funds U_n for advertising is in this case an important feature of the project, see below for more details). Similarly, the risk component is $P_R V_k = P_{Mk} R_0 / \sum_{m=1}^{M} R_m$, where R_m is the risk for the m-th object in the framework of the project, which has to be considered by its participants, m[1; M]; P_{Mk} is probability of the agreement between its participants to hand the completed project for approval, attractiveness (comparative effectiveness) of the project is $X_k = F_k / F_0$. Then FEU will have form:

$$Q_k / Q_0 = P_k \, [P_F (F_k / F_0) + P_{Nk} \, G_0 / \sum_{n=1}^{N} \left(P_{Gn} G_n + P_{Un} U_n\right) + P_{Mk} R_0 / \sum_{m=1}^{M} R_m], \quad (1)$$

Where G_0; F_0; R_0 and Q_0 are the amount of resources, the gain, the risk and the FEU for the base case (with k = 0); they are conditionally accepted as the basis for comparison by decision makers (DM). The scheme for determining FEU in the interest of assessing the effectiveness of the IoT project is illustrated by the scheme in Figure 1.

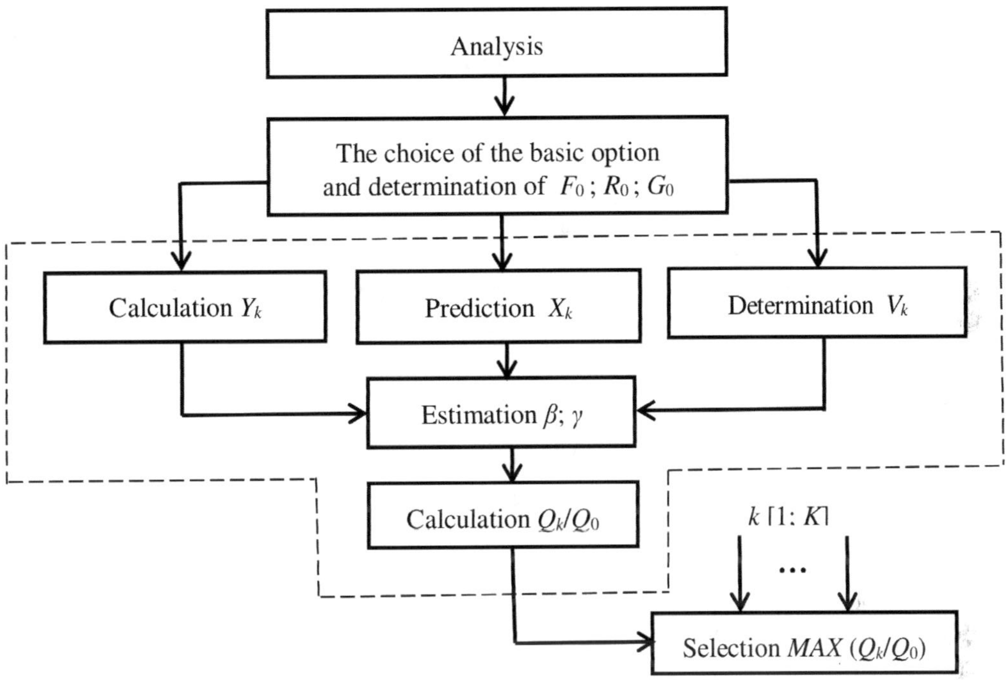

Figure 1. The algorithm for the formation and selection of the maximum FEU

After aggregation elements (1) FEU becomes to a compact form:

$$Q_k/Q_0 = P_k \, (P_F X_k + P_G Y_k + P_R V_k) \quad (2)$$

The most "understandable", in fact, deterministic component of (2) is $Y_k = P_G \, G_0 / G_k$, since both the cost of resources and the probability of their receipt by the decision makers can be accurately predicted heuristically: by cost estimates for each k-th version of the project.

This cannot be said about the components $X_k = F_k / F_0$ and $V_k = R_0 / R_k$, since it is impossible to predict the expected incomes and the change in environmental risk in the same way, and the probabilities $P_F = \exp(-\gamma X_k)$ and $P_R = \exp(-\beta V_k)$ associated with them according to [8–9] can only be specified with accuracy to parameters β [β_{MIN}; β_{MAX}]; γ [γ_{MIN}; γ_{MAX}]. These parameters are formalizing the uncertainty of knowledge of various decision-makers about the properties of the random components of FEU: X_k [X_{kMIN}; X_{kMAX}] and V_k [V_{kMIN}; V_{kMAX}].

Probability is $P_k = 1 - \exp(\alpha X_k)$ and after calibration $Q_k = Q_0 = 1$ with $X_0 = Y_0 = V_0 = 1$, FEU (2) becomes:

$$Q_k / Q_0 = (1 - (\frac{\exp(-\beta) + \exp(-\gamma)}{\exp(-\beta) + \exp(-\gamma) + 1})^{Xk\,Yk\,Vk}) \times (X_k \exp(-\gamma X_k) + Y_k + V_k \exp(-\beta V_k)) \quad (3)$$

In the structure of model (1) - (3), in addition to the determinate component (a possible variable of simulation) Y_k, there are two stochastic variables: X_k and V_k, as well as two secondary random parameters: β and γ. Figure 1 illustrates the algorithm for determining the FEU (2). The initial stages common to all K scenarios are the analysis of the subject area and the choice of the base case. Then, for each k-th option, actions are performed and highlighted by a dashed contour; at the final stage, the largest value of the FEU Q_k/Q_0 is selected for all options.

We adapt the FEU to the conditions of the problem: we mean N REE implementing the IoT project as participants; G_n is amount of resources determined by the costs of the acquisition, installation and operation of the n-th REE; U_n has the same meaning; R_m is risk for the m-th Internet user or any other "biological" object, including the decision-maker involved in the IoT project; F_k is a gain in respect of which so far it can only be said that, according to the forecasts of the decision-maker, it should be multifactorial and radically large [1-2]. The virtual components F_k and R_m significantly "exceed" the real components of the FEU G_n and U_n due the uncertainty of knowledge, which can be accurately estimated using the functional-cost analysis.

What are the gain, resources and risk in (1) - (3)? It depends on the subject area and models of objects in the k-th project. The FEU model helps to understand that promotion of the project to the level of profitability is impossible without active (targeted, creative, etc.) advertising in the media, which should include social networks. This is an objective necessity in "digital" economy, where the investment of resources in equipment $\sum_{n=1}^{N} G_n$ should be supplemented with funds $\sum_{n=1}^{N} U_n$ for intensive advertising of the project, since without it, it's no possible to receive a "bonus" in the form of F_k from consumers, according to (1).

Therefore, the FEU criterion becomes almost the only objective, true and convincing indicator of the effectiveness of the IoT project. The strategic plan of interested decision makers becomes obvious: "invest" in the FEU's component: $\sum_{n=1}^{N} U_n$ and get the consumer to pay for it and for $\sum_{n=1}^{N} G_n$, by intensive formation of F_k as soon as possible. On the way to achieving this goal, all means are good, since it is a matter of survival of the indicated decision-makers, not their ethics or aesthetics. However, the environmental-ergonomic and informational components $\sum_{m=1}^{M} R_m / R_0$ are no less important in the composition of FEU (1) - (3) for consumers of the Internet of Things services.

3. Initial data for simulation

We will consider the features of PCS-technologies using the example of Bluetooth, designed to transmit data over distances of up to 10m (with possibility of increasing range up to 100m), which allows you to connect telephones, computers, and other peripheral devices to the network without providing direct visibility between them. Note that the development of Bluetooth in 1994 by Ericsson Mobile Telecommunication began with a relatively modest goal: to obtain an economical and cheap interface for communication between a cell phone and a headset. However, then the creators of Bluetooth decided to do the same with computers, air conditioners, video cameras, coffee makers, refrigerators and vacuum cleaners, which are directly related to the concept of "Internet of Things".

It's important for IoT, that the Bluetooth protocol supports both point-to-point and point-to-multipoint connections. Using the common channel REE form a PicoNet, in which one of them works as the main (Master), the others work as subordinates (Slave). The piconet can have several active subordinate REEs, while the rest are in standby mode, remaining are synchronized with the main ones. Interacting piconets form a distributed network (Scatter Net). It is believed that Bluetooth continues to distinguish itself from other PCS technologies by its versatility and close to the optimal ratio between price and functionality [5]. The most important parameter of the REEs for security examination of PCS technologies within the framework of the IoT concept is the transmitter power P_A, mW.

By P_A level, Bluetooth devices are divided into three power classes (PC):

- PC-1: maximum power 100 mW, range up to 100 m;
- PC-2: maximum power 2.5 mW, range up to 10 m;
- PC-3: maximum power 1 mW, range u pto 0.1 m;

In Russia, the use of Bluetooth REEs at frequencies of 2400 ... 2483.5 MHz was allowed in 2003 for PC-1 and PC-2, and the PC-1 is the most popular. Examples of other wireless technologies for IoT on the market are shown below [2]:

- ML-Node-Z sensor nodes (they are manufactured by Mesh Logic, Russia), IEEE 802.15.4 standard;
- Zig Bit sensor nodes (Atmel, USA), IEEE 802.15.4 standard;
- Z-Wave modules (ITU ITU G.9959 specification);
- Zig Bee sensors (Zig Bee international alliance), IEEE Std 802.15.4 standard;
- BLE (Bluetooth Low Energy), an energy-efficient development of the Bluetooth Special Interest Group.

The maximum and minimum values P_A of the transmitters of listed devices are shown in Table 1; it can be seen that all these REEs are related to PC-3. A comprehensive security review of the IoT project (even by the SSM method) is the serious challenge. However, the examination of REEs' EMR seems to be an exception, since here one can use the mathematical apparatus of radio engineering systems and the traditional theory of probability [5].

Table 1. Transmitter power of IoT REE

Type	P_{AMIN}, μW	P_{AMAX}, mW
ML-Node-Z	4	1
Zig Bit	1,6	2
Z-Wave	10	1
BLE	10	1
Zig Bee	0,6	1

4. Risk simulation results

Let us consider the methodology and conditions for the SSM experiment of the project "Smart House" in the form of a three-room apartment of 100 m^2, the room is equipped with N ≥ 80 REEs Bluetooth PC-1 ... PC-2. Six PCS network operating modes were simulated:

- mode No. 1: "constant" radiation, N >> 1, Bluetooth PC-2;
- mode No. 2: "alternating" radiation, N >> 1, PC-2 Bluetooth; we need to take into account their separation into four groups according to the values of k, depending on the schedule of daily activity;
- mode No. 3: “constant” radiation, N/2 >> 1, PC-2 Bluetooth and N/2 >> 1, PC-1 Bluetooth;
- mode No. 4: “alternating” radiation, N 2 >> 1, PC-2 Bluetooth and N/2 >> 1, PC-1 Bluetooth; we need to take into account their separation into four groups similar to mode 2;
- mode No. 5: "constant" radiation, N >> 1, PC-1 Bluetooth (which corresponds to the "energy threat" in our terminology);
- mode No. 6: "alternating" radiation, N >> 1, PC-1 Bluetooth, it is similar to mode No. 2.

The SSM program of the situation in the room where the IoT network is deployed provides for M-fold “playing” using the Monte Carlo method of the values V_n [0; 1] and r_n [0,3; 10], these values are accepted equally probable within the specified limits in accordance with the "principle of indifference". Next, we calculate the S-fold total level of energy flux density $EFD = \sum_{i=1}^{4} k_i \sum_{n=1}^{N} EFD_{ni}$ for N = 50...500; M = 10^3...10^4 and S = 10^3...10^4. According to the results of statistical processing of data arrays obtained with data level S, estimates of the predicted average value of the EFD_{avg} are determined, as well as the boundaries of the $EFD_{0,95}$ and $EFD_{0,05}$ of the confidence interval corresponding to a probability of 0.90. Table 2 presents the indicated estimates of EFD depending on the number of REEs N involved in the IoT project as an example. Unsafe EFD levels are shown in bold [6].

Table 2. Values of EFD_{avg}, EFD_{MAX}; boundaries of the confidence interval of the EFD; μW / cm^2 for mode No. 5 of REE PC-1

N	50	100	200	300	400	500
$EFD_{0,05}$	0,1	0,2	2,0	5,2	8,0	**11,0**
EFD_{avg}	1,6	3,2	6,7	**11,3**	**16,4**	**19,0**
$EFD_{0,95}$	5,0	7,2	**13,5**	**19,0**	**27,0**	**28,0**

The recommended way to "improve" the indoor environment is to switch to using PC-3 REEs, since all EFD levels (as shown in table 2) are reduced by two orders of magnitude and the value of EFD is <6 μW / cm^2 even at N = 500. The aforesaid is confirmed by the data in Table 3 for mode No. 2 when using PC-3 REEs; it is clear that the situation with electromagnetic radiation in this case is safe, and the "energy threat" is impossible in principle.

Table 3. Values of EFD_{avg}, EFD_{MAX}; boundaries of the confidence interval of the EFD; μW / cm^2 for mode No. 2 of REE PC-3

N	50	100	200	300	400	500
$EFD_{0,05}$	0,001	0,008	0,030	0,044	0,068	0,084
EFD_{avg}	0,013	0,031	0,058	0,082	0,107	0,139
$EFD_{0,95}$	0,040	0,076	0,110	0,144	0,160	0,204

5. Statistical simulation of FEU

Variables and random parameters (3) are played by the Monte Carlo method according to a uniform distribution within the limits determined by decision makers by an expert method; thus, each combination of the indicated parameters and variables corresponds to one FEU value, and after N_p-fold repetition of this procedure, array of N_p>> 1 FEU values is formed. This array is subjected to standard statistical processing:

- the histogram is built within $(Q_k / G_0)_{MAX}$; $(Q_k / G_0)_{MIN}$ for N_c intervals found by the Sturges formula;
- the expected value (average value), the variance and the boundaries of the 90% confidence interval are determined: $(Q_k/G_0)_{0.95}$ and $(Q_k/G_0)_{0.05}$.

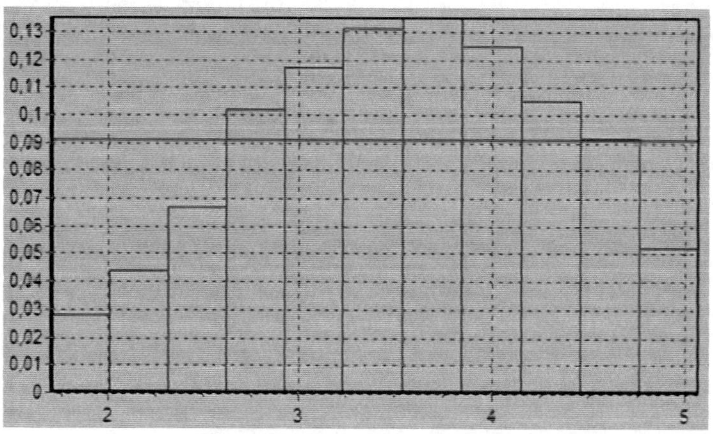

Figure 2. Histogram of Q_k/Q_0

In addition to the initial data, the histogram of FEU and its parameters all $N_p = 10^3$ played values of the FEU are displayed, which can be viewed. Analysis of specific options shows that when the number of similar REEs is reduced to N = 100, resource savings are $Y_k = 5$, risk change is V_k [1; 5], attractiveness reduction of the project is X_k [0.2; 1] and the uncertainty of the knowledge of decision makers about development of the situation is corresponded to β [0;0.1]; γ [0; 0.1]. According to the data in Figure 2, this leads to a noticeable improvement in all the characteristics of FEU despite the negative dynamics of X_k.

Similarly, if replace PC-1 with PC-3 with a five-fold increase in the number of REEs result in Y_k = 2.5; V_k [1; 2.5] and X_k [0.5; 1], then this is also be justified from the point of view of the growth of the parameters of the FEU. In both cases the cause can be considered the prevalence of the resource saving factor Y_k in the structure of the FEU with a relatively stable V_k and an acceptable decrease in X_k.

6. Discussion

The decision-maker can use a computer to get answers to all questions of interest by forming the initial data using any available information about the project. A broad field of activity for virtual research of the effectiveness of various options for implementing the IoT project opens up for the decision maker, since the considered fragment of the SSM model is the open system and it operates in a dialogue mode. The ability to select the initial values of parameters β and γ, which evaluate the most uncertain factor of knowledge uncertainty for analysis, allows decision makers to set them close to zero, and then vary within the necessary limits, checking the response of the FEU to their dynamics. It should be remembered that the SSM model is a "device" in the hands of the decision maker, which performs a significant part of the work for it, but it is not able to completely replace him.

7. Conclusion

A comprehensive examination of the Internet of Things project is possible using the SSM method, which is combining the methods and means of mathematical modeling with probabilistic combinatorics and the FEU criterion. It is necessary to pay attention to both informational [3-4] and environmental-ergonomic safety based on the EMR factor of wireless PCS technologies, on the basis of which PAN and LAN are built when implementing the project [5-6]. The preferred elements of IoT networks based on the Bluetooth protocol are PC-3, on condition that a rational way of deploying radio electronic equipment will compensate the negative consequences of increasing their number compared to more powerful PC-1s'.

References

[1] *Overview of the Internet of things* 2012 (ITU-T Rec) Y.2060.
[2] *Architecting the Internet of Things* 2011 (Springer) 351 p.
[3] *ISO/IES JTS 1. Technical Document of ISO/IES 1 Study Group on Sensor Networks (SGSN). Version 3* 2009 166 p.
[4] *Sensor networks and configuration: fundamentals, standards, platforms and applications* 2007 (Springer) 510 p.
[5] Dargie W, Poelabauer C 2010 *Fundamentals of Wireless Sensor Networks Theory and Practice* (John Wiley & Sons Ltd) 311 p.
[6] Hersent O , Boswarthick D, Elloumi O 2012 *The Internet of Things: Key Application and Protocols* (Willey) 370 p.
[7] Cook D , Das D 2005 *Smart Environments. Technologies, protocols and applications* (Hoboken NY: Wiley-Interscience) 404 p.
[8] Atzori l , Iera A, Morabito G 2010 *The Internet of Things: A survey* (Computer Networks) 54 pp 2787-2805.
[9] Bandyopadhyay D, Sen J 2011 *Internet of Things: Application and Challenges in Technology and Standardization* (Wireless Personal Communications) 58 (1) pp 49-69.
[10] Kuniavsky M 2010 *Smart Things: Ubiquitous Computing User Experience Design* (Morgan Kaufmann) 336 p.
[11] McEwen A , Cassimally H 2013 *Designing the Internet of Things* (John Wiley & Sons) 260 p.
[12] Vasseur J, Dunkels A 2010 *Interconnecting smart objects with IP* (Morgan Kaufmann) 432 p.
[13] Zhou H 2012 *The Internet of Things in the Cloud:A Middleware Perspective* (CRS Press) 365 p.

[14] Evdokimov S , Fabian B and Günther O 2011 *RFID and the Internet of Things: Technology, Application and Security Challenges* (Now Publishers Inc) 88 p.
[15] Von Neumann J and Morgenstern O 2004 *Theory of Games and Economic Behavior: Second Edition (*Princeton: Princeton University Press) 625 p.
[16] Savage L J 1954 *The Foundations of Statistics* (N.Y.: Wiley) 310 p.
[17] Kahneman D and Tversky A 1979 *Prospect Theory: An Analysis of Decision under Risk* (Econometrica XVLII) pp 263-291.
[18] Shoemaker Paul J H 1982 *The Expected Utility Model: Its Variants, Purposes, Evidence and Limitations* (Journal of Economic Literature) XX(2) pp 529-563.
[19] Maslov O N , Rakov A S and Silkin A A 2013 *Statistical Simulation of Random Antennas like Development of the Statistical Theory Antennas* (Proceedings of the IX International Conference on Antenna Theory and Techniques ICATT'13. – IEEE Ukraine, September 16-20 Odessa) pp 53-58.
[20] Maslov O N , Rakov A S and Silkin A A 2013 *Statistical Characteristics of the Field of an Array of Random Aperture Antennas* (Journal of Communications Technology and Electronics) 58(11) pp 1056-1064.
[21] Maslov O N , Rakov A S and Silkin A A 2015 *Statistical Models of the Wave Field of a Random Aperture Antenna* (Journal of Communications Technology and Electronics) 60(6) pp 603-610.
[22] Maslov O N and Shatalov I S 2018 *The Simulation of Equivalent Currents Exciting a Stochastic Aperture Radiator* (Proceedings of SPIE Vol. 10774) pp. 107740S-1 – 107740S-9 doi: 10.1117/12.2317729
[23] Maslov O N and Shatalov I S 2019 *Simulation of Statistical Characteristics of Aperture Random Antenna by Triad-Cluster Method* (Proceedings of SPIE Vol.11146) pp 11146A-1–11146A-13 doi: 10.1117/12.2524835.
[24] Maslov O N 2018 *Electromagnetic Safety of Workstations Equipped with Active Systems for Data Protection* (Journal of Communications Technology and Electronics) 63(2) pp 163-172.

The use of electronic semantization of the cognitive activity manifestations with the aim of detection of intentions of the group of people leading to the destabilization of the digital super system

E L Loginov [1], V V Grigoriev[2], V S Balandin[3], P A Boyko[4], A A Shkuta[5]

[1]Situational Analytic Centre Service of the Ministry of Energy of Russia, Shchepkina, 42, Moscow,107996, Russia

[2] MGIMO (University) MFA of Russia, Vernadskogo Ave. 76, Moscow, 119454, 355017, Russia

[3]Financial University under the Government of the Russian Federation, Leningradsky prospekt, 49, Moscow,125993, Russia

[4]Russian State Humanities University, Miusskaya sq., 6, Moscow, 125993, Russia

[5]Financial University under the Government of the Russian Federation, Leningradsky prospekt, 49, Moscow, 125993, Russia

E-mail evgenloginov@gmail.com *(e-mail of corresponding author)*

Abstract. The article deals with the possibilities of monitoring of information communities and their members' cognitive activity in order to analyze and forecast the actions of these individuals with the use of the method of automated structuring of their real political interests and of detection of inner- and inter-class connections in a digital super system. Analysis is a relevant method to detect tendencies of the political positioning of groups of people that are forming digital information assets as well as of the management of the organizational risks that emerge.

Keywords: digital super system; information communities; cognitive activeness; political interests; digital information assets.

1. Introduction

In last years the new social phenomena are drawing attention, those connected with the realization of information interests by an individual or a group of people that form digital information assets with the use of network technologies [1-3].

The genesis of the active network activity of a vast group of people requires its monitoring, automatic classification and identification the cluster connections in the digital super system because this can make a significant influence on the global socio-politic structure of the various state levels [4-5].

IOP Conf. Series: Materials Science and Engineering **873** (2020) 012002 doi:10.1088/1757-899X/873/1/012002

The development of the strategy and the planning of the development of the political events through the implementation of the observation, structuring and forecasting of the individual's acts on the base of the identification and assessment of the nonlinear regularity of the external facts' impact is a strategically crucial goal for all countries [6]. All state institutes are interested in the provision of their political and social stability [7].

It's necessary develop the basics of the synchronization of the set signals and of the whole complex of the measured parameters of the cognitive activeness in the analyzed information area of a digital super system as well as to detect the following tendencies of the political, social and info-digital reforms.

2. The detection of the taxonomic structure of the information societies and of cognitive activeness for the structuring and forecast of their acts in the digital super system

It's suggested to carry out the analysis of the electronic subject texts that will enable us to visualize the modern society and the processes of the maintenance of the livability of the political system. The assessments are based on the network information possibilities taken as the means of organized acts of an evident or of a latent group of people. This is an effective method of the identification of a taxonomic structure from the concrete area of information societies and its cognitive activeness that forms digital information assets for the structuring and forecast of the individuals' acts in the digital super system by the computing structuring of their real political interests and of the detection of the inner- and inter-cluster connections. Automated analysis is a relevant method of the detection of tendencies of dynamic development of pro-government and anti-government cognitive activity on the base of information available in the digital area.

Automated observation with the use of network technologies of the information assets logistics and of intellectual mobility makes it possible to detect and identify acts of groups of people in an information area which is of a special interest to a researcher, which forms digital information assets.

The visualized expression of the agent-based analysis results for the effective detection of the current level of aggressiveness of groups of people by an automatic classification of the identified network community members and the detection of the cluster connections inside the digital super system is the information structure of the members' messages making it possible to observe an individual's reaction on some certain events and to identify the dependences and the deviations.

On this base it's possible pass on to the formation of identified persons' "aim reflex" towards the operated network resources.

The construction of hierarchic and network connections or of operated network resources is a flexible method of the development of the strategy of management and planning of the development of the political events on the base of identified persons' acts emerging in the especially interesting information interests implementation field which form digital assets [8-9]. The use of the agent-based modeling on this purpose is widely used within the control of those who act in a way that doesn't go along with the official position and in which the state bodies aren't interested as well as with the aim of maintenance of planning of the state body acts in this sphere [10-12].

The analysis of changes in the frequency of the emergence of key concepts in the information networks — terms defining the essence of the subject interest – over a long period of time can deepen the understanding of the process of development of the latent communities. Another contribution to this is made by the identification and visualization of "hotspots", new communication opportunities, the development of newly emerging acts of people in the digital super system on the base of detection and assessment of the nonlinear regularity of the external factors' influence and identification of the evident and latent leaders.

In the computer processing of an analyzable amount of electronic information transactions mediating the expression of information interests and personal contacts in any detectable form we can highlight the information themes or blocks of data taken as the expressions of the dynamism of psychological parameters and conditions detailed by the issues of concern such as parameters of the level of aggressiveness and sociopathy.

The agent based analysis of the distribution of the analyzable amount of electronic information transactions makes it possible to identify the inverse relationship between the response to the advertisement and propaganda elements for the stimulation of digital information asset consumer's behavior in the spheres which affect the political stability of society and state.

For a more deep understanding of the global distribution of events it's necessary to detect the information organizational structure of the members which will make it possible to observe the participation of a concrete person in some certain events and to identify the dependences and deviations on the base of information concerning the connection between individuals. This is the ground of the entire process of identification of those whose political position is latent and doesn't go along with that of the officials that forms the vectors of their acts that contradict the official position, i.e. that in which the state bodies aren't interested.

Every buyer wants to buy some set of goods. Goods are delivered from the warehouse (to firms and shops), which corresponds to a given type of raw materials and goods. Also introduced the concept of value of each end product. It represents the sum of the value of the goods at the point of extraction of raw materials, transportation costs for its delivery first to the storage warehouse, and then to the firm, production costs, costs for delivering goods to the store and storage costs in warehouses [6-7].

3. The formation of a structured field of thematic chains for the construction of the conditions suitable for the positive interpretation of information and virtual programming of the social reality in the form of acts under the influence of information signals

The structured field of thematic chains on the information and social networks (the ground of the event sequences) can become a powerful instrument of the maintenance of taking important decisions of great responsibility by the management at an early stage of the aggressive trends deployment as well as under the information attacks. The substantive actions of a political or a social group of people can be detected by the analysis of thematic data injection and of semantic characteristics of the amounts of information messages and their sources by the use of inverse relationship signals that reflect a person's actions in information and social networks (as the ground of event sequences) in regards to the most important spheres of the government's interests.

Thus, we can get a systematic understanding of groups of people including the modern level of processes of the maintenance of livability of the political system and the activities of organizational clusters in the identified sectors of substantial acts that form digital information assets that belong to the sphere of government's interests in the organizational field of the maintenance of stability and order which is an issue of concern for them [13]. But because the projects of this kind provide a more broad overview in comparison to the assessment of information themes (blocks of data), they are used in marketing researches, political developments and the assessments of information situations with the aim of linkage between the information strategy and the general aims of the government [14].

The results of the formation of a structured field of thematic chains can be used for the construction of the conditions for the positive interpretation of information or for the virtual programming of the social reality in the form of actions that can be used for offensive and defensive purposes [15]. Depending on the final destination of thematic injections and semantic characteristics of the amounts of information messages and their sources with the use of inverse relationship signals, that describe an individual's activity in information and social networks, the political strategies of the detected groups can vary significantly according to their resources and focusing [16].

The contextual flexibility and the permissible breadth of coverage of information by the information, virtual game based and communication technologies entails both advantages and disadvantages. They can lead to the optimization of the information injections and semantic characteristics of the amounts of information messages (taken as the ground of the event sequences) as well as to the overloading of people with redundant information, to the dispersion of aims and the reduction in the coordination of the received results. The transformation of the sets of data by the information, virtual game based and communication technologies is aimed at the optimization of thematic injections and semantic

IOP Conf. Series: Materials Science and Engineering **873** (2020) 012002 doi:10.1088/1757-899X/873/1/012002

characteristics of the amounts of information messages and is focused on the political positioning and management of political risks.

Governmental bodies are interested in the monitoring of the actions of people with the set up space-, frequency- and temporal characteristics of the expression of activeness in the communication networks and in the streets with the aim of detection of the organized actions of a large political or social group [17]. The government is interested in the structured field of thematic chains for the construction of the conditions suitable for the positive interpretation of information and virtual programming of the social reality as the set of acts under the influence of information signals that enable us to identify the strategy of the organized groups [18]. Usually the organization of thematic injections is characterized by a broad coverage of the organizational area [19].

4. The detection of the linkage between the information interests and the cognitive activeness of individuals or groups

The use of the strategy of detection of the linkage between the information interests and the cognitive activeness of individuals or groups has a number of advantages when used by the government in the construction of the instruments of information and social networks monitoring:

— the connections between people are reflected visually helping the analytic to assess the organizational structure in general;

— the analysis of the structured field of thematic chains in the information and social networks (taken as the ground of the event sequences) is carried out taking into account the numbers of various key concepts and words determining the essence of the substantive interest which provides as a result getting more reliable indicators;

— the analysis of the thematic chains within the information and social networks reduces the time and expenses involved because the initial random set of electronic information messages is transformed into the represented visually structured data;

— the configuration of the information field received with the use of advertisement and propaganda elements for the stimulation of the digital information asset consumer's behavior makes it possible to give the direction to the acts of people with the use of computing structuring of their real political interests and the identification of the inner- and inter-cluster connections through time and from a group of people to the nucleus of the information messages overcoming thematic barriers;

— the strategy of the construction of the monitoring mechanisms in the global networks will make it more easy to analyze the secondary substantive events, the influence of concrete identifiable groups of people, temps of dynamic development of the expressions of pro-government and anti-government cognitive activity.

The state agencies are assessing and analyzing the available amount of the electronic and information messages observed in the analyzed information area of the digital super system with the detailing by the issues of concern such as the parameters of the level of aggressiveness and sociopathy. The identification of individuals in the interesting field of information interests and the measuring of the level of aggressiveness of such groups of people making stable acts leading to the "gliding" of the super system in a risky direction makes it possible to determine the measures of neutralization of their threats to government and society.

5. Conclusion

Having identified the general trends and the active participants with the use of automatic classification of the identified members of the network communities and having detected the cluster connections in the digital super system we can pass on to the qualitative analysis of the role structure of the organizational members which makes possible to observe an individual's response to some certain events, to distinguish dependences and deviations. For this purpose we'll have to enlarge the quantitative assessments received earlier with the analysis of the contents of texts, relevant descriptions of the relevant themes of blocks of information data or network interactions as the base of the organized actions of an evident or a latent group of people.

To identify the nucleus of the active members it's recommended to focus the concentration on the particularly significant substantive areas of expressions of cognitive activity and the network interactions of the detected and identified group of people. On base of the study of analyzable amount of electronic information transactions we find out the inverse relationship of the response to the advertisement and propaganda elements for the stimulation of the digital information asset consumer's behavior in any available form. Thus we forecast their position on the general substantive structure of political views, build the diagram of the dynamic development of the expressions of the pro-government and anti-government cognitive activity which reflects the positions of all the relevant groups of people making stable acts leading to the mutual activation or "extinguishment" of the aggressive acts of the dispersed members of network communications in the digital super system.

6. The indicators of the organization of groups of people making stable information acts (implementing information strategies)

The political institutes of the state that don't possess the information about the substantive trends of the development of network cognitive activeness are losing ground in the political system of the society. To assess the cognitive activeness it's necessary to identify the groups of people making stable aggressive acts and inter alia the leaders of the dispersed members of the network communications that maintain the management of the organized network groups. The indicators of the organization of groups of people making stable acts enable us to assess more accurately the threat of the destabilization of the super system (network information attacks, the calls for participation in the riots etc.).

The indicators of organization of groups of people as dispersed members of network communications include:

– the number of the information messages to enter the information area of a certain political or social groups of people;

– the average number of the network information contacts mediating the information injections;

– the average age of the members of organized groups;

– the number of the information themes they use;

– the size of the information messages in the network;

– key words used for the description of information themes (or blocks of the used words);

– the availability of a structure of the information interests as the ground of the network information contacts with the previous or the following information injections from unknown authors;

– the dispersion of the information leaders as the basis of the network groups over clusters and groups.

With the use of indicators of the organization of groups we identify the level of structuring of the information area controlled by groups of people, the thematic coverage of the individuals' acts in the digital super system, their origin and development. The latter is measured with the use of retrospective and forecast analysis of the linkage between the information interests and cognitive activity of individuals and groups. The final assessment of the organization is carried out with the use of data about the enlarged information themes, the timing of the updating of the information assets, reciprocal links, the size of messages and detailing of the description of information themes. The model of the organization of groups of people making stable acts leading to the "gliding" of the super system in a risky, built in this way, is used for the assessment of their political potential.

The assessment methods used are based on the counting of information messages as the base for the organizational structuring of the acts of people in the digital super system with the set space-, frequency- and temporal characteristics of the expression of the activeness in the communication networks for the detection of their coordinated actions. The advantages of these methods consists in the possibility of the use of the electronic information messages of a political or social group of people available in the Internet, and the disadvantage consists in the lack of the direct correlation between the interests of the financial and political groups. Qualitative assessments are most often used in the interests of management of the information portfolio of the electronic Mass Media cluster under the control of the government. They are also used by the ranging of the groups of people making stable acts or by comparing them to the groups of people from the detected sectors of the substantive actions that form digital information assets.

7. Acknowledgment

The article is written with the financial support of the Russian Fundamental Research Fund (project № 19-07-01066 "The creation of an artificial intelligence system in the form of a digital platform component for the monitoring of the behavioral activeness of large groups of people on the base of the application of the methods of analysis of large semi-structured data, of the construction of thematic models with cognitive and multi-parameter semantic interpretation, of the exploration research and of collaborative filtration with a convergent management").

References

[1] Ageev A and Loginov E 2018 *Convergent monitoring and programming of a person as an instrument of operation of the intellectual dynamism of large groups of people* (Economic strategies) 2(152) pp 70-87.
[2] Ibragimov L 2015 *The technologies of the internet manipulations in the network space of politics* (Problem analysis and public administration design) 2 pp 34-40.
[3] Petuhov A 2014 *Chain reactions in the complex political and social systems* (Vector Society Togliatti State University. Series: Pedagogy, Psychology) 3(29) pp 207-210.
[4] Lepskiy V and Avdeeva Z 2018 *Disruptive situation management in digital economy* (Proceedings of 2018 11th International Conference "Management of Large-Scale System Development") MLSD pp 8551949.
[5] Loginov E and Grigoriev V 2019 *The use of artificial intelligence's elements to block the manifestations of individuals' behavioral activity going beyond the quasi-stable states* (IOP Conference Series: Materials Science and Engineering) pp 012028.
[6] Loginov E and Rajkov A 2018 *The use of neurotechnologies in programming of cognitive and behavioral stereotypes of person's acts for the stability of the systems controlling the society* (Neurocomputers: development, application) 9 pp 34-45.
[7] Bakshutova E 2018 *The problem of qualitative measurements of the latent psychology of large groups* (Vector Society Togliatti State University. Series: Pedagogy, Psychology) 3(34) pp 42-50.
[8] Dokuka S 2014 *The use of network based approach for the analysis of the structure and the operation of communities in the social networks* (Innovation and investment) 1 pp 179-182.
[9] Nazarov A and Galushkin A 2016 *The models of risks and the criteria of the information confrontation in the social networks* (T-Comm: Telecommunications and Transport) 7 pp. 81-86.
[10] Gorlushkina N and Ivanov S 2019 *The method of the generalized centrality for the analysis of the network cyber-spac*e (Cybernetics and programming) 2 pp 55-64.
[11] Skuratov A 2016 *The leadership in the local internet communities: practice of the socially stratified analysis* (Bulletin of the South Russian State Technical University (Novocherkassk Polytechnic Institute). Series: Socio-Economic Societies) 3 pp 75-79.
[12] Sergeev N and Celyh A 2017 *The methodological aspects of the modelling of social systems for the analysis and management of conflict*s (Political Internet electronic scientific journal of the Kuban State Agrarian University) 133 pp. 936-959.
[13] Loginov E and Grigoriev V 2019 *Intelligent monitoring, modeling and regulation information traffic to specify the trajectories of the behavior of organizational agents in the context of receipt of difficult-interpreted information* (IOP Conference Series: Materials Science and Engineering) pp 012015.
[14] Kravchenko Y 2018 *The model of the knowledge filtration for the aims of the semantic identification* (News SFU. Technical science) 4(198) pp. 166-174.
[15] Afanaseva T and Sibirev I 2018 *The application of unclear models in the analysis of the processes in the organizational and technical systems* (Automation of management processes) 1(51) pp. 83-

90.
[16] Tereshonok M and Rautkin Y 2018 *The assessment and forecast of the actions of mass events members with the use of intellectual analysis of the parameters of mobile network traffic* (Cybersecurity issues) 3(27) pp. 70-76.
[17] Sokolov I 2019 *The theory and practice of the AI-methods application* (Bulletin of the Russian Academy of Sciences) 4 pp. 365-370.
[18] Bahtizin A and Bortalevich V 2019 *Using artificial intelligence to optimize intermodal networking of organizational agents within the digital economy* (Journal of Physics: Conference Series) pp 12042.
[19] Raikov A 2019 *Uncaused semantic interpretation of cognitive models in networked decision support systems* (11th IEEE International Conference on Application of Information and Communication Technologies) AICT 2017 Proceedings 11 pp 8686968.

IOP Conf. Series: Materials Science and Engineering **873** (2020) 012003 doi:10.1088/1757-899X/873/1/012003

EEG Neuro - processing for the development of neurointerfaces

N I Chervyakov, P A Lyakhov, M R Kiladze, A A Kovalets and U A Lyakhova

North-Caucasus Federal University, Department of Applied Mathematics and Mathematical Modelling, 1, Pushkin Street, Stavropol 355017

E mail ljahov@mail.ru

Abstract. Fast and accurate processing of electroencephalogram (EEG) signals is necessary for correct realization of neurointerfaces. Neural network is one of the ways of its processing. In this article the method of EEG processing with the help of multilayer perceptron trained by the algorithm of error backpropagation is considered. The results of modeling during training of the neural network showed the minimum value of the Mean Squared error.

Keywords: neuro-interface; electroencephalogram; neural network; backpropagation; perceptron

1. Introduction

For the first time, the idea of human communication with the machine was proposed by J. Vidal, describing the laboratory for the analysis of the electroencephalogram (EEG) signal [1]. Later, the universal architecture of brain-computer interface (BCI) was described in [2]. Classification of BCI schemes was suggested in [3]. BCI is divided by the way the user presents information into active ones, in which the user initiates a command, e.g. a hand movement, reactive ones, where the user initiates a command in response to the system, e.g. a hand movement in response to the screen signal, and passive ones, where the system reads and analyzes the user's state without initiating a command.

BCIs are classified by the way the signal is received into invasive, e.g. implanted electrodes, and non invasive, e.g. EEG, ECG, etc. BCI has found its main application as a tool for rehabilitation of patients with various motor and neurological disorders, but recently such an interface has also been used by healthy people, for example, to control cognitive condition through biofeedback, games and entertainment.

In this article we suggest using multilayer perceptron for EEG processing and will show the result of modeling such a neural network.

2. Application of EEG for neurointerfaces

A device for exchanging information between the brain and the computer is called a neuro-interface or BCI. BCIs can be invasive and partially invasive, as well as they differ in their application and area of use [2], [3].

The architecture of the BCI is shown in figure 1. It contains "Signal Acquisition", "Preprocessing", "Feature Extraction", "Translation/Classification", "Commands/Control". The points " Signal Acquisition " and "Commands/Control " are implemented by the device. Items " Preprocessing " and " Feature Extraction " implement artifact selection and motion features selection, respectively. Artefacts of EEG are conditionally divided into physical artifacts arising due to influence of various physical or technical hindrances on EEG equipment, and on physiological artifacts which are shown at various biological processes in an organism of the investigated person [4]. The item " Translation/Classification " determines which movement is performed.

EEG of the brain is a method for registering electrical signals of brain origin by placing electrodes (figure 2) on the scalp surface as a result of electrical summation and filtration of elementary processes in neurons [5]. The EEG has the form of a graphical image, which shows the gradient oscillations of somatodendrite potentials corresponding to the excitation post-synaptic potential (ESP) and inhibitory post-synaptic potential (PSP) [5].

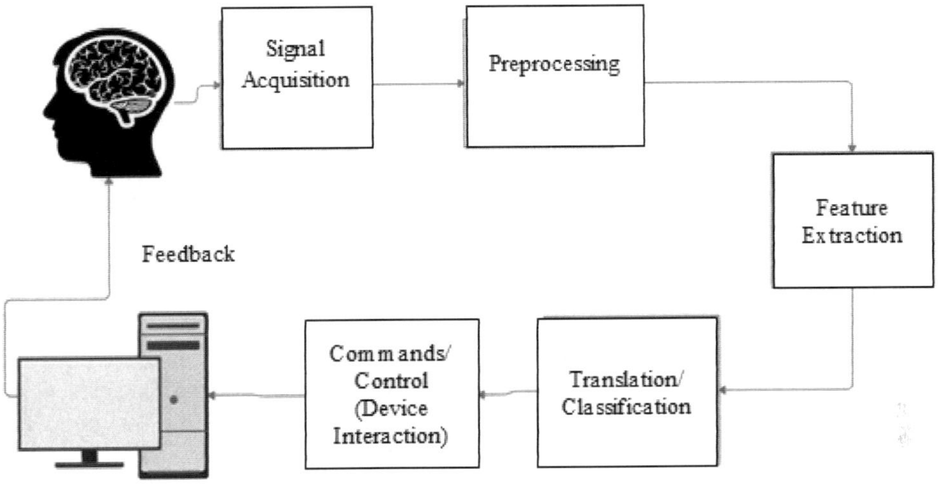

Figure 1. Architecture of the BCI

Electrode designations include the first letter of the Latin name of the mounting area of the electrode and a number indicating the side and location. The electrodes are placed on the scalp surface according to the standard international system "10-20%" recommended by the International Federation of Electroencephalography and Clinical Neurophysiology. According to this system, the line connecting the nose bridge (Nasion) and occipital mound (Inion) is accurately measured and divided into 10 equal segments. The first and last electrode are placed at a distance equal to 10% of the total long line from the nose bridge (Fp1/Fp2) or occipital tuber (O1/O2). The first electrode is followed by the next one at a distance equal to 20% of the total line length. The second main line, which connects the left and right ear cavities, is also divided into equal sections, followed by mounting the electrodes at a distance of 10% of the ear canals (T3/T4) and 20% of each other (C3, Cz, C4). The third line represents the arrangement of the electrodes along the head circumference from the occipital electrodes (O1/O2) to the frontal electrodes (Fp1/Fp2) at a distance of 20% from each other. Indifferent electrodes (A1/A2) are located at the earlobes. In this way, 21 electrodes are applied over the entire head surface [6].

The excitatory postsynaptic potential allows the neuron to become more excitable and thus increases the probability of generating the action potential, while the inhibitory postsynaptic potential leads to a decrease in neuron activity and decreases the probability of generating the action potential [7]. For generation of action potential by neuron to be transferred to other neurons or effectors it is necessary that the level of its excitation reaches a certain threshold value. In its turn the level of excitation is determined by the sum of excitatory and inhibitory influences which appear on the neuron by means of synapses at the moment. As soon as the sum of excitatory actions exceeds the sum of inhibitory actions by a value higher than the threshold level, the neuron generates a nerve impulse transmitted to the axon [5].

The most significant characteristics when estimating the EEG can be called average frequency of oscillations, maximum amplitude, phase of oscillations and estimation of curves difference on different channels with their dynamics in time. The frequency indicator is estimated by a fixed number of wave oscillations per second, calculated from the number of waves on the taken 4-5 sections of the record and expressed in Hertz (Hz). The amplitude gives an idea of the range of wave oscillations of eclectic potential and is measured by the distance between the wave peaks in opposite phases, expressed in microvolts (μV). The oscillation phase value estimates the current state of the process occurring in the human brain and determines its vector changes. After the end, the physical and pathological interpretation of the obtained materials is made and on its basis the EEG conclusion is formulated.

IOP Conf. Series: Materials Science and Engineering **873** (2020) 012003 doi:10.1088/1757-899X/873/1/012003

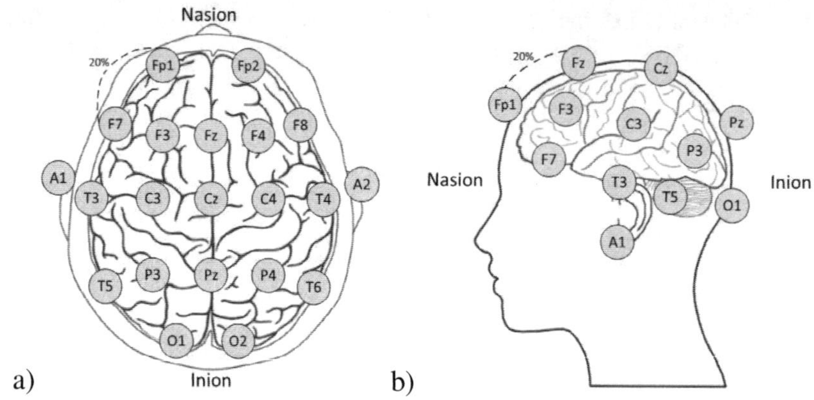

Figure 2. Scheme of electrodes location on the head surface according to the international system "10-20%". (a) is top view, (b) is side view

The activity of the adult brain, which can be recorded by an EEG in a waking state, is divided into the following types of rhythms. The α-rhythm is fixed at a frequency of 8-14 Hz, present at rest with the examiner with his eyes closed, is determined at the occipital region. β-rhythm. Rhythm with wave frequency in the interval of 13-30 Hz in the active state of the brain is diagnosed in frontal lobes [8]. The γ-rhythm is fixed with frequency of fluctuations from 30 to 180 Hz, it is defined at the solution of tasks and the situations requiring a high level of concentration and attention. κ-rhythm is characterized by frequency in 8-12 Hz and is observed in a temporal lobe of a brain at mental processes. The rhythm has a low frequency of 4-5 Hz and is determined in the occipital region when the visual receptors are used. μ-rhythm has a frequency of 8-13 Hz and is fixed in the occipital lobe of the brain in a calm state.

In sleep, a separate category of rhythm types is recorded [9]. δ-rhythm characterizes the phase of deep sleep as well as the coma. θ-rhythm has an oscillation frequency within 4-8 Hz. Waves of the given rhythm are launched by the hippocampus and appear in dreams. σ-rhythm differs in frequency from 10 to 16 Hz and is considered to be the rhythm of oscillations that occur during natural sleep in the initial stage.

By the end of XX century this method of research was applied in BCI, which allowed a person to communicate with the outside world in real time by means of changes in the activity of his own brain [10]. Today, the EEG is the leading method for studying the stages of sleep, as it allows tracking large changes in brain activity. This study also shows rhythms of brain electrical activity, consequences of brain operations, brain tumor processes and their influence on functional activity, presence or absence of foci of increased convulsive readiness and their localization, etc. [11].

The method is highly accurate, efficient, safe and painless for humans, allowing the use of EEGs to receive brain signals and their subsequent use in BCIs.

3. Application of a neural network for the analysis of EEG

The neural network can be used to decrypt the EEG. A neural network is a sequence of neurons connected to each other by synapses. The structure of a neural network is a machine interpretation of the human brain, which contains millions of neurons transmitting information in the form of electrical impulses.

Neural networks are not programmed, they are trained. The possibility of learning is the main advantage of neural networks. Learning a network is about finding the relationship coefficients between neurons. In the training process, the neural network detects complex relationships between input and output data and summarizes them. If the training is successful, the network can get the result if there is no or partial data distortion [12].

A neuron is a unit of information processing in a neural network. Figure 3 shows the biological (a) and mathematical (b) models of the neuron underlying artificial neural networks [13]. Dendrite, presented in biological neuron model, is a prototype of input signals in mathematical model (x_1 , x_2 , ..., x_k). Axon is the prototype of the output signals, in a mathematical model it is "Activation", i.e. the result. Interaction between neurons is realized in synapses, in the mathematical model they correspond to weight coefficients (w_{1j}, w_{2j}, ..., w_{kj}), which can take positive and negative values.

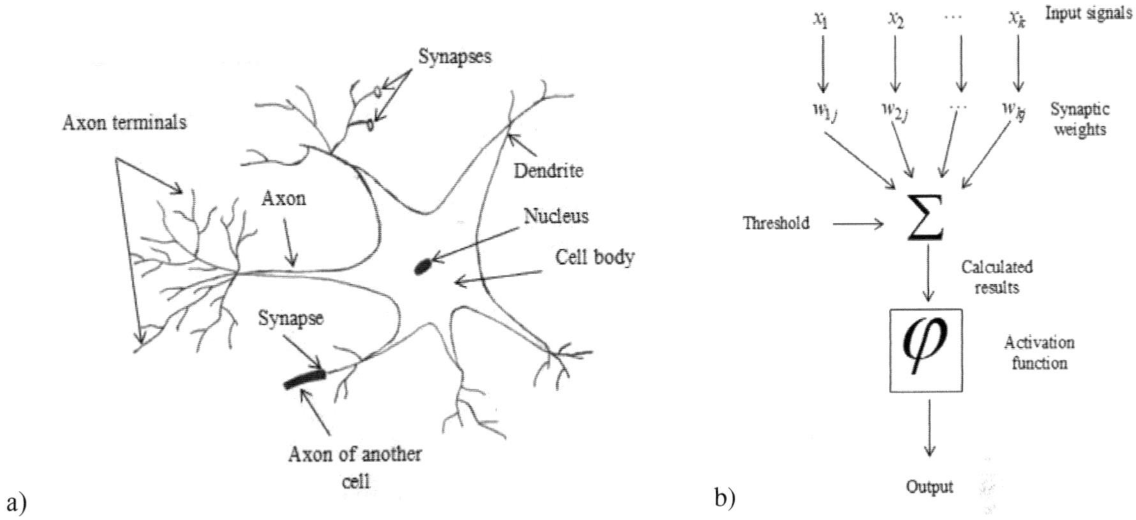

Figure 3. Neuron model. a) biological model of the neuron; b) mathematical model of the neuron

The nucleus and body of the cell in the biological model correspond to the adder and activation function of the mathematical model. The adder adds up the input signals, this operation is a linear combination. The activation function limits the amplitude of the neuron output signal. This function is also called the compression function. Normally, the neuron output amplitude range is in the range [0.1] or [-1.1] [13].

There are several models of neural network. We will consider one model of perceptron, which is the McCullock model. Its nonlinear activation function is a discrete step-type function, where the neuron output can be only two values 0 or 1 according to the rule $y_k(v_k) = \begin{cases} 1 & for \quad v \geq 0 \\ 0 & for \quad v < 0 \end{cases}$. In this expression u_k the adder output signal $v_k = \sum_{j=0}^{N} w_{kj} x_k$. Perceptron teaching implies a teacher and consists in selecting weights so that the output value is as close as possible to the specified value [14].

In our proposed method, we will use a multilayer Perceptron with at least one hidden layer. The multilayer perseptron was chosen by us as the basis for its simplicity and versatility.

The output signal of the j-type neuron of the hidden layer can be described by the function $v_j = f\left(\sum_{j=0}^{N} w_{kj}^{(1)} x_j \right)$, provided that a single polarization signal is considered as one of the components of the x input vector $y_k = f\left(\sum_{i=0}^{K} w_{ki}^{(2)} v_j \right) = f\left(\sum_{i=0}^{K} w_{ki}^{(2)} f\left(\sum_{i=0}^{K} w_{ki}^{(1)} x_j \right) \right)$. It follows that the signals generated in the hidden layer do not depend on the weights of the output layer.

In the proposed method, an algorithm of backpropagation was used to teach the described neural network. This method of teaching a multilayer neural network is called a generalized delta rule. The method was proposed in 1986 by Rumelhart, McCleland and Williams [15]. This algorithm is the first and main practically applicable for training multilayer neural networks.

For the output layer, the correction of weights is understandable, but for hidden layers the algorithm has not been known for a long time. Weights of the hidden neuron will change directly proportional to the error of those neurons with which the given neuron is connected. Therefore, the reverse distribution allows you to correctly adjust the weights of links between all layers. In this case the error function value is reduced and the network is trained [16].

The main relations of the method of reverse propagation of the error are obtained with the following designations - function error for the image p, - desired neuron output j for the image p, - active neuron output j for the image p, - weighted sum of outputs of bound neurons of the previous layer on the weight of the bond (inactive state of the neuron), - weight of the bond between i and j neurons.

The main ratios of the error propagation method are obtained with the following designations. E_p is function error for image, t_{pj} is desired neuron output j for image p, y_{pj} is active neuron output j for image p, s_{pj} is weighted sum of yields of bound neurons of the previous layer on the weight of the bond, w_{pj} is the weight of the connection between i and j neurons.

The value of the error is determined by formula (1)

$$E_p = \frac{1}{2}\sum_j (t_{pj} - y_{pj})^2 \tag{1}$$

The inactive state of each neuron j for image p is recorded as a weighted sum using formula 2

$$s_{pj} = \sum_i w_{ij} y_{pj} \tag{2}$$

The output of each neuron j is the value of the activation function f_j that switches the neuron to the activated state [5]. Any continuously differentiable monotonic function can be used as an activation function. The active state of a neuron is calculated using formula 3:

$$y_{pj} = f_j(s_{pj}) \tag{3}$$

Gradient descent method is used as a method to minimize the error. To find the minimum, the movement must be in the direction of the antigradient.

The gradient of the loss function is a vector of private derivatives calculated by formula 4, where $\nabla E(W)$ is the gradient of the loss function

$$\nabla E(W) = \left[\frac{dE}{dw_1}, \ldots, \frac{dE}{dw_n} \right] \tag{4}$$

A derivative error function for a specific image can be written using formula 5

$$\frac{\partial E}{\partial w_{ij}} = \frac{\partial E}{\partial y_j} \frac{\partial y_j}{\partial s_j} \frac{\partial s_j}{\partial w_{ij}} \tag{5}$$

The neuron error $\frac{\partial E}{\partial y_j}$ is usually written as an δ (delta) symbol. Let us determine the error of the output layer [5]. To do this, let's calculate a derivative of formula 1. We will get $t - y$ what is the difference between the desired and the obtained output.

Error δ for the hidden layer is calculated using formula 6, where δ_i - neuron error i of the hidden layer; δ_j - neuron error j of the next layer

$$\delta_i = \frac{\partial y_i}{\partial s_i} \sum_j \delta_j w_{ij} \tag{6}$$

The algorithm of error propagation is reduced to the following stages: direct signal propagation through the network, neuron state calculation, error value δ calculation for the output layer, reverse error propagation, update of network weights [12], [16].

IOP Conf. Series: Materials Science and Engineering **873** (2020) 012003 doi:10.1088/1757-899X/873/1/012003

4. Simulation of EEG processing using a neural network

The method we proposed was implemented in MATLAB R2017b. We faced the task of determining hand movement on the basis of EEG, which we solved by teaching the neural network the inverse gradient distribution of error. Figure 4 shows the first 1200 EEG signals performed using a standard 10-20% protocol with 24 sensors the first 4 sensors.

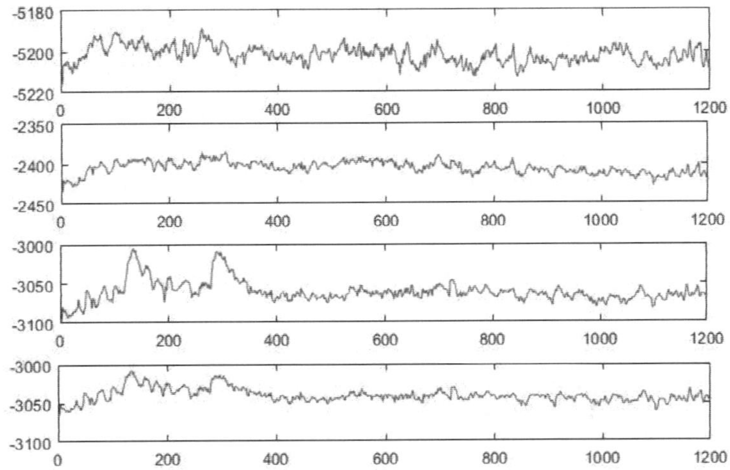

Figure 4. The first 1200 EEG signals: sensors T5, T3, F7, F3

To train the neural network, vectors were created to denote hand movements. The first vector (1,0,0) denotes movement of the right hand, the second (0,1,0) - movement of the left hand, and the vector (0,0,1) - state of calm. Training is structured in such a way that some part of data is used by the neural network for self-study and the other part is used for its evaluation. In figure 5 shows the result of neural network training conducted in the mathematical environment MATLAB R2017b.

We trained the neural network with 1000 iterations and stopped training after the network made 6 consecutive errors. To evaluate the correctness of the network training, you must make sure that the root-mean-square error decreases with each iteration.

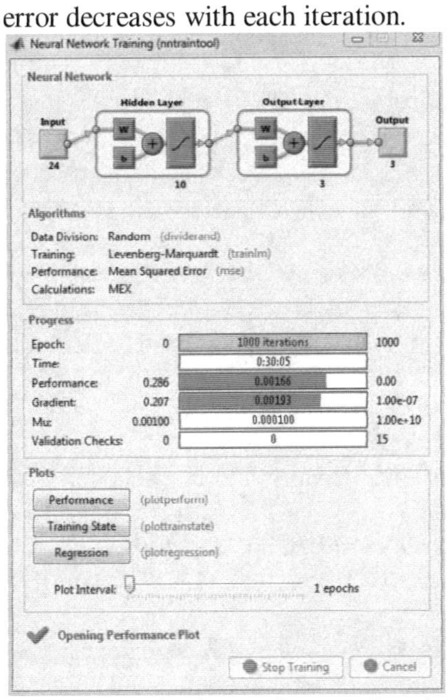

Figure 5. The result of neural network training in Matlab

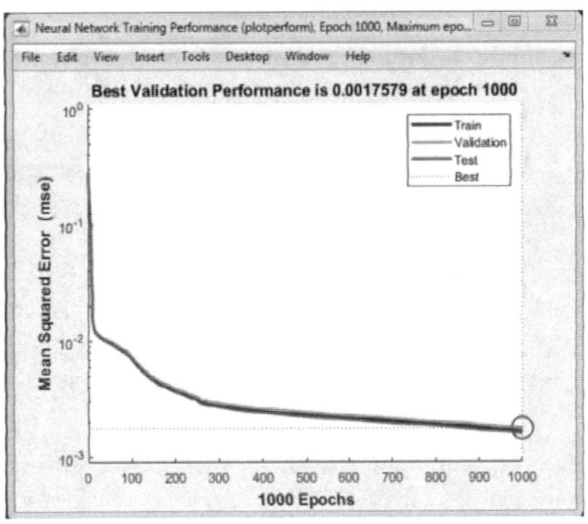

Figure 6. The MSE during training of neural network.

The figure 6 shows a graph of the process of training the neural network with 1000 iterations and reduction of the standard error with each iteration, which allows to draw a conclusion about correct training of the network.

5. Conclusion

The article shows the application of neural network training by the algorithm backpropagation for EEG data processing. It is shown that such training can be used for operation of BCI. A good result of the training is obtained with $MSE \approx 10^{-3}$. The BCI can be applied in human life, for example, in medicine to create the human exoskeleton or bionic prosthetics.

Acknowledgements
This work was supported by the Russian Foundation for Basic Research (projects no. 19-07-00130 A and no. 18-37-20059 mol-a-ved), and by the Presidential Grant of the Russian Federation (projects SP-2245.2018.5 and SP-126.2019.5).

References

[1] Jacques J Vidal 1973 “Toward direct brain-computer Communication“ *Annual Review of Biophysics and Bioengineering* Vol 2 (USA: Annual Reviews) pp 157-180
[2] Wolpaw J R, Birbaumer N, McFarland D J, Pfurtscheller G and Vaughan T M 2002 Brain-computer interfaces for communication and control *Clinical Neurophysiology* (Netherlands: Elsevier) pp 767–791
[3] Zander T O; Kothe C 2011 Towards passive brain-computer interfaces: Applying brain-computer interface technology to human-machine systems in general *Journal of Neural Engineering* (United Kingdom - SIR icon SIR Ranking of United Kingdom)
[4] Daly I, Scherer R, Billinger M, Müller-Putz G 2015 FORCe: Fully Online and Automated Artifact Removal for Brain-Computer Interfacing *IEEE Trans Neural Syst Rehabil Eng*. Sep 23(5) pp 725-736
[5] Alfaro M, Argüelles A, Yañez C and Chairez I, 2012 "Continuous Neural Networks for Electroencephalography Waveform Classification," *VI Andean Region International Conference* (Spain: Cuenca) pp 153-156
[6] Jasper H H 1958 The ten-twenty electrode system of the International Federation *Clinical Neurophysiology* (Netherlands: Elsevier) p 375
[7] Ramzan M and Dawn S 2019 "Temporal measures for analysis of emotional states from human electroencephalography signals *Twelfth International Conference on Contemporary Computing* (India: Noida) pp 1-6
[8] Jensen O 2005 On the human sensorimotor-cortex beta rhythm: Sources and modeling *NeuroImage* (Amsterdam, Netherlands: Elsevier) vol 26 pp 347-355
[9] Karimzadeh F, Boostani R, Seraj E and Sameni R 2018 "A Distributed Classification Procedure for Automatic Sleep Stage Scoring Based on Instantaneous Electroencephalogram Phase and Envelope Features," *IEEE Transactions on Neural Systems and Rehabilitation Engineering*, vol 26 pp 362-370
[10] Wu Q, Zhao W and Qiu T 2018 "Using Human Electroencephalography to Determine Word Interpretation via an Artificial Neural Network," *IEEE International Conference on Systems, Man and Cybernetics (SMC)* (Japan: Miyazaki) pp 3620-3624
[11] Jinwon An and Sungzoon Cho, "Hand motion identification of grasp-and-lift task from electroencephalography recordings using recurrent neural networks," 2016 International Conference on Big Data and Smart Computing (BigComp), Hong Kong, 2016, pp. 427-429.
[12] Haykin S 1999 *Neural Networks: A Comprehensive Foundation (2nd Edition)* (USA: Prentice Hall International)
[13] Kandel E R, Schwartz J H and Jessell T M 2000 *Principles of neural science* (USA: Appleton & Lange)

[14] Barron A R 1994 Approximation and estimation bounds for artificial neural networks *Journal of Machine Learning Research* (USA: JMLR, Inc., MIT Press, Microtome Publishing) vol 14 pp 115–133
[15] Rumelhart D E, Hinton G E, Williams R J 1986 Learning internal re presentations by error propagation *Parallel distributed processing* (USA: Fullerton) vol 1. p 31 - 362.
[16] Gonzalez R C, Woods R E 2008 *Digital Image Processing 3rd Edition* (USA: Prentice Hall International)

Digitalization as a driver of development of domestic animal breeding

T E Marinchenko

Russian Research Institute of Information and Feasibility Study on Engineering Support of Agribusiness, the Federal State Budgetary Scientific Institution (Rosinformagrotekh FSBSI), U60 st. Lesnaya, Pravdinsky Township, Pushkinsky District, Moscow Region, 141261, Russia

E-mail 9419428@mail.ru

Abstract. The aim of the study is to analyze the level of digitalization of livestock production, determine the prospects of the process against the background of their existing strategic documents and the level of domestic developments. The study was based on the program documents, scientific publications, and data on Russian developments. The study used the monographic methods, comparative and system analysis methods, idealization and mental modeling methods, as well as a logical approach. Since the government has posed a challenge to ensure the entry of the Russian Federation in the five largest economies in the world, the transformation the agriculture through the introduction of digital technologies and platform solutions, the issue of digitalization of agriculture and livestock as less advanced in matters of digitalization is very relevant.

Keywords: agriculture, production, digitalization, digital technologies, market, state programs, trends, prospects

1. Introduction

Currently, digital data is becoming the key factor in production. This data comes from various devices and sensors located in a field, at a farm, in equipment, from weather stations, drones, satellites, and from external systems throughout the production chain. The data acquired and processed allows obtaining information of a new quality, finding patterns, forecasting while using modern big data processing techniques , minimizing risks, optimizing production and logistics costs, creating added value for all participants in the production chain, and ensuring the sale of products.

The modern model of technological development provides for the accelerated development and modernization of the Russian economy through the widespread development of the basic innovations of the fifth and accelerated transition to the sixth technological structure, the core of which is nano-, bio- and information and telecommunication (digital) technologies.

The “digital economy” term means an activity in which the key production factors are digital data, its processing and use in the modeling of production processes. The data can significantly improve the efficiency, quality and productivity of production, storage, sale, delivery and consumption [1].

To accelerate the penetration of digitalization in production, a national program entitled "Digital Russian economy" is implemented, which aims to promote the transformation of priority sectors of economy including agriculture through the introduction of digital technology and platform solutions, to accelerate technological development, to increase the number of organizations implementing technological innovations up to 50 % of their total amount, to ensure the entry of the Russian Federation in the five largest economies in the world, to create in the basic sectors of the economy, particularly in the manufacturing industry and the agricultural sector, a highly efficient export-oriented sector that would use modern technology and provided with highly qualified personnel [2].

Efficiency of the domestic livestock is significantly lower than that in the developed countries, therefore the increase in production efficiency through the introduction of modern digitalization technologies becomes operational need that allows increasing the competitiveness of agricultural producers

A considerable amount of works of domestic and foreign authors are devoted to various aspects directly or indirectly aimed at solving the problems of agriculture digitalization; the issues of the public support, positive implementation experience, and producers' challenges are widely covered [3-14].

The purpose of the study is to analyze the level of digitalization penetration in the domestic animal husbandry, to identify the prospects for the process in the framework of the tasks set by the government and the presence of domestic developments.

2. Materials and Methods

A base for the study was the departmental program titled "Digital Agriculture", scientific publications on the problems of digitalization of agricultural sector, data regarding Russian developments in the field of digitization of livestock. The study used the monographic methods, comparative and system analysis methods, idealization and mental modeling methods, as well as a logical approach.

3. Results

Russia occupies the 15th place in the world in the general level of digitalization and the 45th place in the world in the level of the information technology penetration in agribusiness. Based on estimates, only 13-15 % of producers are able to implement the digital e technology and commercialize the scientific and technical inventions.

The effectiveness of domestic agriculture is noticeably inferior to the largest economies. According to expert forecasts, 25 % of the global economy will switch to the implementation of digitalization technologies by 2020. Smart farming and precision farming programs are available in dozens of countries. The pace of implementation of "artificial intelligence" technologies in the agricultural sector is growing by 22.5 % annually. According to the MarketsandMarkets research company data, the size of this market will be $2.6 billion by 2025 [15-17].

Upon an initiative of the Russian Ministry of Agriculture, a framework of a national platform titled "Digital Agriculture" has been developed along with a preliminarily comprehensive examination of the current state of the level of digitalization of basic processes, and an algorithm for the digitalization of the state information resources has been developed too.

Digital data to be processed by modern software tools is considered as one of the key assets of the industry. The national platform called Digital Agriculture to be created, which is integrated with platforms of related industries, will accumulate information, industry best practices and models, as well as provide access and data processing services, which will significantly speed up the process and allow achieving a synergistic effect. The platform should become the basis for building an ecosystem of digital services and services in the agricultural sector [16].

A platform, integrated with sub-platforms of the regional and municipal levels, should provide a technological breakthrough and double labor productivity by 2024 [3].

The analytical center created for the project should determine the most effective project parameters, develop standard projects and stimulate their implementation, for which an information platform is being formed that aggregates 13,000 indicators for agricultural enterprises. The interaction of the federal system with 16 regional IT systems has been worked out.

The market for information technologies in the agribusiness is estimated at more than 360 billion rubles. The digitalization level is projected to grow five-fold by 2026, including through start-ups. The ultimate goal of the Digital Agriculture program is to develop and launch replicable integrated innovative projects of end-to-end intelligent systems called “Country”, “Region”, “Agricultural enterprise”, “Field (Farm)” based on domestic methods, algorithms, digital technologies, and samples of systems and devices [3, 16]

The Ministry of Agriculture of Russia involved the following databases in the implementation of the project: TsIAS SKh (Central Information and Analytical System of Agriculture) and EFIS ZSN (Unified Federal Information System of Agriculture).

TsIAS SKh is a data bank integrated with the information systems of the Ministry of Agriculture of Russia, Rosstat, Federal Customs Service, Roshydromet that has a functionality of the analysis for operational monitoring of the status and development of agribusiness facilities.

EFSIS ZSN has formed the basis of the “effective hectare” area based on BigData, is integrated with the databases of Federal Agency for State Registration, Cadastre, and Cartography, and Roscosmos, which provides a map of agricultural land with a high level of verification. The service will allow, by 2021, to introduce intelligent industry planning in the entities, which involves differentiated cultivation using adapted technologies of the most profitable crops for specific areas of agricultural land and the rational use of logistics infrastructure. The “effective hectare” will allow real-time monitoring and modeling the export flows of agricultural raw materials taking into account many factors. Integration with the databases of Roshydromet and Agrochemical Centers will make it possible to forecast harvests and harvesting dates, which will form the basis of the export area called “from the Field to the Port”. Such a model will take into account the forecast flows of raw materials considering the yields and the transport shoulder, that is to say, the rolling stock of Russian Railways. By 2024, the export of products will be accompanied by a paperless system "from the Field to the Port."

The Smart Contracts project provides a system of intellectual measures of state support and personal accounts of subsidy recipients with electronic document management and package solutions for agribusiness (subsidy + loan + insurance). Integration with the databases of Roshydromet and EMERCOM will allow adjusting subsidies in connection with emergencies. The contracts are expected to be available in the SMART mode Since 2021.

It is planned to implement the project of “smart” distribution of food production and regional differentiation, taking into account the “health map” of the population titled “Industrial FOODNET” based on data on the lack of vitamins, minerals, antioxidants in the framework of public-private partnership with transnational corporations.

The “Digital Agriculture” will be implemented through pilot regional scalable and replicable projects of end-to-end scientific and production cooperation in the concept of “Digital Agriculture” based on domestic integrated digital agricultural solutions: “Smart Farm”, “Smart Field”, “Smart herd”, “Smart greenhouse”, “Smart processing”, “Smart warehouse” and “Smart agricultural office”, which will contribute to the economic, social and technological development of the regions. It is planned that, by 2024, all agricultural data will be consolidated on a single platform, all agricultural producers will become users of digital services, and agricultural product traceability will reach 80 %. [2, 16].

The objective of the departmental project is the digital transformation of agriculture through the introduction of digital technologies and platform solutions to ensure a technological breakthrough in the agricultural sector and to achieve a two-fold increase in productivity at digital agricultural enterprises by 2024 (table 1).

Table 1. Target indicators of the departmental project titled "Digital Agriculture" [2]

Indicator	Years					
Share of data on agricultural resource facilities included in the "Digital Agriculture " digital platform, %:	2019	2020	2021	2022	2023	2024
agricultural land (of the total area of agricultural land)	50	75	9 0	100	100	100
working and productive livestock (from the total number of livestock in the category)	25	35	5 0	75	90	100
farm machinery (the total number of units)	45	60	75	90	100	100
Growth rate of labor productivity (for enterprises that have introduced digital agricultural solutions), %	105	125	150	175	190	200
Ratio investment in digital products and technology of the total investment, %	1	3	7	10	15	25
Including that in domestic development, %	0.5	1.5	5	7	10	20
Share of SMART contracts with recipients of subsidies, %	5	25	50	75	100	100
Share of regions that have implemented planning based on the Digital Agriculture platform, %	0	6	29	59	100	100
Share of material costs in the cost of production (for enterprises that have introduced digital agricultural solutions), % of the cost	60	55	50	47	45	43
Share of specialists in the field of the digital economy, % (of the total number of specialists, on an accrual basis), %	10	15	20	30	40	50

The total economic effect from the switch of farms to digitalization-based business models can amount to more than 4.8 trillion rubles. in annual terms, the growth of labor productivity can be increased three to five times, and the growth in the consumption of information technology due to the digitalization of the agricultural sector can reach 22 % [18].

The level and rate of digitalization in livestock and crop production is markedly different. So, in 2017, the structure of investments in IT equipment included the share in livestock production amounted to only 7 % of the total costs in the agricultural sector, while in crop production was 51 % [19].

The main consumers of digital technologies in animal husbandry are large manufacturers. Demand for automated milking, feed preparation and herd management systems is growing annually. All digital innovations in animal husbandry to be offered on the world market are available in our country. In addition to robotic milking machines, there is a great demand for electronic heat period detection systems.

Almost all components of digitalization of management in livestock are integrated into herd management systems. The main requirements for such systems are switching the visual monitoring to digital one through the measured parameters; minimization of the human factor when performing production operations and entering information; transition from reactive to preventive management; minimization of the impact of individual objects of management on production indicators. The creation of "intelligent farms" envisages the development and implementation of automated centralized management through process control subsystems (feed production and feeding, herd reproduction, animal veterinary service, milking, microclimate, manure removal, etc.); automation of most operations of specialists (veterinarians, livestock specialists, engineers); information and analytical units for product quality assessment, logistics planning, etc. [20].

An example of digitalization at the state level is the creation of the Mercury system projected by the Federal Service for Veterinary and Phytosanitary Surveillance to provide electronic veterinary accompanying documentation. This system was launched on July 1, 2018. Currently only part of the planned "from field to counter" chain concerning raw materials is working The system will work in full after it has spread to finished products.

It is now under development the integration of information systems of the Federal Service for Veterinary and Phytosanitary Surveillance and the veterinary services in a digital platform for interfacing the monitoring and supervisory systems in the business management systems for business entities for the purpose of identification and traceability of animals and for inclusion of the complete livestock production cycle in the digital end-to-end chains.

The use of digital, automated production and management technologies in pig and poultry farming allowed for a dynamic growth of production volumes in difficult economic conditions and with increasing veterinary risks.

In the future, digital identification systems and sensors for identifying the physiological state of animals will be integrated into domestic Selex systems and will become tools for valuation work along with data processing and provision in electronic form, monitoring of the physiological state and treatment of animals, automated control of the quality of milk in the stream at milking plants (protein, fat, somatic, electrical conductivity, etc.), and non-contact remote monitoring of animal behavior.

Significant progress in the field of significance of digitalization in the selection of farm animals in the world is, in recent decades, associated with the development and implementation of technologies of gene and genomic selection. Creating DNA passports of animals requires the development of methods and test systems that allow for the genetic differentiation of breeds, types and lines of animals with high accuracy [21].

The process of digitalization of selection in dairy cattle breeding is actively ongoing in the Orenburg, Volgograd, Bryansk, Sverdlovsk, Chelyabinsk, Tyumen, Kurgan, Novosibirsk, Samara, Rostov, Astrakhan regions, the Stavropol Territory, the Altai Republic, and in the Udmurt Republic.

The studies that have being performed over the years have allowed for the development in the Russian Federation of national systems for the genetic identification of animal species compatible with the systems of pedigree cattle importing countries.

Digitalization in livestock farming will allow agricultural producers to integrate into the global space using world standards for meeting the requirements for product quality and traceability. Specialists predict the fragmentation and specialization of farms in the production of milk, reproduction of the herd, cultivation of livestock, fodder storage, slaughter, transportation and sale of products, etc. [22, 23].

4. Discussion

Digitalization of processes related to the maintenance and reproduction of animals, milk production, information collection and database generation, the development and prompt refinement of algorithms and digital models provide objective, complete and efficient "digital" display of facilities and their management. The need for constant visual observation and control is minimized. Management adjustments are promptly made.

The production and economic results, that is to say, calf yield and milk production, are increased, feed conversion is improved, insemination costs, losses due to culling and the cost of treating animals are reduced.

In livestock farming, the rate of use of digital technologies is much lower; however, there is a positive experience of farms using information from all stages of complex production processes based on a single “digital platform”.

5. Conclusions

The digital transformation of the agribusiness has become an object of increased attention of both the state and manufacturers; the formation of a high-tech infrastructure with access to all market participants is ongoing.

The technology of "precision" production in agriculture and livestock breeding is becoming a strategic necessity, the state supports its implementation, software products and platform solutions have been created, including domestic ones that have proved their effectiveness.

In the short term, the transition to advanced digital, intelligent manufacturing technologies and robotic systems should be considered as a development priority. The project developed by the Russian Ministry of Agriculture provides for the end-to-end digitalization of all data, allows simplifying the collection of statistical information, monitoring the system and promptly applying preventive measures against the system destabilization, predicting more accurately and simulating raw material and commodity flows with a multiple reduction in logistics costs, speeding up communication between government bodies and manufacturers, and also simplifying the receipt of government support measures. The system must be dynamic and transparent for all platform participants.

The basis of "digital agriculture" is not only statistical data and information from sensors, but ready-made mathematical models of production and marketing processes, which include the entire value chain of goods, allow planning volumes, quality, and profitability in geographical and time dimensions.

References

[1] Buklagin D 2017 *Fifth technological structure: place of the Russian agribusiness* (Economic analysis: theory and practice) V16 (Issue 1) pp 19-35.

[2] *Digital transformation of agriculture in Russia* 2019 (official publication, Moscow: Rosinformagrotekh) p 80

[3] *Departmental project "Digital Agriculture"* 2019 (official publication, Moscow: Rosinformagrotekh) p 48

[4] Ognivtsev S 2019 *Digitalization of the economy and the economy of agribusiness* (International Agricultural Journal) No 2 (368) pp 77-80 DOI: 10.24411/2587-6740-2019-12034/

[5] Guzueva E R, Vezirov T G, Beybalaeva D K., Batukaev A A and Chaplaev Kh G 2020 *The impact of automation of agriculture on the digital economy* (IOP Conference Series: Earth and Environmental Science) Vol 421 Issue 2 022047 DOI: 10.1088/1755-1315/421/2/022047

[6] Sokolova A P, Litvinenko G N 2020 *Innovation as a source of agribusiness development* (IOP Conference Series: Earth and Environmental Science) 421(2) 02205 DOI: 10.1088/1755-1315/421/2/022053

[7] Panov A, Panova N, Malofeev A, Nemkina E 2019 I*nteraction of regional agribusiness entities in the transition to a digital economy* (IOP Conference Series: Earth and Environmental Science) 403(1) 012138 DOI: 10.1088/1755-1315/403/1/012138

[8] Ivanov P A, Kornilova L 2018 *Financing innovation in agriculture: the basis of its digitalization* (*Accounting, analysis and audit in a digital economy:*Proceedings of All-Russia Scientific and Practical Conference) pp 255-261

[9] Khabarov V, Volegzhanina I 2019 *Knowledge management system of an industry-specific research and education complex* (IOP Conference Series: Earth and Environmental Science). 403(1), 012197 DOI: 10.1088/1755-1315/403/1/012197

[10] Serbulova N, Kanurny S, Gorodnyanskaya A and Persiyanova A. 2019 *Sustainable food systems and agriculture: The role of information and communication technologies* (IOP Conference Series: Earth and Environmental Science) 403(1) 012127 DOI: 10.1088/1755-1315/403/1/012127

[11] Boev V U, Ermolenko O D, Bogdanova R M, Mironova O A and Yaroshenko S G 2020 *Digitalization of Agro-Industrial Complex as a Basis for Building Organizational-Economic Mechanism of Sustainable Development: Foreign Experience and Perspectives in Russia* (Lecture Notes in Networks and Systems) 87 pp 960-968 DOI: 10.1007/978-3-030-29586-8_109

[12] Gravshina I N, Denisova N I, Kuzmin V N 2019 *On the issue of improving the competitiveness of agricultural products of the Ryazan Region in the conditions of digital transformation. (*Agribusiness: Economics, Management) No 11 pp 77-83

[13] Semkin A .G 2019 *Strategic areas of the development of the regional agribusiness management*

system (Economics of agriculture in Russia) No 2 pp 22-27
[14] Butyrin V V, Butyrina Yu A 2019 *Areas of digital transformation of agriculture* (Economics of agriculture of Russia) No 6 pp 9-14
[15] Marinchenko T E 2019 *Digital transformation of agriculture (Modern agrarian economy: problems and prospects in the development of digital technologies:* Proceedings of All-Russia Scientific and Practical Conference) pp 69-73
[16] *Departmental project: "Digital Agriculture"* [Electronic resource] URL: www.agro.cap.com/usercontent/minselhoz/news/2018_12/03/4fb5c435-356c-4e3a-9279-bfca943c459c/mcx.pdf
[17] Marinchenko T E 2019 *Monitoring of innovation activities in the agricultural sector* (Machinery and equipment for rural area) No 1 pp 40-46.
[18] *Automation of the Russian agricultural sector: realities and prospects* [Electronic resource] URL: https://foodretail.com/news/avtomatizatsiya-rossiyskogo-agrosektora-realii-i-perspektivi-401250
[19] Presentation of the Ministry of Agriculture of Russia (2018). URL: https://www.sas.com/content/dam/SAS/ru_ru/doc/Events/Presentation/agro-bb-2018/2-gerasimov-digitalization-of-the-agroindustrial-complex.pdf
[20] *The concept of "Scientific and technological development of digital agriculture. Digital Agriculture"* [Electronic resource] URL: https://www.coursehero.com/file/49273324/97d2448548e047b0952c3b9a1b10eddepdf/
[21] IoT in Russia: what do experts think? [Electronic resource]. http://www.csr-nw.ru/files/publications/iot_in_russia.pdf.
[22] Marinchenko T E 2019 *Digitalization as a driver of the agribusiness technological development* (Status and prospects of the agribusiness development, Proceedings of the XII Int. Scientific and Practical Conf) pp 30-34
[23] Salnikov S G *Actual areas of the Russian agribusiness digital transformation* [Electronic resource] URL: http://www.viapi.ru/

Expert assessment of agrotechnologies with positioning key components

N N Krupina
Saint-Petersburg State Agrarian University, Faculty Rural Development and Management, 2 Petersburg highway, St. Petersburg, Pushkin, 196601, Russia
E-mail Krupina_n17@mail.ru

Abstract. Fuzzy expert assessment of the current state and prospects for changing technologies is based on knowledge of the subject area of research, understanding of foresight priorities, and the choice of components whose loss leads to maximum damage to functionality. The relevance of the structural analysis of technologies is justified, the basic components and their weights are highlighted (by the example of poultry farming), and the positions of agricultural technologies are defined in the coordinates «level of innovation – impact on cost» are determined. Basic components of the technology are input and output flows; volume-space parameters of objects; technological mode of operations, materials, reagents, biologics; machines, equipment, vehicles; means of automation and computerization of quality control and management; staff competencies. A graphic technique of technology choice in the conditions of market uncertainty, constant risk and instability of agricultural production conditions is considered. The results of the study can serve as a working tool to facilitate the construction of the expert's chain of reasoning.

Keywords: input and output flows, technological mode, spatial parameters, machines and equipment, materials, reagents, biologics, technological innovations, costs, technological analysis.

1.Introduction

A fuzzy expert system uses the representation of knowledge in the form of linguistic variables, so the expert should have a detailed idea of the essential nature, value, orientation and degree of influence of technology components on the results of activities [1]. It is important to identify the components, the loss of which leads to the maximum loss of functionality, and then clearly connect the linguistic values of the ratings with specific numbers on the rating scale. In addition, the structuring of the object allows you to build more reliable aggregated assessments and develop fairly universal assessment programs. In fuzzy expertise, knowledge of the research subject area and understanding of foresight priorities are especially relevant, which inevitably increases the importance of technology expertise [2].

The author's goal is the development of theoretical ideas about the structure of agricultural technologies and the identification of components that form specific relationships and impacts, which can be useful in the work of specialists to achieve a coordinated aggregate expert assessment.

2.Informal statement of the problem

Let us formulate the basic concept: agricultural technology is a highly organized system of means of production and staff competencies in conjunction with consciously implemented physical, mechanical and biological-chemical methods of exposure when growing and/or processing specific natural («living») labor objects (land, plants, animals, fish, birds, bees) reproduced systematically and sequentially in time and space in the prescribed manner (streams and cycles). It is customary to distinguish three levels of agricultural technology - extensive, basic and intensive.

The innovation process determines competitive advantages and the growth of agricultural efficiency in Russia, but its activity remains quite low (Table 1).

Table 1. Innovation in the agricultural sector of the Russian economy.

	Share of Innovations in the Issue, %	The Share of Organizations Implementing, %	
		Innovation	Technological innovation
Plant growing	1,5	4,2	3,8
Livestock	1,7	4,3	3.4
Mixed farming	1,2	1,5	1,8
Auxiliary operations	1,3	2,1	1,7
Total for the Russian economy	7,8	8,5	7,4

Out of the total number of progressive applied scientific and technical developments, less than 5% is realized, and selection is carried out based on the results of an expert assessment of the current state and prospects of technology change. These estimates are constructed as a chain of reasoning (rules) from the knowledge base. The constant risk, the uncertainty of the market environment, the instability of agricultural production conditions, the transition of the economy to the best available technologies (BAT, «best available techniques») set the task of increasing the reliability of the assessment.

The prerequisites of this study were the ideas of other scientists:

- about the «materialization» of technology as its distinctive feature, which ensures the reduction of risks and costs of enterprises [3];
- on the continuous improvement of the structural components of industry technologies [4];
- about the possibility of maximizing the potential of the commercializability of the technology through analysis of the level of satisfaction of the properties of the components with the requirements imposed by investors, owners, consumers, regulatory authorities, etc.;
- on management selection as a process of systematization of structural components to be analyzed; determining the range of their permissible changes and their impact on the indicators of the operating cycle [5];
- on the development of an assessment methodology through the introduction of technology output indicators (productivity, cost, energy intensity) and a simplified modular structure of the technology, which avoids listing a large number of options for used machines, pesticides and fertilizers in order to reduce inefficient methods [6];
- on the possibility of reducing energy costs , achieving profitability of production not less than 25%, forming competitive advantages and providing leadership in the market with the introduction of advanced agricultural technologies [7];
- about adaptive resource-saving agricultural technologies based on the maximum use of the soil-climatic potential of the area and the assessment of their energy efficiency, as the ratio of the energy intensity of the grown crop to the total energy costs of the technology [8].

3.Study results

3.1. Structure-forming components of agricultural technology

The study of the accumulated volume of scientific and practical results of adaptive agricultural technologies allowed us to identify eight structural components (Figure 1).

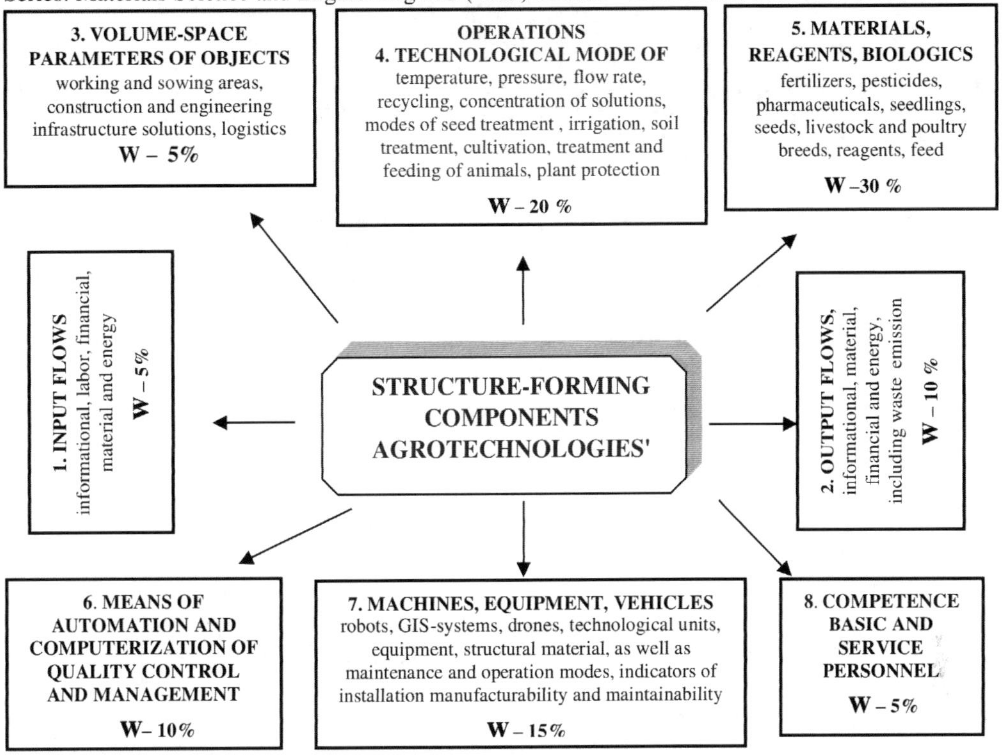

Figure 1. Structure-forming components of agricultural technology
(weight (W) is indicated by the example of intensive poultry farming)

Briefly we describe these components:

1. ***Input streams***. Through orientation and power, they mediate the temporal and spatial organization of technological operations, set the dynamics of the production process, determine energy consumption, material consumption, waste, safety and production efficiency. The material intensity of the technology determines its labor intensity and energy intensity, but sometimes less material-intensive products may have less useful effect (for example, they may be perishable or short- lived). This makes it necessary to evaluate the material-saving level of technology in each specific production by measuring the ratio of the mass of processed raw materials, the resulting product, the useful result and the negative external effect [9]. The use of combined units for soil treatment allows you to combine several operating flows in a single pass of equipment. Thus, the efficiency of dairy production is largely determined by the rational flow of harvesting, grinding, applying preservatives and distributing feed using a computerized multi-functional feed distributor. This organization of the cycle reduces the specific material consumption by 1.7 times and the specific energy consumption by 1.5-2 times [10].

2. ***Output stream***. Specify the output parameters of the main technology (yield, cost, energy intensity) and the specific values of emissions, discharges and waste generation that are achievable in the associated environmental technology («commercial product + externalities»). At the initial stages of using new technologies, information on input and output streams can be unstable, contradictory, insufficiently verified, and conflicting, which increases the demand for fuzzy logic that gives the expert a greater degree of freedom [11].

3. ***Volumetric-space parameters of objects.*** Complex biological processes are conservative, they determine the multistructure of the industry technological cycle and the presence of the limit of the possible maximum productivity of animals and crop yields. Volumetric-space characteristics of the technology are recognized as an industry feature of the reproduction cycle in agriculture. The placement of fields, gardens, greenhouses, livestock complexes and other structures and communications depends on the prevailing technological order, which often does not allow developers-innovators to focus on

minimizing the size of the area, reducing the density of development and optimizing transport routes. The space determines the sequence of operations, equipping the production with equipment, devices, tools, dosing and control means, forms of labor organization. So, according to the required nutritional area, there are crops that require a small up to 0.005 m^2 (legumes and grains), an average 0,1-0,2 m^2 (corn, sunflower) and a large 1,0-2,5 m^2 land area. According to the degree of development of the root system, there are cultures with a highly developed deeper than 2,5 m (corn, sugar beets), moderately developed up to 1.5 m (potatoes, rye) and an underdeveloped 0.7-0.9 m system.

4. ***Technological mode.*** These are the best working conditions for organizing the processing of objects of labor and the transformation of raw materials flows - temperature, pressure, acidity, aggregate state of the medium, volumetric and linear flow rates, concentration of the working substance, quality and frequency of seed treatment, animal feeding rates, composition, doses of fertilizers, pesticides, biological products, types of solvents, fillers, duration of working strokes and other. In real conditions, it is impossible to single out one significant parameter; therefore, so it is necessary to normalize different types of costs, develop a single criterion for evaluating efficiency, and select the initial set of technological factors at each stage of the cycle.

5. ***Materials, reagents, biological products.*** The use of highly effective herbicides, fertilizers, meliorants, and biological products is not an alternative within the framework of basic and intensive agricultural technologies in large- scale production: Complete exclusion, replacement with organic products, or «strict» restriction is possible in the production of a limited amount of environmentally friendly products.

6. ***Automation and computerization tools.*** The digital economy is a strategic guideline for the development of the industry. Automated systems make agricultural technologies self-organizing, ultra-flexible and realize the synergistic potential of innovative development, reduce risks. «Smart greenhouses» and «smart firms» demonstrate high labor productivity and return on intellectual capital, increasing the quality and safety of production in General [12]. The availability of reference databases makes it easier to select experts.

7. ***Machines, equipment, vehicles***. This is the most important multi-aspect and multi-level component of agricultural technology, forming its «core», often having a contradictory effect on production indicators and requiring a responsible approach from experts. Minimization of costs, capital productivity, cycle times, labor productivity, product quality are ensured by achieving the greatest conformity of the machine design to optimal conditions and production volumes in specific given circumstances. A few arguments:

- increasing the volume of the grain harvester's hopper increases the utilization rate of the machine, but a filled hopper increases the energy consumption for transporting grain, and when the soil is wet, it dramatically increases the technogenic impact of the running system, destroying the structure, reducing the bunker's occupancy by 25-35% and soil fertility. But placing the hopper directly to the crawler frame truck leads to synergy effects - the reduced weight of the harvester, enhanced stability, improved permeability, increasing storage tank capacity, reduces the grinding of grain and technogenic load on the soil [13];
- quality soil treatment depends on the design of the disks (flat, knife-shaped, wavy, spherical, spherical cut). The Great Plains company (USA) uses a spiral knife-shaped roller Turbo-chopper on soil processing units, the design of which differs in the angle of inclination of the knives along the radius of the roller, which improves the penetrating ability during the cutting of plant residues and prevents them from being pressed into the soil [14]. When processing high-stem crops, their splitting is achieved and the time of decomposition of plant residues in the soil is reduced, which prevents the spread of diseases and pests. The spiral shape of the rollers ensures constant contact with the soil, prevents its compaction, improves cutting action, reduces vibration and increases productivity [15]. However, for low-stem crops (up to 15 cm) or stony soil, these effects are not typical due to damage to the roller knife and clogging the space between the knives with plant residues and the soil;
- «precision farming» technologies minimize negative environmental effects by tightening the requirements for design and performance indicators of equipment, taking into account the yield, energy

and labor costs, quality of work, and uniformity of the load on the machine. The benchmark are units for making organic fertilizer into the soil , competitive positions which are determined by the types of grinding , dosing and distributing of the working bodies (the angle of the helix, the speed of rotation of the drum, the angle of sharpening a knife, the feed rate of the mass of fertiliser, working width, etc.).

Another mandatory aspect of the examination is the reliability of the equipment and the stability of the process as a whole. Experts evaluate three types of equipment defects that can be eliminated: on the move (without stopping the equipment); when the equipment is stopped, but without immediate stopping, and those that require immediate stopping. Knowledge of production information about possible failures, breakdowns, failures in operation is the necessary competence of the expert, allowing him to calculate the probability of failure-free operation of the equipment, its «throughput», therefore, the volume of sales and the final financial result.

Thus, the optimal geometric shapes of the equipment are characterized by a rational design, minimal consumption of material and energy while maintaining operational properties. Their high cost will be offset by lower operating costs and resource savings. Modern software tools allow for the design of spherical, cylindrical or other shaped containers to develop structural and functional maps indicating the element structure of the product (weight, volume, material, etc.) and its various functions (main secondary, main, auxiliary, useless), which is in demand by experts.

8. ***The competence of the staff.*** Technological modernization increases intellectual capital, changes the requirements not only for personnel qualifications, but also for the design of the production environment, workplaces, and work clothes, and strengthens the role of organizational culture. The priority is the unique ability of the intellect to penetrate deeply into the fundamental scientific basis of processes and to understand the physical, chemical, biological, social, psychological, organizational, and managerial patterns of technological development.

The indicator determining the weight (importance) of these components is of fundamental importance. Depending on the type of agricultural technology, the actual soil and climate conditions of its application , and the resource capabilities of the organization, experts use various algorithms for establishing weight coefficients (pairwise comparison, ranking, Delphi method, etc.). On the example of a large poultry farm (Stavropol territory, 2017) that implements intensive broiler farming technology, the weight coefficients of the technology components are calculated using the cost method (Figure 1). The total weight at the level of 50% is the technological mode of poultry cultivation and biological products used in breeding, feeding, and veterinary services. The technology does not require large areas and special training of personnel, but involves a certain cycle of drying, processing and disposal of chicken manure.

3.2. The dynamics of costs

So, advanced agricultural technologies are aimed at minimizing resource consumption, maximizing the reduction of negative biosphere impact, maintaining the reproduction of the natural productivity of biological assets and land fertility. These goals are not always achievable in different economies and territories. Therefore, experts conduct a comprehensive assessment of the entire chain of technological operations, for example, in agriculture, from preparing the soil for sowing to storing grown products, and Supplement the functional approach with an analysis of costs and efficiency. The selection of the essential components of the technology, the determination of the range and limits of their permissible changes are necessarily carried out according to the efficiency criterion the boundary of which is considered the allowable operating costs per unit of output for a particular operation. The calculation is based on the determination of the proportion of the complexity of the operation in the total cost of labor prices and the minimum level of profitability.

Experts can use graphical methods to justify the decision. Let's assume that we compare three corn growing technologies that differ in soil and climate conditions, the type of equipment , the mode of soil treatment, the quality of fertilizers used, and, as a result, the level of variable and fixed costs. The yield versus cost curves help you identify the optimal application area for each technology, as a segment bounded by segments that correspond to the lowest costs (Figure 2).

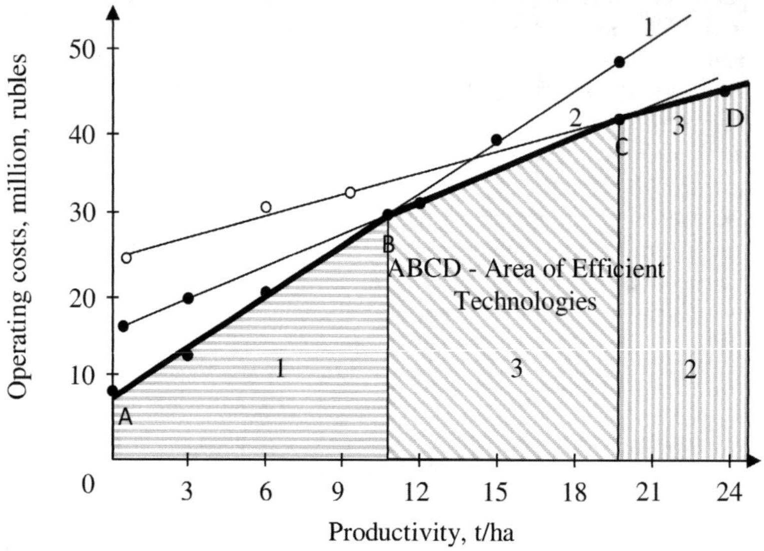

1, 2, 3 – alternative technologies

Figure 2. The dynamics of costs as yield increases under different regimes of tillage

The transition from one mode to another is determined by the goal setting, and technological operations serve as a tool for managing the process to achieve the planned yield and product quality while ensuring environmental safety and certain economic efficiency. For example, the cultivation of organic products due to the refusal from the use of chemicals and other methods of intensification of production, which reduces the productivity and directs technology1, under certain market conditions, it is advisable to limit the yield on average 12 – 20 kg/ha, applying technology 3 technology, and in terms of intensification of production – technology 2.

3.3. The matrix map

The portfolio technology analysis uses a two-dimensional technology matrix map built in the coordinates «technology attractiveness – market position of an organization» that implements a significant number of technologies. When building it, the ordinate axis shows the importance of the technology (relative efficiency, productivity compared to the accepted standard), the abscissa axis shows the degree of use of the technology compared to competitors and its impact on costs. All technologies are distributed in four quadrants:

1 - the most important and attractive with a very low degree of use;

2 - the most important and attractive with a high degree of use, determining the market value of the organization;

3 - quite old, but important and unattractive technologies that occupy a strong and stable position in the market, do not require significant investment and are characterized by high commercial returns;

4-technologies that have a weak commercial appeal and a weak position in the market, recommended for exclusion from the company's portfolio.

For the tasks of expert assessment, the known distribution should be supplemented, the list of possible types of technologies expanded, and the dominant component specified (Figure 3). In the agricultural sector, there are simultaneously livestock production with outdated technological systems («museum» and «forced» technologies) and a plant growing company equipped with modern technology with space navigation («smart field»).

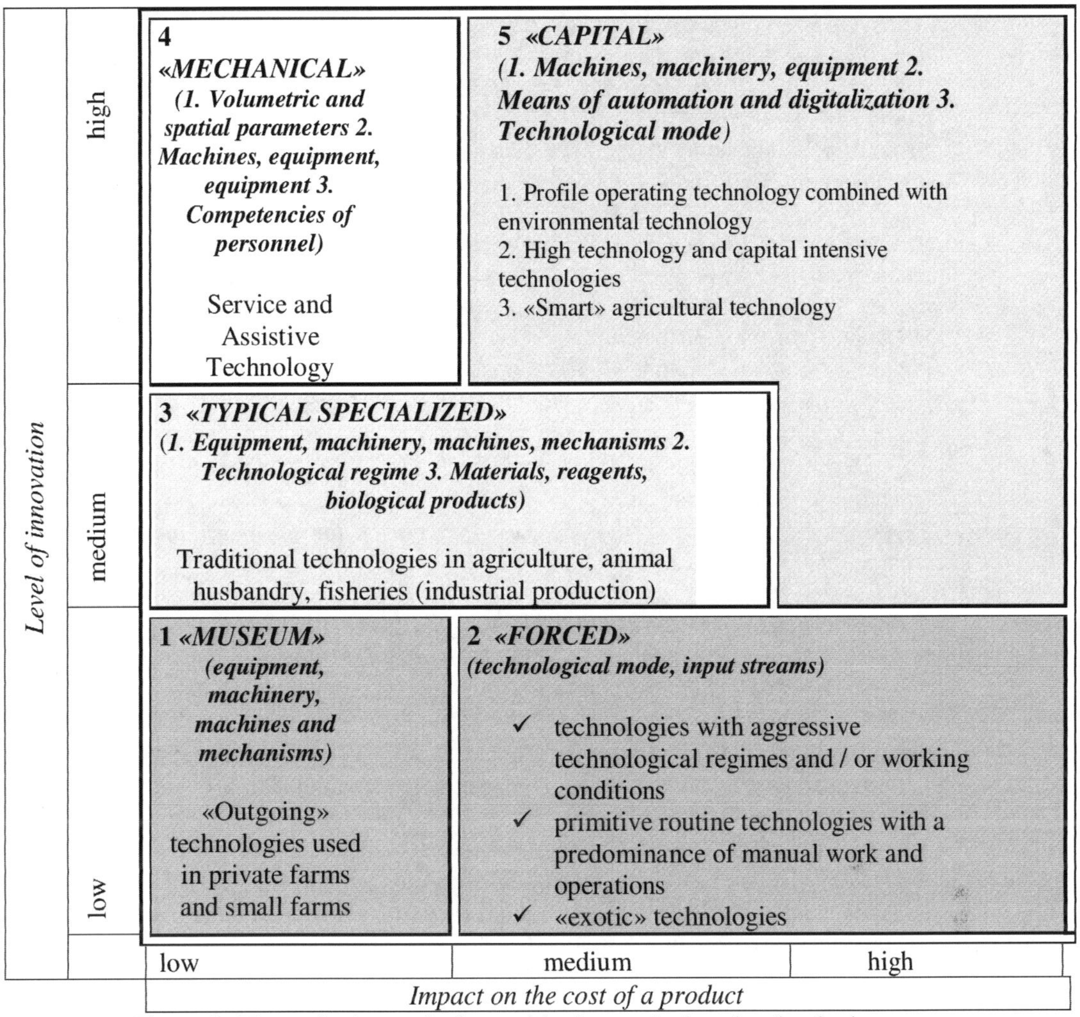

Figure 3. The author's matrix for positioning agricultural technologies

4. Conclusion

The presented structure of agricultural technologies is not exhaustive and comprehensive, it can be supplemented, and the components do not always clearly manifest and mutually penetrate each other. It illustrates the opinion of the outstanding thinker Manuel Castelis about technology as a method of using scientific knowledge to determine how to make things in a reproducible manner. It is the unconditional reproduction of components in each new modification of agricultural technology that allows them to be positioned as objects of comparison, which reflect the purpose and stability of the technological process, the adaptability of the technology to changing environmental conditions, resource provision, and safety. Understanding their essence makes it easier to build a generalized agreed point of view of experts about weighting factors; about factors that have a significant impact on the effectiveness of the technology.

References

[1] Zadeh L A 2008 *A Is there a need for fuzzy logic?* (Information Sciences Rlsevier) 178 pp 2751-2779.
[2] Maslennikov M 2017 *The Technologiical innovations and their Impact on the Economy*

(Economy of Region) 13 (4) pp 1221-1235.
[3] Denisyuk V and Markov F 2008 *On the formation of the market of advanced production technologies in the CIS* (Innovations) 7 pp. 11-17.
[4] Ivanova E 2012 *Technological capital and its system characteristics* (Economic analysis: theory and practice) 30 (285) pp 2-8.
[5] Brom A and Shuteev V 2012 *Multivariate approach to choose optimal technological support for machine-building complex* (News of Higher Educational Institutions) 12 pp 75-80.
[6] Cherkasov G and Pykhtin I 2014 *The urgency of creating the registers of cultivation technologies for oilseed crops* (Dostizheniya nauki i tekhniki APK) Vol. 28 (12) pp. 3-4.
[7] Kondratyeva O and Fedorov A 2017 *Domestic innovative machinery is the important factor of milk production efficiency increasing in the import substitution conditions* (Journal of VNIIMZH) 2 (26) pp 91-95.
[8] Pykhtin I and Gostev A 2016 *Theoretical Basis for Effective Application of Modern Resource-Saving Technologies of Grain Crops Cultivation* (Zemledelie) 6 pp 16-19.
[9] Voronkina D 2014 *Economic problems of the productive use of structural materials and methods for their solution* (Izvestiya of the Tula state University. Economic and legal Sciences) 5 pp 359-379.
[10] Skorkin V.K. and Karpov V 2012 *Research results for the multifunctional feed chopper* (Journal of VNIIMZH) 2 (6) pp 156-162.
[11] Herva M and Roca E 2013 *Review of combined approaches and multi-criteria analysis for corporate environmental evaluation* (Journal of Cleaner Production) vol. 39 pp 355–371.
[12] Kuznetsov B and Kuznetsova S 2016 *Technological Management in the Conditions of Scientific-Technological Revolutions of the XXIst Century* (Upravlenets/The Manager) 3 (61) pp 2-7.
[13] Candelya M and Ryabchenko V 2012 *The development of harvesters combines with new configuration of technological part* (Far Eastern Agrarian Herald) 3 (23) pp 72-75.
[14] Raper R and Simionescu P 2004 *Reducing Vibration while Maintaining Efficacy of Rollers to Terminate Cover Crops* (Applied Engineering in Agriculture: American Society of Agricultural Engineers) Vol. 20 (5) pp 581–584.
[15] Holland H and Debbeler A 2016 *Marktübersicht: So kriegen Sie Zwischenfrüchte klein* (Agrarmagazin) Vol. 134 (2) pp 99–101.

The methodology for calculating the interval of the shortwave radio link frequency correlation with the sphericity and small-scale inhomogeneities of the ionosphere

V P Pashintsev [1], N N Gakhova [2], A D Skorik[1], D V Alekseev[1], M A Senokosov[1]

[1] North Caucasus Federal University, 1, Pushkin Street, Stavropol 355009, Department of information security of automated systems
E-mail pashintsevp@mail.ru
[2] Belgorod State University, 85, Pobedy St., Belgorod, 308015, Department of applied informatics and information technologies, Russia
E-mail gahova@bsu.edu.ru

Abstract.The paper suggests the methodology for calculating the interval of the fading frequency correlation in the shortwave radio link with one discrete beam and its diffuse scattering. This methodology takes into account the effect of the sphericity and small-scale inhomogeneities of the ionosphere reflecting layer.

Keywords: short-wave coupling, ionosphere, sphericity, small-scale inhomogeneities, diffuse multipath, fading, frequency correlation interval.

1. Introduction

It is known [1] that in a shortwave (SW) radio link there can be one path (discrete beam) of the wave propagation in the reception point. In this case, the signal at the receiver input is almost always subject to the interference fading (amplitude and phase fluctuations) due to the diffuse scattering of a beam (a single wave) within small-scale inhomogeneities of the ionosphere reflecting layer F and the diffuse multipathing with the maximum relative delay time $\Delta\tau_{i\,\max} \approx 50\ldots200$ mcsec [1]. That is why the interval of the fading frequency correlation (or the band of undistorted transmission) of the single-beam SW radio link is constrained by the values $F_{\kappa} \approx 1/\Delta\tau_i$ =20…5 kHz. If there are several (2-3) paths (discrete beams) of wave propagation in the reception point, their relative delay time within the range of SW communication R =1500…4000 km is $\Delta t_i \approx 1\ldots3$ msec [1-3]. And the interval of the SW radio link frequency correlation with the discrete multipathing gets narrow up to $F_{\kappa} \approx 1/\Delta t_i$ =1…0,3 kHz. Since it is usually assumed [1-3] that the frequency correlation interval in the SW radio link is $F_{\kappa} \approx 0{,}3\ldots1$ kHz, the transmission rate $c_{\text{т}} = T_{\text{c}}^{-1}$ of simple binary signals (that is with the base $B_{\text{c}} = T_{\text{c}}F_0 = 1$, where T_{c} and $F_0 = 1/T_{\text{c}}$ are the signal length and width) does not usually exceed tens to hundreds bit/sec due to the necessity to meet the requirements of simultaneous elimination of the frequency selective fading (FSF) $F_{\kappa}/F_0 >> 1$ and the intersymbol interference (ISI) $F_{\kappa}/c_{\text{т}} = T_{\text{c}}F_{\kappa} = F_{\kappa}/F_0 >> 1$ [4, 5].

Using the known method [2] of the separation (radiation) of one discrete beam through narrowly directed receiving (transmitting) antennas enables providing the frequency correlation interval $F_к \approx 5...20$ kHz in the SW radio link. Moreover, according to [2, 3] the percentage of the lifetime of the discrete-single-beam propagation models in the SW radio link is quite high on long-distance radio paths (85% when the communication range $R = 3000$ km, 64 % when $R = 4000$ km and 31 % when $R = 1500$ km). That is why in single-beam SW radio links, the signal transmission rate can significantly increase (by one order of magnitude and greater) if there are no FSF and ISI ($F_к/F_0 = F_к/c_т >> 1$). To quantify the maximum available values $c_т$ in the single-beam SW radio link, it is necessary to determine the value of the interval of the fading frequency correlation $F_к$.

The interval of the fading frequency correlation $F_к$ in SW radio links with multipathing (both discrete and diffuse) is usually determined experimentally [1, 4]. There is the method [5] that enables estimating analytically the value of the frequency correlation interval $F_к$ in the single-beam SW radio link considering the plane ionosphere reflecting layer (with diffuse scattering) but with the Earth sphericity. The impact of the ionosphere sphericity on the various paths of the wave propagation is considered in works [7, 8], which is of great significance for long-distance SW radio paths. However, the method suggested in [7, 8] does not enable quantifying all the parameters of the wave path in the single-beam SW radio link that is necessary to determine $F_к$. Therefore, the known methods [6 – 8] require generalizing and specifying.

The goal of the paper is to develop the methodology to calculate the interval of the fading frequency correlation in the shortwave radio link with one discrete beam with the diffuse multipathing which considers the impact of sphericity and small-scale inhomogeneities of the ionosphere reflecting layer F.

2. Method for calculating the frequency correlation interval of a short-wave radio line

According to [6], the interval of fading frequency correlation in the single-beam SW radio link with the diffuse multipathing is described by the expression

$$F_к = f_0 / \left(\sigma_\varphi \sqrt{2 + d_1^2}\right), \tag{1}$$

where f_0 is the radio carrier (operating frequency) of the SW radio link [Hz]; σ_φ is the mean-square deviation of fluctuations (distortions) of the wave phase front at the output of the ionosphere reflecting layer F with small-scale inhomogeneities [rad]; $d_1^2 \geq 1$ is the diffraction parameter characterizing the rise in diffraction effects when the wave front removes from the reflecting layer of the inhomogeneous ionosphere to the reception point.

The mean-square deviation (MSD) of the wave phase front fluctuations at the output of the ionosphere layer F in the SW range is described by the standard expression common to transionospheric radio links with small-scale inhomogeneities of any frequency ranges [6, 9-11]

$$\sigma_\varphi = (80{,}8\,\pi/c\,f_0)\,\beta\,\overline{N(h_{re})}\sqrt{2\,l_0\,L_{eq}}, \tag{2}$$

where $c = 3 \cdot 10^8$ [m/sec] is the speed of light in vacuum; coefficient 80.8 is expressed in [m^3/sec^2] in SI system; $\beta \approx 10^{-3}...10^{-2}$ is the rate of small-scale ionospheric inhomogeneities; l_0 is the largest length of ionospheric inhomogeneities [m]; L_{eq} is the equivalent homogeneous wave propagation path in the ionosphere reflecting layer F [m] which corresponds to the length of the imaginary curved path of wave propagation in the "tube" from the input to the output of the ionosphere reflecting layer F with average electron concentration $\overline{N(h)} = const$ of constant (homogeneous) height (h) that corresponds to its maximum value at the height $h = h_{re}$ of wave reflection $\overline{N(h_{re})}$ [m^{-3}].

The diffraction parameter included in (1) is determined by the expression [6, 9,10]

$$d_{\mathrm{i}}^2 = \frac{3\,L^2 - 3\,L\,L_{eq} + L_{eq}^2}{6\left(2\,\pi\,f_0/c\right)^2}\left(8\,l_0^2\,l_{\mathrm{i}}^2\right)^{-1}. \tag{3}$$

Here, $L = L_{eq} + L_{fs}$ is the sum of the equivalent homogenous propagation path in the reflecting layer F and the wave propagation path in the free space L_{fs} [m] behind the ionosphere, that is from the output of the reflecting layer to the reception point; l_{i} is the smallest length of ionospheric inhomogeneities [m].

The wave operating frequency, that is at the angle φ_0 to the lower limit $h = h_0$ of the spherical reflecting layer F of the ionosphere, is linked to the frequency of the equivalent wave directed vertically $f_v = \left[80{,}8\,\overline{N(h_{re})}\right]^{0,5}$ by the known ratio [3, 6]

$$f_0 = f_v K_s \sec\varphi_0 = \left[80{,}8\,\overline{N(h_{re})}\right]^{0,5} K_s \sec\varphi_0, \tag{4}$$

where $K_s \le 1$ is the coefficient of the Earth and ionosphere sphericity.

If (4) is substituted in (2), the expression for the MSD of phase fluctuations at the wave front at the output of the spherical reflecting layer F of the ionosphere with small-scale inhomogeneities can be written as [6, 12-16]

$$\sigma_{\varphi} = \pi\beta f_0 \sqrt{2\,l_0\,L_{eq}}\,/c\,K_s^2 \sec^2\varphi_0\,. \tag{5}$$

According to (5), the MSD of the fluctuations of the wave phase front at the ionosphere output in the single-beam SW radio link rises with the increase of the operating frequency (and approaches to its maximum usable frequency), the rate of the ionosphere small-scale inhomogeneities β (which characterizes the degree of the ionosphere diffuseness [17] and can increase to the values $\beta \approx 10^{-1}$), the equivalent homogenous path of the wave propagation in the ionosphere reflecting layer L_{eq} and with the decrease of the Earth and ionosphere sphericity coefficient K_s and the angle of the wave incidence φ_0 to the lower limit of the spherical reflecting layer of the ionosphere.

Thus, to calculate the interval F_{κ} of the frequency correlation of the SW radio link with one discrete beam taking into consideration the sphericity of the reflecting layer F and diffuse scattering on the small-scale inhomogeneities of the ionosphere according to the expressions (1-5), the values K_s, L_{eq}, L_{fs} and φ_0 should be determined.

The expressions to calculate the coefficient of the sphericity and equivalent homogenous path of the wave propagation in the spherical reflecting layer F of the ionosphere of the single-beam SW radio link are given in [7]:

$$K_s = \left\{1 + \frac{2\,f_{\mathrm{cr}}^2\,C_1(1-C_1)}{f_0^2\cos^2\varphi_0}\left[1 - \left(1 - \frac{f_0^2\cos^2\varphi_0}{f_{\mathrm{cr}}^2 C_1^2}\right)^{\frac{1}{2}}\right]\right\}^{-\frac{1}{2}}; \tag{6}$$

$$L_{eq} = \frac{L_{\mathrm{F}}}{2}\left[1 + \frac{f_{\mathrm{cr}}^2 C_1(4\text{-}3C_1)}{f_0^2\cos^2\varphi_0} - \frac{2Z_{\mathrm{m}}(4\text{-}3C_1)}{l_g\cos\varphi_0}\right]K_s^2. \tag{7}$$

Here, $f_{\mathrm{cr}} = \left[80{,}8\,\overline{N(h_{\mathrm{m}})}\right]^{0,5}$ is the critical frequency of the ionosphere reflecting layer [Hz] at the height $h_{\mathrm{m}} = h_0 + z_{\mathrm{m}}$ with maximum mean value of the electron concentration $\overline{N(h_{\mathrm{m}})}$, where Z_{m} is the half-thickness of the ionosphere reflecting layer F [m];

IOP Conf. Series: Materials Science and Engineering **873** (2020) 012006 doi:10.1088/1757-899X/873/1/012006

$$C_1 = 1 - \frac{f_0^2}{f_{cr}^2}\frac{Z_m \sin^2\varphi_0}{R_E + h_0} \le 1 \tag{8}$$

is the coefficient that characterizes the rate of decline f_{cr} in the spherical reflecting layer F of the ionosphere as compared to the plane one; $R_E \approx 6370 \cdot 10^3$ [m] is the radius of the Earth;

$$L_F \approx L_g - \frac{L_g}{4\sec^2\varphi_0} - \frac{L_g}{4}\frac{f_{cr}^2}{f_0^2}C_1(4\text{-}3C_1) + \frac{Z_m(4\text{-}3C_1)}{2\sec\varphi_0}; \tag{9}$$

$$L_g = Z_m \frac{f_0}{f_{cr}} \ln\frac{1 + f_0\cos\varphi_0 / f_{cr}C_1}{1 - f_0\cos\varphi_0 / f_{cr}C_1} \tag{10}$$

is the actual (L_F) and group (L_g) paths of the wave propagation in the spherical reflecting layer F of the ionosphere [m].

Considering the geometry of the wave propagation in the free space behind the ionosphere, its path L_{fs} from the output point from the reflecting layer F to the reception point can be determined as

$$L_{fs} = \frac{R_E}{\sin\varphi_0}\sin\gamma = \frac{R_E}{\sin\varphi_0}\sin\left\{\arcsin\left[\sin\varphi_0\left(1 + \frac{h_0}{R_E}\right)\right] - \varphi_0\right\}, \tag{11}$$

where γ is the geocentrical angle [rad] that corresponds to the wave straight path from the reflecting layer output to the reception point of L_{fs} length.

To calculate the values C_1 (8), K_s(6), L_g (10), L_F (9), L_{eq} (7), L_{fs} (11), d_1^2 (3) and σ_φ(5) which determine the interval F_κ (1) of the frequency correlation of the SW radio link with one discrete beam with the sphericity and diffuse scattering of the reflecting layer F, it is necessary to determine the wave incidence angle φ_0 onto the lower limit (h_0) of the ionosphere reflecting layer F. To obtain the analytic expression for φ_0 considering the Earth and ionosphere sphericity is rather a difficult task. However, according to [18], the angle φ_0 can be determined if the functional dependence $R = \Psi(\varphi_0)$ of the SW communication range R (over the Earth surface) on the angle φ_0 for the known values of the operating frequency f_0 as well as the parameters of the ionosphere reflecting layer (h_0, Z_m, f_{cr}) and the specified range of communication $R = R_{pre}$ is previously found.

According to [18], the range of SW communication over the Earth R involves the sections R_1 and R_2 that correspond to the ionosphere and extra-ionosphere parts of its path and is determined as

$$R = R_1 + R_2 \approx \frac{R_E}{R_E + h_0}\sin\varphi_0 \frac{f_0}{f_{cr}} Z_m \ln\left[\left(1 - \frac{f_0^2}{f_{cr}^2}\frac{Z_m \sin^2\varphi_0}{R_E + h_0} + \frac{f_0}{f_{cr}}\cos\varphi_0\right) \Big/ \left(1 - \frac{f_0^2}{f_{cr}^2}\frac{Z_m \sin^2\varphi_0}{R_E + h_0} - \frac{f_0}{f_{cr}}\cos\varphi_0\right)\right] +$$

$$+ 2R_E\left[\operatorname{ctg}\varphi_0 - \left(\operatorname{ctg}^2\varphi_0 - 2h_0/R_E\right)^{\frac{1}{2}}\right]. \tag{12}$$

According the functional dependence (12) $R = \Psi(\varphi_0)$, when the operating frequency f_0 is given and the ionosphere parameters (h_0, Z_m, f_{cr}) are known, it is possible to determine (either in graphically or numerically) the wave incidence angle φ_0 onto the spherical ionosphere, which corresponds to the specified range of the SW communication $R = R_{pre}$.

Let us see how the interval of the frequency correlation F_κ (1) of the single-beam SW radio link for various values of the operating frequency f_0 is calculated based on the definition of the wave incidence angle φ_0 onto the lower limit of the ionosphere reflecting layer F according to the dependence (12) $R = \Psi(\varphi_0)$ when the communication range $R = R_{cr}$ is given and according to the expressions C_1 (8), L_g (10), L_F (9), K_s (6), L_{eq} (7), L_{fs} (11), d_1^2 (3) and σ_φ (5).

For the communication range $R_{pre} = 3000$ km and standard parameters [2, 3, 9-14, 16, 18] of the inhomogeneous layer of the ionosphere F2 ($h_0 = 300$ km, $Z_m = 200$ km, $f_{cr} = 5,7$ MHz, $l_0 = 390$ m, $l_i = 10$ m, $\beta = 10^{-2}$), the variations of the specified features of the wave in the single-beam SW radio link with various values f_0 are given in Table 1.

Table 1. variations features of the wave in the single-beam SW radio link with various values

f_0, MHz	φ_0, degrees	C_1	L_g, km	L_F, km	K_s	L_{eq}, km	L_{fs}, km	d_1^2	σ_φ, rad	γ^2	$F_к$, kHz
12.3	69.15	0.878	1168	1104	0.918	799	1061	$1.4\cdot10^5$	4.84	$7\cdot10^{-11}$	6.91
11.13	70.86	0.898	698	668	0.939	465	1254	$1.7\cdot10^5$	2.71	$6\cdot10^{-4}$	10.03
9.89	71.41	0.919	483	465	0.954	319	1346	$2.2\cdot10^5$	1.82	0.04	11.63
8.66	71.71	0.938	341	329	0.966	223	1408	$2.9\cdot10^5$	1.26	0.26	12.76
7.42	71.91	0.954	236	228	0.976	154	1454	$4\cdot10^5$	0.86	0.92	13.62
6.18	72.03	0.968	156	151	0.983	102	1490	$5.8\cdot10^5$	0.57	2.6	14.28

The analysis of the results given in Table 1 shows the following. As the operating frequency in the single-beam SW radio link drops against the value of the maximum usable frequency (MUF) $f_{MUF} = 12,37$ MHz, within the range from f_0=12,3…6,18 MHz there happens an insignificant rise of the angle φ_0, and in this case the specified communication range $R_{pre} = 3000$ km is ensured. And the values of coefficients C_1 (8) and K_s (6) increase approaching to 1 (this can be explained by the fact that the height of the wave reflection h_{re} decreases as well as the degree of the impact of the sphericity of the ionosphere reflecting layer). The lengths of the group and phase paths of the wave propagation in the ionosphere reflecting layer F L_g (10) and L_F (9) decrease and, as a result, the equivalent homogenous path (7) significantly decreases (from L_{eq}=799 km to 102 km). The decrease of values f_0 and $\sqrt{L_{eq}}$ in the numerator and the increases of K_s^2 and $\sec^2\varphi_0$ in the denominator of the expression (5) are responsible for the significant decrease of MSD of the phase fluctuations at the wave front at the output of the inhomogeneous reflecting layer of the ionosphere. The decrease of σ_φ with the operating frequency f_0 lowering ensures expanding the interval of the frequency correlation of the single-beam SW radio link (1) $F_к \sim f_0/\sigma_\varphi$ despite some increase of diffraction effects at the front of the received wave (that is the increase of d_1^2 (3) due to the rise of L_{fs} (12)).

The diagrams of the dependences of the MSD of phase fluctuations at the wave front at the output of the ionosphere $\sigma_\varphi = \psi(f_0)$ and the interval of the frequency correlation of the single-beam SW radio link $F_к = \psi(f_0)$ on the selection of the operating frequency f_0 are presented in Figure1.

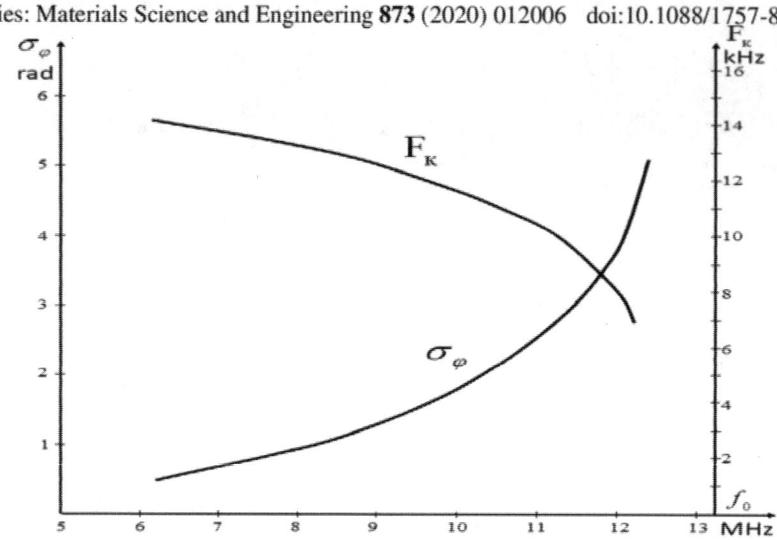

Figure 1. Dependences of the phase fluctuations at the wave front at the inhomogeneous ionosphere output and the interval of the frequency correlation of the single-beam SW radio link on the operating frequency selection

The analysis of Figure 1 shows that as the operating frequency of the single-beam SW radio link approaches to MUF, phase fluctuations at the front of the output wave increase from $\sigma_\varphi \approx 0{,}57$ to 4.8 rad. This results in increasing the parameter of the Rician fading of received signals, which is determined as [6, 12, 14-16]

$$\gamma^2 = \left[\exp(\sigma_\varphi^2) - 1\right]^{-1} \geq 0, \tag{13}$$

from $\gamma^2 \approx 2.6$ to $\gamma^2 \approx 7 \cdot 10^{-11}$, which practically corresponds to the case when general Rayleigh fading (when $\gamma^2 = 0$) occurs. Moreover, the interval of the frequency correlation of the single-beam SW radio link becomes twofold narrower – from $F_\kappa \approx 14.3$ kHz to 6.9 kHz.

The reliability of the results is confirmed by the fact that the calculated values of the frequency correlation interval of the single-beam SW radio link ($F_\kappa \approx 6{,}91 \ldots 14{,}28$ kHz), constrained by the diffuse multipathing, approximately correspond to the known experimental data [1] ($F_\kappa \approx 5 \ldots 20$ kHz) for standard parameters of the spherical inhomogeneous reflecting layer of the ionosphere and various values f_0.

Thus, the methodology is developed to calculate the interval F_κ of the frequency correlation of the single-beam SW radio link with the sphericity of the inhomogeneous reflecting layer of the ionosphere under the diffuse multipathing based on the definition of the φ_0 angle (according to (12)) and a set of analytical expressions (8, 10, 9, 6, 7, 11, 3, 5, 1).

This methodology includes the following stages:

1. Based on the measurement of the frequency-height characteristics (FHC) of the ionosphere layer F (f_{cr}, h_0, Z_m) for the selected values of the operating frequencies of the single-beam SW radio link by the ionosphere vertical sounding (IVS) station, a number of functional dependencies can be established $R = \Psi(\varphi_0, f_0)$.

2. For the selected operating frequencies f_0, it is possible to determine graphically or numerically the wave incidence angle φ_0 onto the lower limit of the ionosphere reflecting layer F that corresponds to the specified range of the SW communication $R = R_{pre}$.

3. For the found value of the wave incidence angle φ_0 with the frequency f_0 onto the lower limit of the ionosphere reflecting layer F with the known parameters (f_{cr}, h_0, Z_m), which ensures the specified range of the SW communication $R = R_{pre}$, the coefficient that characterizes the rate of decline f_{cr} within the spherical reflecting layer F of the ionosphere as compared to the plane one (8) $C_1 = \psi(f_0, \varphi_0, f_{cr}, h_0, Z_m)$ is determined.

4. Taking into account $C_1 = \psi(f_0, \varphi_0, f_{cr}, h_0, Z_m)$, the group (10) $L_g = \psi(f_0, \varphi_0, f_{cr}, h_0, Z_m, C_1)$ and actual (9) $L_F = \psi(L_g, f_0, \varphi_0, f_{cr}, h_0, Z_m)$ paths of the wave propagation within the spherical reflecting layer of the ionosphere F are determined as well as the coefficient of the Earth and ionosphere sphericity (6) $K_s = \psi(f_0, \varphi_0, f_{cr}, h_0, Z_m, C_1)$.

5. Taking into account these dependences, the length of the equivalent homogenous path of the wave propagation within the spherical reflecting layer of the ionosphere of the single-beam SW radio link (7) $L_E = \psi(L_F, L_g, f_0, \varphi_0, f_{cr}, h_0, Z_m, C_1, K_s)$ is determined .

6. According to the found wave incidence angle φ_0 onto the lower limit of the ionosphere layer F and the height of this limit h_0, the wave path in the free space behind the ionosphere from the output point from the layer to the reception point (11) $L_{fs} = \psi(\varphi_0, h_0)$ is determined .

7. According to the found values of the equivalent homogenous path of the wave propagation with the frequency f_0 in the spherical reflecting layer F of the ionosphere of the single-beam SW radio link (7) $L_{eq} = \psi(L_F, L_g, f_0, \varphi_0, f_{cr}, h_0, Z_m, C_1, K_s)$ and the sum of this path $L = L_{eq} + L_{fs}$ and the length of the wave propagation in the free space behind the ionosphere (11) $L_{fs} = \psi(\varphi_0, h_0)$, taking into account the known lengths (the smallest l_i and the greatest l_0) of the ionosphere inhomogeneities, the diffraction parameter (3) $d_1^2 = \psi(f_0, L_{eq}, L_s, l_i, l_0)$ is determined .

8. For the selected operating frequency f_0 and the found wave incidence angle φ_0 onto the lower limit of the ionosphere reflecting layer F, the coefficient of the Earth and ionosphere sphericity (6) $K_s = \psi(f_0, \varphi_0, f_{cr}, h_0, Z_m, C_1)$ and the equivalent homogenous path of the wave propagation with the frequency f_0 in the spherical reflecting layer F of the ionosphere (7) $L_{eq} = \psi(L_F, L_g, f_0, \varphi_0, f_{cr}, h_0, Z_m, C_1, K_s)$, it is possible to determine the MSD of the phase fluctuation at the wave front at the output of the spherical reflceting layer F of the ionosphere with small-scale inhomogeneities (with the known values of their maximum length l_0 and rate β) according to (5) $\sigma_\varphi = \psi(f_0, \varphi_0, K_s, L_{eq}, l_0, \beta)$.

9. For the selected operating frequency f_0 and found values of the MSD of the phase fluctuations at the wave front at the output of the spherical reflecting layer F of the ionosphere with small-scale inhomogeneities (5) $\sigma_\varphi = \psi(f_0, \varphi_0, K_s, L_{eq}, l_0, \beta)$ and the diffraction parameter (3) $d_1^2 = \psi(f_0, L_{eq}, L_{fs}, l_i, l_0)$, the interval of the fading frequency correlation in the single-beam SW radio link with the diffuse multipathing (1) $F_\kappa = \psi(f_0, \sigma_\varphi, d_1^2)$ is determined.

3. Conclusion.

The application of the developed methodology allows you to (selection) of one discrete beam by the data of sounding the parameters of the ionosphere layer F ($f_{cr}, h_0, Z_m, l_i, l_0, \beta$), the application of this methodology enables selecting (lowering) the operating frequency f_0 that ensures the significant decrease of the phase fluctuations at the wave front at the output of the spherical reflecting layer F of the ionosphere with small-scale inhomogeneities (5) $\sigma_\varphi \sim f_0 \beta \sqrt{L_{eq}} / K_s^2 \sec^2 \varphi_0$, which leads to the decrease in the depth of interference fading (that is the increase in the Rice parameter (13) $\gamma^2 \sim 1/\sigma_\varphi^2$) and the expansion of the interval of the fading frequency correlation (1) $F_\kappa \sim f_0/\sigma_\varphi$ in the single-beam SW radio link.

Acknowledgments
The work was supported by the Russian Foundation for Basic Research within the project No. 18-07-01020.

References

[1] Stein S Jones D 1971 *Principles of modern communication theory and their application to the transmission of discrete messages* (Moscow: Svyaz)

[2] Khmelnitsky E A 1975 *Evaluation of the real noise immunity of receiving signals in the HF range* (Moscow: Svyaz)
[3] Cherenkova E L Chernyshov O V 1984 *Propagation of radio waves* (Moscow: Radio and communications)
[4] Volkov L N Nemirovsky M S Shinakov Yu S 2005 *Digital radio communication Systems: basic methods and characteristics* (Moscow: EKO-Trenz)
[5] Pashintsev V P Borovlev I I 2003 Messages transmission opportunity rise in one-path decameter radio link because of choice of optimal signals transmissions rate *Izvestiya Vysshikh Uchebnykh Zavedenij Radioelektronika* No 46(11) pp 26-33
[6] Pashintsev V P Kolosov L V Tishkin S A Antonov V V 1996 Application of the phase-screen theory for developing a model of a one-hop decameter communication link *Journal of Communications Technology and Electronics* No 41(1) pp 16-21
[7] Pashintsev V P Tishkin S A Smirnov A A Borovlev I I 2001 Equivalent path of decameter wave propagation in the spherical-layered ionosphere *Journal of Radioelectronics* No 8
[8] Pashintsev V P Solchatov M E Kondrashin A Ye Senokosova A V 2005 Maximal frequency of reflection of a decameter wave from the spherically stratified ionosphere *Radioelektronics and Communications Systems* No 48(5) pp 8-14
[9] Knepp D L 1983 Multiple phase-screen calculation of the temporal behavior of stochastic waves *Proc of the IEEE* Vol 71 No 6 pp 722-737
[10] Maslov O N Pashintsev V P 2006 Models of transionospheric radio channels and interference stability of space communication systems *The information and communication technologies* (Samara: pgati) Vol 4 p 358
[11] Pashintsev V P Kolosov L V Tishkin S A Smirnov A A 1999 Influence of the Ionosphere on Signal Detection in Space Communications Systems *Journal of Communications Technology and Electronics* No 44(2) pp 132-139
[12] Pashintsev V P Tishkin S A Ivannikov A I Borovlev I I 2001 Calculating the fading depth parameter in single-beam decameter radio link *Izvestiya Vysshikh Uchebnykh Zavedenij Radioelektronika* No 44(12) pp 57-65
[13] Pashintsev V P Koval S A Potyagov D A Spirin A M 2010 Estimation of noise error when measuring virtual height during diffusivity of ionospheric F layer *Radioelektronics and Communications Systems* Vol 53 No 7 pp 348-355
[14] Pashintsev V P Tishkin S A Ivannikov A I Solchatov M E 2001 Determining the optimal working and least applicable frequency of the DCM radio line with the depth of fast fading *Elektrosvyaz* No 12 pp 16-19
[15] Pashintsev V P Chipiga A F Shevchenko V A Kiselev D P 2016 Influence of ion-sphere diffusivity on the optimal operating frequency of a decameter radio line *Izvestiya Samar-go scientific center of the Russian Academy of Sciences* Vol 18 No 2(3) pp 946-951.
[16] Pashintsev V P Skorik A D Koval C A Kiselev D P Senokosov M A 2019 Dependence of communication reliability in a decameter radio line on the choice of the operating frequency with consideration of signal-interference situation and ionosphere diffusivity *Management communication and security systems* No 4 pp 300-322
[17] Pashintsev V P Omelchuk A V Koval S A Galushko Yu I 2009 Method for determining the magnitude of the intensity of inhomogeneities based on ionospheric sounding data *Dual technologies* No 1 pp 38-42
[18] Davis K 1973 *Radio Waves in the ionosphere* (Moscow: Mir)

IOP Conf. Series: Materials Science and Engineering **873** (2020) 012007 doi:10.1088/1757-899X/873/1/012007

System of automatic video stream images evaluation and preprocessing for ADAS

A A Egorchev

Kazan Federal University, 18 Kremlyovskaya str, Kazan, 420008, Russia

E-mail eanton090@gmail.com

Abstract. In this article, the algorithm of preliminary filtering of images of a video stream based on assessment and the subsequent correction of such qualitative characteristics of images as contrast, noise level, sharpness and existence of illumination is considered.

Keywords: image preprocessing; image filtering; noise reduction; unsharp mask; histogram correction; ADAS

1. Introduction

One of the most important tasks of ADAS systems is to classify the environment using computer vision algorithms. Using unprocessed video streams and images in the algorithms often fails to achieve sufficient accuracy because of the quality of the visual image settings (low sharpness, contrast, noise, etc.), and also increases the processing time due to the presence of large amounts of unused data (e.g., color channels of the image or high resolution image), which makes impossible the use such approaches in systems operating in real-time.

To assess the quality of the input image, the criteria corresponding to the intervals of feature values, as well as functions for evaluating each criterion for the image flow will be defined. The functions for evaluation of criteria were tested on various types of video (with different levels of sharpness, contrast, time of day, varying degree of noise).

The video stream of images, previously converted to 256 grayscale mode (8bit grayscale) using colorimetric conversion with preservation of brightness, is used as input data [1]. (The input data is a video stream of images previously converted to 256 grayscale (8bit grayscale) mode using a colorimetric transformation with preservation of brightness [1].) The content of the images is the results of shooting video cameras installed inside the cabin on the windshield of various cars.

In this field of science multiple authors offer a number of solutions for related problems. Ralph Gross and Vladimir Brajovic propose a new image preprocessing algorithm that compensates for illumination variations in images. From a single brightness image, the algorithm first estimates the illumination field and then compensates for it to mostly recover the scene reflectance. Unlike previously proposed approaches for illumination compensation, our algorithm does not require any training steps, knowledge of 3D models or reflective surface models.[2] The authors use an image preprocessing solution to improve the quality for facial recognition. Jones M. in their scientific work described the operational system for the precision preprocessing of imagery. The operational system performs three basic functions: image data acquisition; amplitude preprocessing, including such functions as interchannel registration, line-phase correction, and fine gain adjustment; and navigation, involving the production of accurate image deformation models [3]. Siddhartha Bhattacharyya in thouse article presents a brief survey of the trends in color image enhancement and segmentation: methods in this direction involve Markov random models, vector directional filters and statistical mixture models like Gaussian and Dirichlet mixtures, the non-classical approaches comprising the neuro-fuzzy-genetic paradigm or its variants are bestowed with features for real time applications [4].

Solving specific tasks in narrowly defined areas requires different approaches to image preprocessing, since the criteria for image preprocessing differ. In particular, a system of automatic video stream images evaluation and preprocessing was developed for ADAS. The research results presented in this article are unique and were obtained as a result of many practical experiments.

2. Criteria for evaluating images of video stream

2.1. Noise level

Noise estimation is important in many image processing and analysis algorithms. Quantifying the noise component makes it possible to adapt the preprocessing algorithms to the amount of noise instead of using fixed threshold values.

The model for noisy images is as follows(formula 1):

$$I\,(x,y) = f\,(x,y) + n\,(x,y), \tag{1}$$

where f is the "perfect" image, n is the noise, I is the observed image.

One of the main problems of noise estimation is the measurement of deviations I from f, which may contain significant structural information, such as edges and texture features.

[5] presents a fast algorithm of assessment of image noise dispersion. It is used as method for noise determination in our task. In the offered method the operator with a zero average value, which is almost insensitive to the image structure is used. The dispersion of the output values is taken as an estimate of the noise variance. Since the structural elements of images, such as edges, correspond to the differential components of the second order, the noise estimation must be resistant to the Laplacian of the image.

The advantage of this method is that it uses the Laplace operator, which is insensitive to the image structure, and its result depends only on the noise level in the image.

When using the ESTIMATE_NOISE function that implements this method, the values of the NL noise level in the image belong to intervals $NL \in (0,3) \cup [3, +\infty)$. The value of $NL = 3$ is considered a threshold: images with the value of NL from 3 and above are considered noisy and require additional processing.

The image is pre-translated into Grayscale. At high noise levels above $NL = 3$, the median noise reduction operation is applied, at lower noise levels, the image does not need additional processing.

The median filter is a nonlinear digital filtering technique that is often used to remove noise from an image or signal.

Median filtering is one of the ways to smooth the image, equally with linear Gaussian filtering. All smoothing methods are effective at removing noise in smooth areas without structural features, but adversely affect the edges. When choosing smoothing filtering algorithms, it is important to be guided by the degree to which the filter allows you to preserve edges that bear significant visual information about the image. In this respect, for small or medium levels at a given fixed window size, the median filter is significantly better than the Gaussian blurring.

2.2. Level of sharpness

The sharpness of the image describes the subjective visibility of details, the degree of clarity of the border between the two sections of the photographic image that have received different exposures. The sharpness value is related to the amplitude of the brightness derivative in spatial variables.

In the most general sense, the sharpness of the image S at any point can be taken as the brightness gradient I (formula 2).

$$S = \nabla I. \tag{2}$$

Thus, the sharpness of the image is a vector field.

For the input image using the ESTIMATE_SHARPNESS function, the sharpness of the SL is measured (method [6], based on the calculation of the image gradient modulus). SL values are less than 2 correspond to the image of low sharpness, from 2 to 4.5 – the average, more than 4.5 – high sharpness (see examples in figure 1).

Figure 1. (a) low sharpness image ($SL = 1.7$), (b) high sharpness image ($SL = 6.3$)

Images with a high level of sharpness do not need to be pre-processed. For images with sharpness values lower than 4.5, an unsharp masking filter is applied.

The unsharp masking filter applies a blur to a copy of the original image. If, when comparing the mask with the original, the differences exceed a certain threshold, the image is subtracted. The threshold setting is necessary to avoid amplifying unwanted details such as noise in a digital photo or grain on a film.

A typical scheme of applying an unsharp masking filter is as follows (formula 3):

$$sharpened = original + (original - blurred_r) \times amount \tag{3}$$

where $original$ is the input image, $blurred_r$ is the image with the blur filter applied (usually Gaussian) with radius r, which affects the degree of blurriness of the image copy and the width of the zone in which the change in tone will appear, $amount$ is an Unsharp masking parameter that determines how much the areas of the image that are on the boundaries of the contour are darkened or lightened.

When using the unsharp masking filter, the $threshold$ parameter is also used, which determines the minimum difference in tone at which the unsharp masking is accomplished. For a noisy or grainy original, it is set to a higher value to prevent these unwanted details from underlining. Pre-noise reduction allows this parameter to be fixed ($threshold = 0$).

For sharpness values 4.5 and above, no filtering is applied to the image. If $SL \in (0, 2)$ an unsharp masking filter is applied with the parameters $amount = 0.5$, $radius = 10$. If $SL \in [2{,}4.5)$ an unsharp masking filter is applied with parameters $amount = 0.5$, $radius = 20$.

2.3. The presence of illumination

The illumination on the image is characterized by the presence of overexposed regions in the area of excessive sunlight on the matrix, the presence of sharp shadows, as well as areas of low contrast, which has an extremely negative impact on the subsequent recognition of the image objects.

The FLASH_TEST function is used to determine the presence of FL illumination.

Since the input image was converted to grayscale mode calorimetrically, the value of each pixel retains the illumination parameters. To determine the presence of illuminated parts in the image, 3 windows (W1, W2, W3) are selected, located in two corners and center of the image. Then the average pixel value (brightness) in each window is measured: $Y_{av}(W_1)$, $Y_{av}(W_2)$, $Y_{av}(W_3)$ (see examples in figure 2). If threshold $th = 200$ is exceeded in at least one of the windows, the values of $Y_{av}(W_i)$ are compared. If there is a difference in brightness values of more than 100%, then there is illumination and the value of FL is set to 1, otherwise $FL = 0$.

Figure 2. An image with illumination ($FL = 1$). Brightness levels Y_{av} of the image illumination

If there is an illumination in the image($FL = 1$), an adaptive histogram alignment filter is applied to the image (see examples in figure 3).

Histogram alignment [4] is used in image processing to increase the contrast level. This method typically increases the global contrast of many images, especially when pixel-by-pixel image contrast data is represented by similar values. By using histogram alignment, the intensity values can be better distributed across the histogram. This allows you to increase the contrast of areas of low local contrast ratio. Histogram alignment allows you to efficiently distribute the most common intensity values in an image.

The disadvantage of the method is its non-selectivity and the fact that its application can increase the contrast of background noise of the image. Therefore, noise reduction is used as a pre-processing for noisy images.

Adaptive histogram equalization (AHE) [8] is an image processing technique used to improve global contrast. Ordinary histogram equalization uses the same transformation derived from the image histogram. This method works well when the distribution of pixel values is the same throughout the entire image. However, when an image contains areas that are significantly lighter or darker than the rest of the image, the contrast in these regions when applied to the image of a regular histogram alignment will not be sufficiently enhanced.

The adaptive method calculates several histograms, each corresponding to a separate part of the image, and uses them to redistribute brightness values of the image. Therefore, it is suitable for improving local contrast and edge refinement in all areas of the image.

Contrast Limited Adaptive histogram equalization (CLAHE) [9] is a version of adaptive histogram equalization in which contrast enhancement is limited to avoid the occurrence or amplification noise in the image. In the case of the illuminated image, adaptive histogram alignment helps to smooth out the contrast of the image in both over - and underexposed areas, and to get rid of sharp shadows.

Figure 3. (a) Image with the presence of illumination on the left ($FL = 1$) in the grayscale mode (b) Image with an adaptive histogram alignment filter

2.4. Contrast Level

Contrast measures the difference in brightness or color that makes objects in an image distinguishable. The contrast level is related to the intensity gradient. In addition, contrast can be defined locally or globally. When an image consists of textured areas, it is more appropriate to perform local image analysis and determine local contrast. It is well known that the level of local contrast can be increased by calculating the differences between the intensity value in each pixel of the image and its surrounding pixels and the subsequent increasing these differences [10].

For the input image, the contrast level CL is measured using the method of A. Beghdadi and A. L. Negrate [11]. This algorithm is based on the Gordon's method [7], which directly determines the contrast function and the function of improving the contrast of the image without involving the histogram of the gray levels of the input image. To improve the operation of the method in conditions of high noise level and image detail, algorithms for edge selection are implemented in this method.

The algorithm is implemented in the ESTIMATE_SHARPNESS function.

For the current image, the contrast is measured and recorded in CL. At values of CL from 0 to 70 the image is considered low contrast. When $CL > 70$, the image is high contrast. See examples in figure 4.

Figure 4. (a) low contrast image ($CL = 32$), (b) high contrast image ($CL = 110$)

For high-contrast images $CL \geq 70$ filtering is not applied (figure 5(a)). At $CL \in (0, 70)$ the filter CLAHE is applied (figure 5(b)).

Figure 5. (a) low contrast image ($CL = 32$) in grayscale mode (b) image with histogram alignment filter applied to it ($CL = 67$)

3. Algorithm for automatic evaluation and filtration of images

Figure 6 shows the algorithm of automatic filtering of the image in a video stream. In order to avoid excessive computational costs, as well as to optimize the operating time of the algorithm, the frequency of application is determined for each criteria evaluation function, based on the nature of the image properties change in the video stream, as well as the requirements of subsequent recognition algorithms. Table 1 shows the order and frequency of application of criteria evaluation functions.

IOP Conf. Series: Materials Science and Engineering **873** (2020) 012007 doi:10.1088/1757-899X/873/1/012007

Table 1. Frequency of application of functions.

Function	Frequency of application (number of applications per second)
ESTIMATE_ NOISE	6
ESTIMATE_ SHARPNESS	3
FLASH_TEST	18
ESTIMATE_ CONTRAST	3

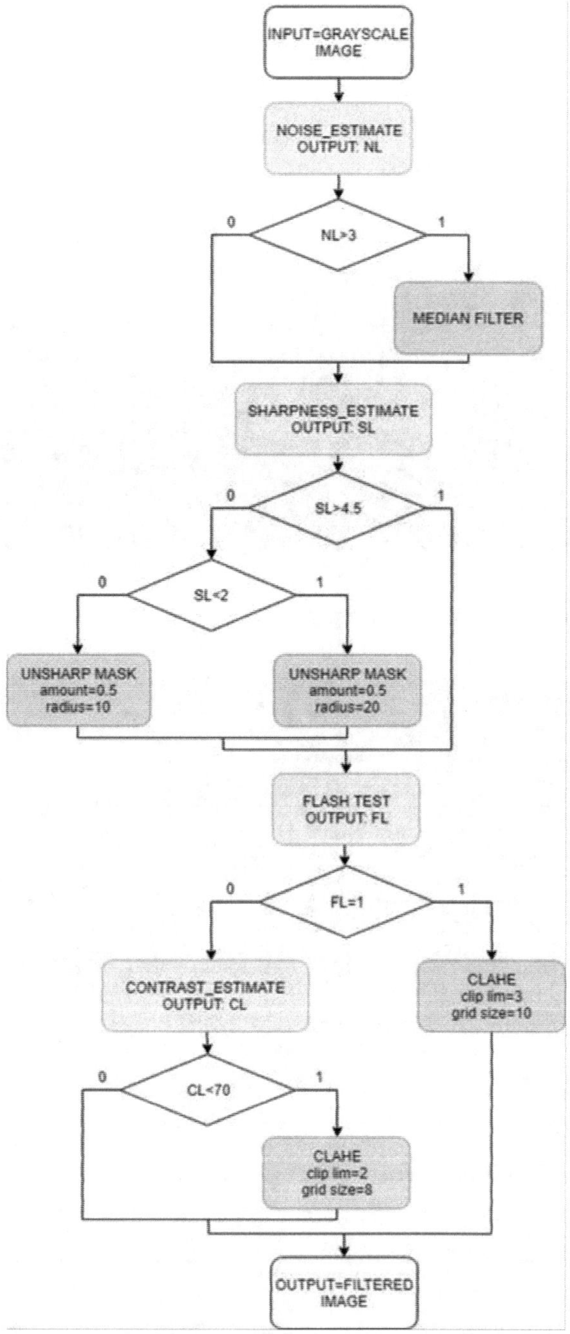

Figure 6. Algorithm of automatic evaluation and filtering

IOP Conf. Series: Materials Science and Engineering **873** (2020) 012007 doi:10.1088/1757-899X/873/1/012007

4. Software implementation

Jetson TX2, an NVidia industrial product, is used for the automatic evaluation and image preprocessing system. The characteristics of the Jetson TX2 are listed in Table 2.

Table 2. Jetson TX2 Features.

Feature Specification	Jetson TX2
Processor	NVIDIA Pascal architecture with 256 CUDA cores
Computing Micromodule	HPM Denver 2/2 ARM A57/2
Video	Encoding 4K x 2K 60Hz (HEVC) 4K x 2K 60Hz decoding (12 bit support)
Memory	8 GB memory
Display	2x DSI, 2x DP 1.2 / HDMI 2.0 / eDP 1.4
CSI	Up to 6 cameras (2 channels) CSI2 D-PHY 1.2 (2,5 ГБ\s per channel)
PCIE	Gen 2 | 1x4 + 1x1 or 2x1 + 1x2
Data storage	32 ГБ eMMC, SDIO, SATA
Other	CAN, UART, SPI, I2C, I2S, GPIOs
USB	USB 3.0 + USB 2.0
Connection	1 Gigabit Ethernet, 802.11ac WLAN, Bluetooth

The location of the Jetson TX2 computer in the logic of the ADAS system is shown in Figure 7.

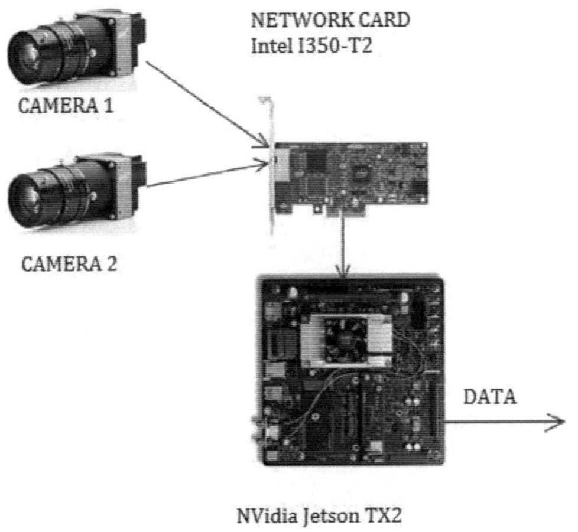

Figure 7. Interaction Diagram

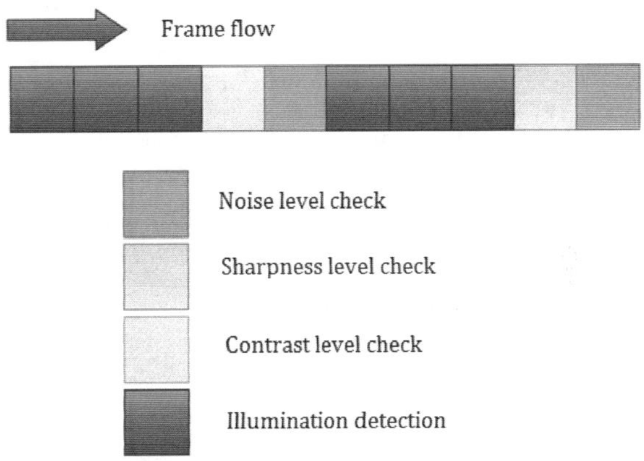

Figure 8. Software Block diagram

Running applications on this product involves running resource-intensive parts of the program on the GPU. In connection with this requirement, computer vision algorithms were developed using NVIDIA's CUDA software-hardware parallel computing architecture, which the program to increase its speed by using computational micromodules.

The algorithmic base of the developed system was implemented using OpenCV and VisionWorks machine vision libraries based on CUDA technology. The generalized block diagram of software modules interaction is presented in figure 8.

To ensure real-time operation of the software, the calculation of the image characteristics was distributed over time intervals. During the processing of one frame, one image characteristic is calculated, the image is filtered at each step based on the last measurements.

The operating time of the algorithms on Jetson TX2 is presented in table 3. The tests were carried out on images in HD resolution.

Table 3. Jetson TX2 Computer Specifications.

Algorithm Name	Operating time, ms
Median	14
Unsharp	23
CLAHE	12
Estimate noise	15
Estimate sharpness	40
Estimate contrast	50
Flash test	14

An example of the results of the algorithm is presented in Figure 9.

Figure 9. (a) original image (b) filtering result

5. Conclusion

The considered algorithm for automatic filtering of video stream images uses evaluation of characteristics for video stream images and can be used to improve the quality of detection and classification of various types of objects in ADAS systems.

6. Acknowledgment

The work is performed according to the Russian Government Program of Competitive Growth of Kazan Federal University.

References

[1] Poyton C A 1998 Rehabilitation of gamma *Proc. SPIE 3299 Human Vision and Electronic Imaging III* ed B E Rogowitz and T N Pappas (San Jose: SPIE) pp 232-249 https://doi.org/10.1117/12.320126

[2] Gross R and Brajovic V 2003 An image preprocessing algorithm for illumination invariant face recognition *Lecture Notes in Computer Science* **2688** 10-18

[3] Jones M, Lauque P and Haemmer R 1980 An introduction to the meteosat image preprocessing and image navigation system *ESA Journal* **4** 211-226

[4] Bhattacharyya S 2011 A brief survey of color image preprocessing and segmentation techniques *Journal of Pattern Recognition Research* **6** 120-129

[5] Immerkaer J 1996 Fast noise variance estimation *Computer Vision and Image Understanding* **64** 300-302

[6] Birdal T 2019 Sharpness estimation from image gradients *MATLAB Central File Exchange* https://www.mathworks.com/matlabcentral/fileexchange/32397-sharpness-estimation-from-image-gradients

[7] Gordon R and Rangayyan R M 1984 Feature enhancement of film mammograms using fixed and adaptive neighborhoods *Applied Optics* **23** 560-564

[8] Hummel R 1977 Image enhancement by histogram transformation *Computer Vision, Graphics and Image Processing* **6** 184-195

IOP Conf. Series: Materials Science and Engineering **873** (2020) 012007 doi:10.1088/1757-899X/873/1/012007

[9] Pizer S M, Amburn E P and Austin J D 1987 Adaptive histogram equalization and its variations *Computer Vision, Graphics and Image Processing* **6** 355-368

[10] Kim. V and Yaroslavsky L P 1986 Rank algorithms for picture processing *Computer Vision, Graphics and Image Processing* **6** 234-258

[11] Beghdadi A and Negrate A L 1989 Contrast enhancement technique based on local detection of edges *Computer Vision, Graphics and Image Processing* **46** 162-174

Role of innovative digital technologies in transformation of Agricultural and Industrial Complex

E A Strizhakova, Yu N Nazarova, O A Zayats, Yu V Zadorozhneva, R I Shumakova

Volgograd state agrarian university, Ecological and Land-Improvement Faculty, 26 University prospect, Volgograd, 400002, Russia

E-mail StrizhaEl@gmail.com

Abstract. The conditions of the digital economy development are studied in the article taking into account the application of innovative technologies. The authors considered features and the role of their influence on the digital transformation of the Agricultural and Industrial Complex. The review of the main government programs to intensify the industry development by using digital technologies is presented here.

Keywords: digital economics, digital transformation, digitalization of agriculture, innovation, informatization of economics

1. Introduction

The informatization process led to the development of digitalization processes in the economy as a whole, as well as its individual industries. To do this, a single information space is being created, including data warehouses, the Federal Geographic Information System (GIS), regional customers (regional GIS centers), the site of the main computing center and other components.

The existing information environment in the Agricultural and Industrial Complex sector is inefficient. Transactional costs in production are rising. The comparison of foreign analogues with domestic products reflects the low level of domestic food products financial affordability, as well as their low competitiveness.

2. Enterprises functioning in the digital economy features

Enforcing the accelerated application of digital technologies in the economy and social sphere is one of the goals in national development of Russia for the period until 2024 [1]. The national program “Digital Economy of the Russian Federation” was developed to achieve this goal. Within its framework in 2024 it is necessary to:

- increase domestic costs for the development of the digital economy at the expense of all sources at least 3 times compared with 2017 (in terms of the share in the country's GDP);
- create a stable and secure information and telecommunications infrastructure for the large amounts of data (accessible to all households and organizations) high-speed transmission, processing and storage;
- ensure the use of predominantly domestic software by organizations, state bodies and local governments.

Favorable conditions were developed in Russia to solve a number of problems aimed at achieving the above targets. Between 2014 and 2018, the gross added value in the ICT sector increased by 13.7%. The relative share of the ICT sector for the period of 2014-2018 ranged from 2.6 to 2.8%. In 2018, the share of the ICT sector in the business sector added value amounted to 3.2%; however, it is 3 times lower than in the Republic of Korea (9.6%) and 1.5-2 times less than in most developed countries. In 2018, the investments in fixed assets increased significantly - by 26% compared to 2017. The increase in the number of organizations in IT industry was by 6.3%, the decrease in the number of employees was by 7.4%, which indicates such negative trends as job cuts.

During 2018, the share of profitable organizations in the information and communication field amounted to 75.2% (7.1% more than in 2017), including 76% in the IT field. The business sector organizations have widely mastered basic and relatively simple digital technologies, but only a small part of them conducted a deep automation and restructured business processes according to advanced digital technologies. The individual technologies intensity use is distinguished by a significant differentiation. Thus, more than 80% of organizations use broadband Internet and more than 60% of them uses electronic data exchange technologies. The share of organizations using more sophisticated technologies remains quite low: less than a quarter of organizations use cloud services, slightly more than 20% of them uses ERP-systems, and less than 10% uses RFID-technologies. Less than half of the organizations in the business sector have a website, and cloud services are used primarily for e-mail purposes. This is another negative aspect of the digital economy development in Russia. The highest levels of digitalization have organizations in such types of economic activities as "Telecommunications", "IT industry", "Manufacturing industry" and "Whole sale and retail trade". As to the software tools, organizations most often use electronic document flow systems, software for financial calculations, as well as for solving organizational, managerial and economic problems [2].

International comparisons on the totality of factors show Russia's promising positions in terms of the digital technologies development and implementation main indicators, since it is included in the top 50 major international ratings of digital development. Thus, in terms of the e-government development index, Russia lay in the 32-nd place in 2018 (35 in 2016); in terms of ICT development index in 2017 – 45-th place (in 2016 - 43). By the value of the global cyber security index, in 2017 Russia lay in 10-th place, but in 2018 it dropped to 26.

The Institute of Statistic Studies and Economics of Knowledge National Research University Higher School of Economics has developed a Business Digitalization Index to measure the speed of business adaptation to digital transformation. That index characterizes the level of broadband Internet, cloud services, ERP-systems, RFID-technologies use and the involvement of business sector organizations in electronic commerce. The value of this index for Russia in 2017 amounted to 28.4%, that is 1.6-1.8 times lower than in the leading countries (Finland, Belgium, Denmark, the Republic of Korea) [3]. By the intensity of the digital technologies use in the business sector, Russia is located next to Bulgaria, Hungary and Romania. In terms of the digital technologies spread, leading positions, in addition to telecommunications (more than 40%), are occupied by organizations in whole sale and retail trade, manufacturing and IT (35%) [4].

According to the Index “Digital Russia” [5], the Volgograd Region for 2018 lies in 39-th place out of 85 entities, that is 2 points lower than for the previous year. The Governor approved the Program “Digital Development of the Volgograd Region” by the resolution No. 204 of April 23, 2019 in order to strengthen the position of the region and successfully implement of the national program “Digital Economy of the Russian Federation” on its territory.

The Volgograd region has all the necessary prerequisites for the further digital technologies application and the Program implementation. The share of organizations in the Volgograd Region that used special software in 2018 amounted to 82.5%, it is 4.5% higher than that level in 2014. The most common are software tools for solving organizational, managerial and economic problems, for making financial payments electronically, as well as legal reference systems (they are used by more than half of the organizations surveyed). Programs for providing access to databases are less in demand (30% of organizations). Less than 20% of organizations use training programs, programs for managing automated production and/or individual technical means and technological processes, design software, as well as CRM, ERP and SCM systems. However, according to these types of software in recent years, there has been a steady tendency to increase the share of organizations which use them. This is especially true for ERP, CRM and SCM systems: the annual increase in the share of organizations using them amounted to almost 2% [6].

3. Agricultural and Industrial Complex digital transformation based on innovative technologies

Traditional technologies have already exhausted their resources, and the digitalization of agricultural production requires costs, and sometimes significant ones, the change farmers' mentality, and acquiring new skills and abilities. The challenges facing the digital economy are solved by using: relevant knowledge bases; information resources; interactive communities; integration business platforms; digital environment; and personnel who are able to perform work in new conditions.

This integrated approach creates a digital ecosystem. Each participant in this system simultaneously acts as a client and a data server. All this is the basis for the formation of subject-oriented clusters. Inside these clusters, the functioning of agricultural digital ecosystems occurs [7, 8].

The core of such a cluster includes agricultural producers, terminals and reception centers, processing enterprises, logistics centers, distributors, etc.

The sectors of coordination and management, scientific research and technology, personnel training and retraining, transport and technological services, financial services and lending, insurance, information and consulting services, technical supply, etc. are the cluster's satellites.

The authors propose to create a digital economy ecosystems global platform on the basis of the integration and interaction of large corporations, educational structures - training classes (higher education institution, vocational education institution, distance teaching system, etc.).

The effectiveness of the operation through the use of digital technologies in the Agricultural and Industrial Complex sector lies in the growth rate of productivity and profitability in the agricultural industry, which is directly related to the use of digital technologies [9].

According to the head of the Project Office for "Digital Economy of the Russian Federation" program application Vladimir Mesropyan's opinion, the digital platform is "a key tool for the traditional industries and markets digital transformation, the central concept of the global digital agenda that distinguishes between digitalization (digital automation) and digital transformation strategies" [10, 11, 12].

It should be noted that digital platforms appeared in the 90s of the XX century due to the information and communication technologies active use in various sectors of the economy, including the widespread use of the Internet. Thanks to ICT, software and hardware conditions and opportunities appeared for the platform business-model formation and development.

The digital platform belongs to the category of disruptive innovations and represents an "integrated information system that provides multilateral user interactions for the exchange of information and values, leading to lower overall transaction costs, streamlining business processes, and increasing the efficiency of the chain of goods and services supply" [10].

According to experts at the working group led by B. M. Glazkov, the vice president of public joint-stock company "Rostelecom" and the head of the Competence Center "Information Infrastructure" of the Russian Federation Digital Economy Development Program [13], the digital platform is "the system of algorithmic mutually beneficial relationships of the significant number of independent industry participants in economics (or areas of activity) implemented in the single information environment, and leading to lower transaction costs due to the digital technologies package use for work with data and changes in the division of labor system" [14]. The authors identified five key criteria for digital platforms:

1) the algorithmization of the platform participants interaction;
2) the mutually beneficial relationship of the platform participants (the principle "win-win");
3) the significance of the participants number in the activity (scale) using the platform for interaction;

4) the presence of a single information environment where the participants' interactions are carried out, and the corresponding information technology infrastructure;

5) the presence of the effect in the form of lower transaction costs during the interaction of various platform participants - compared with the same interaction without a platform.

An appropriate material base is needed to implement measures aimed at the industry digitalizing. For example, the introduction of precision farming is impossible with outdated equipment. Special GIS-driven tractors, combines, trailed devices are needed for this. In the course of the government program No. 1432 application, "our domestic agricultural machinery has become a competitor to the foreign production most modern models" [15]. However, according to the experts' opinion, the amount of domestic equipment produced is clearly not enough to stop the mechanization dropout level in the village [15]. According to experts' estimation, the aging of agricultural machinery is slightly ahead of the new machines purchase. The quantity of tractors older than 10 years in Russia exceeds 50 percent, the combine harvesters -about 45%. As the result, losses when harvesting with the use of old combines make up 10 percent of the grown crop, and when using new equipment it is about one.

To fulfil the tasks of the Agricultural and Industrial Complex technical modernization, farmers take advantage of the soft lending possibilities, and also receive discounts on the so-called "1432" Program. Leasing is becoming more widespread: in 2019, in cooperation with "Rosagroleasing" JSC, farms in the region acquired equipment worth 425 million rubles, while 80 percent of buyers are small and medium-sized businesses representatives.

For example, the Volgograd region is annually recognized as one of the most active regions in the field of Agricultural and Industrial Complex technical modernization. In 2014, farmers purchased equipment for a maximum of 2.5 billion rubles, in recent years the annual volume has been approaching to six billion [16].

The use of computer systems for automating the management of business processes in animal husbandry allows to work individually with each animal, taking into account its current condition and real needs, and provides detailed data on milk yield, fertility, health and nutrition of the whole herd. The use of such technologies made it possible, in particular, to increase milk yield by 30-40 percent in the farm enterprise "Donskoye".

The information system of veterinary control "Mercury" is successfully used in the Volgograd region. It helps to control the number and condition of farm animals, and makes it easier to obtain veterinary certificates. Through this system, about 28 million documents have already been issued.

Chipping of farm animals is another area of digitalization. In ten live stock districts at the Volgograd region in 2019, the stock of cattle was fully microchipped. 20 million rubles were allocated from the regional budget for these purposes.

The farm land monitoring information system won the title of laureate at the All-Russian competition of regional and municipal informatization "PROF-IT.2019".

The system covers the entire territory of the region and uses satellite imagery data. It contains digitized information about the location, condition and users of more than 57 thousand land plots. With the help of the agricultural portal, accounting and control of farm land is carried out, accurate information on crop rotation and field work is collected [17].

So in 2017, unused lands were identified during the satellite monitoring and inventory of agricultural lands. The result was a 84.6 ha reduction in uncultivated arable land.

It should be noted that at present, there is a Unified Federal Information System on Agricultural Lands (UFIS AL) in our country, which so far is the main information resource for digital land management.

Its main task is to quickly provide the Ministry of Agriculture and its subordinate organizations with relevant and reliable information about the boundaries and areas of agricultural lands and crops, land users, soil quality indicators, negative processes, objects and land reclamation facilities, etc.

However, at present, this information is not enough to form an effective land management system, because the system of updating data has not been fully developed, there is no quality control of the information provided, the system does not contain decision-making blocks [18, 19].

For example, according to the Analytical Center data, 22 regions out of 83 regions with agricultural lands did not provide data to the Unified Federal Information System on Agricultural Lands. More than half of the regions that provided information do not have reliable information in their information databases. In 42 subjects of the Federation, the land user database fill rate is less than 50%. There are also significant errors in cartographic support [20, 21].

It should be emphasized that nowadays the full implementation of the project "smart land management" is impossible due to the following reasons:

- the lack of a unified information resource on agricultural lands [22];
- obsolete cartographic materials (the latest update on agricultural lands was conducted in the period from 1987 to 1990);
- the lack of competent specialists in the field of cadaster and land management (In the course of previous researches, the authors found out that there are a large number of errors in the work of the currently available specialists. For example, according to the cadastral valuation results of the agricultural land in the Volgograd region, the estimated land area is 14.9 million hectares, it is 3.6 million hectares more than the entire area of the region (11.3)) [23];
- the lack of the necessary coordinate description for half of the land plots (about 29 million), and, as the result, the lack of the full register of rights to the land plots;
- the lack of specialized equipment and software.

4. Conclusion

The digital economy contains elements that allow performing land monitoring, optimization of payments between producer and buyer, rationalization of credit systems, investment, the use of marketing techniques, opposition to monopoly, electronic interaction between the digital economy participants.

The purpose of technological digital platforms (elements of an innovative infrastructure) is to ensure efficient and effective communication, stimulating the agricultural producers' interaction with other interested parties.

The initial stage of digital transformation (digitalization) includes the creation of a centralized information space by integrating existing data warehouses, using distributed data storage and tools for managing information and business intelligence, implementing a network-centric approach.

Digitalization, therefore, in the Russian Agricultural and Industrial Complex is at the initial stage. A detailed study of programs for informational support of the industry and the economy as a whole is required. Basically, it is proposed to automate what has already become fixed, but how to make this global transformation is still unknown. It is designed to enable the intelligent systems construction, the use of great data technologies and machine learning technologies, the introduction of technologies that have shown effectiveness in piloting framework. The search for a solution and its execution are carried out by the system.

An optimal digital ecosystem should be created by developing a digital platforms network in all areas of the Agricultural and Industrial Complex sector.

References

[1] On the national goals and strategic objectives of the Russian Federation development for the period until 2024: Decree of the President of the Russian Federation of 05/07/2018 № 204. URL: http://kremlin.ru/acts/bank/43027.

[2] Razvan S 2020 *Several contemporary economy features, consequences of internet expansion and i.c.t. innovations in the world* (Studies in Business and Economics) 14 (3) pp. 175-181.

[3] Mugge P, Abbu H, Michaelis TL, Kwiatkowski A, Gudergan G. 2020 *Patterns of Digitization: A Practical Guide to Digital Transformation* (Research Technology Management) 63 (2) pp. 27-35.

[4] Abdrakhmanova GI [et al.] 2019 Indicators of the Digital Economy 2019 (Statistical Digest of the National Research University "Higher School of Economics") 248 p.

[5] Index "Digital Russia": Report (Moscow School of Management SKOLKOVO Center for

Financial Innovation and Cashless Economy). URL: https://finance.skolkovo.ru/downloads/documents/FinChair/Research_Reports/SKOLKOVO_Digital_Russia_Report_Full_2019-04_ru.pdf.
[6] Statistical Yearbook of Volgograd Region 2018: collected volume (Territorial authority of the Federal State Statistics Service of Volgograd Region) Volgograd: Volgogradstat, 2019. 768 p.
[7] Aletdinova A A 2017 *Innovative development of the agricultural sector based on digitalization and the creation of technological platforms* (Innovation: electronic scientific journal) 4 (33) P. 22.
[8] Ghobakhloo M. 2020 *Industry 4.0, digitization, and opportunities for sustainability* (Journal of Cleaner Production) 252 article № 119869.
[9] Yurchenko I 2019 *Development and improvement of technologies on automated control for the formation of amelioration mode of agro-ecosystems* (Proceedings of nizhnevolzskiy agrouniversity complex: science and higher vocational education) 2 (54) pp. 354-363
[10] Mesropyan V 2018 *Digital platforms are a new market power*. URL: https://www.econ.msu.ru/sys/raw.php?o=46781&p=attachment
[11] Project office for the implementation of the program "Digital Economy of the Russian Federation". URL: https://data-economy.ru/projectoffice.
[12] Ivanov V and Ovchinnikov A and Kochetkova O 2019 *Conceptual basis of digital transformation of the agro-industrial complex of the Volgograd region* (Proceedings of nizhnevolzskiy agrouniversity complex: science and higher vocational education) 2 (54) pp. 18-25
[13] Direction "Information Infrastructure" for the implementation of the program "Digital Economy of the Russian Federation". URL: https://data-economy.ru/infrastructure
[14] Digital platforms: approaches to the definition and typification of Rostelecom, 2018. URL: http://d-russia.ru/wp-content/uploads/2018/04/digital_platforms.pdf.
[15] Dissenting opinion. Agricultural business magazine "My Siberia". URL: http://mysibir.ru/osoboe-mnenie/.
[16] Yurchenko I 2019 *Digital technologies as factor of the competitive ability of enterprises of the meliorative sector of the economy* (Proceedings of nizhnevolzskiy agrouniversity complex: science and higher vocational education) 1 (53) pp. 313-320
[17] Triantafyllou A, Tsouros D C, Sarigiannidis P, Bibi S 2019 *An architecture model for smart farming* (Proceedings - 15th Annual International Conference on Distributed Computing in Sensor Systems, DCOSS 2019) article № 8804834 pp. 385-392
[18] Kozubenko I 2019 *Analysis of the current state and prospects for the development of digital agriculture in Russia* (IOP Conference Series: Earth and Environmental Science) 274 (1) article № 012110
[19] Borodychev V, Lytov M 2018 *System «Analysis – data visualization – decision-making» consisting of irrigation management geographic information system* (Proceedings of nizhnevolzskiy agrouniversity complex: science and higher vocational education) 2 (50) pp. 37-43
[20] Volkov S N, Shapovalov D A 2019 *Digital land management - problems and prospects* (Interexpo Geo-Siberia) № 2. URL: https://cyberleninka.ru/article/n/tsifrovoe-zemleustroystvo-problemy-i-perspektivy
[21] Koshelev A 2018 *Digital mapping of soils with the use of SRTM data* (Proceedings of nizhnevolzskiy agrouniversity complex: science and higher vocational education) 4 (52) pp. 159-166
[22] National report on the progress and results of the "State program for the agriculture development and regulation of the market for agricultural products, raw materials and food for the period 2013-2020" implementation in 2017 (table 7.3.1). Ministry of Agriculture Moscow, 2018. URL: http://mcx.ru.
[23] Vorobyov A V, Smykov A V 2015 Land registration system perfection during the state cadastral valuation of agricultural land in the Volgograd region (Bulletin of the Nizhnevolzhskiy agro-university complex) 3 pp. 237-241.

The study of the applicability of catastrophe theory in predicting the systemic risks of managing the digital economy and Internet of things systems

E E Tikhonov, D V Samoilenko, U N Kocherov, V V Kuzmenko, E N Melnikova

Nevinnomyssk technological Institute (branch) North-Caucasian Federal University, Department of information systems, electric drive and automation, 1 Gagarina street, Nevinnomyssk, 357100, Russia

E-mail igwt@mail.ru

Abstract. The article discusses the applicability of the theory of disasters and nonlinear dynamics in predicting the systemic risks of managing the digital economy and Internet of things systems. One of the approaches may be modeling of nonlinear processes based on the use of the concepts of catastrophe theory. Studies have shown a key indicator of such modeling is to determine the essential features of the disaster. The study showed that the widely used models of the economic processes of the digital economy, as well as the software and technical components of the internet of things, can be transformed into the canonical equation of «assembly», which subsequently leads to the emergence of such an area in the space of phase development variables in which jumps can occur or for the better, or vice versa, such as a collapse. Currently, in the digital economy, the study of such phenomena seems to be very relevant, since an uncontrolled «spasmodic» change in the functioning parameters of such systems can lead to disastrous consequences for the functioning of all interconnected subsystems. The collapse or unstable behavior of one of the components of such systems can lead to the collapse of the entire system. Such phenomena are not allowed in the digital economy and can trigger not only economic consequences, but also social ones.

Keywords: Digital economy, Internet of things, forecasting, catastrophe theory

1. Introduction

The effective functioning of the economy has become dependent on the digital environment so research subject is topical. At the same time more and more noticeable the growing number of uncertainties inherent in the digital space. Digital threats have become wider, which often leads to significant financial, reputational, time costs [1].

As you know, during the implementation and functioning of software and hardware environments that support the work of the entire digital economy, there are three key problematic issues, the study of the behavior and reactions of which using the theory of disasters and nonlinear dynamics can be considered fundamentally significant [2]:

1. Issues related to the information security function of all of the digital economy.
2. Problems of data transmission and processing of large volumes of unsystematized information.

IOP Conf. Series: Materials Science and Engineering **873** (2020) 012009 doi:10.1088/1757-899X/873/1/012009

3. Problems of electric power safety and the stability of the functioning of technical systems of the digital economy [3].

These issues of the functioning of the digital economy are interrelated, and failures and critical changes that can be described, analyzed and modeled using catastrophe theory relate to the systemic risks of the functioning of the entire digital economy.

Also, systemic risks should include both fundamentally economic changes in the rules for the functioning of the elements of the digital economy, as well as software and technical risks, in particular those related to both data transfer systems and the problems of protecting information and organizing external attacks on such systems.

In any economic system, and even more so, based on software and hardware systems, which are influenced by various factors, not only smooth, but also abrupt, spasmodic changes occur. The modern digital economy is a complex system, the sudden changes of which are also difficult to analyze and predict. As you know, "disasters" in mathematics are called abrupt changes that occur in the form of a sudden response of the system to a smooth change in external conditions. If we consider the complex elements of the digital economy as the system under study, it should be noted that its elements are not only new economic relations between entities, but also many complex information technology environments and realizations that, in the aggregate of multifactorial elements of such systems, may be subject to unpredictable and drastic «catastrophic» changes.

Recent studies have shown that traditional forecasting methods are not applicable to real-life problems of the digital economy, such as forecasting system overloads, predicting the appearance of atypical «anomalies» in the operation of controlled systems, etc. In this connection, the study of the applicability of catastrophe theory and methods of nonlinear dynamics in the above questions is an urgent scientific problem of a fundamental nature. As a result of the work, algorithms and models will be obtained that describe and predict the behavior of complex systems, in particular the digital economy. Mechanisms have been developed for early prediction of the phenomena of the digital economy, which can lead to "catastrophic" consequences of the entire system.

2. Informal statement of the problem

The aim of the study is to investigate the possibility of using models of catastrophe theory and nonlinear dynamics to predict the systemic risk management of the digital economy, including systems based on Internet of Things technologies. In particular, basic research is aimed at solving interrelated issues that allow the entire system to function effectively. These are issues of integrated information security and the study using catastrophe theory models of the predictability of instability as a result of hacker attacks, system hacks, etc. The study of these issues will allow not only to assess the impact and consequences of such attacks in time, but also during the development process, specific software and hardware solutions based on a system of residual classes will be proposed, the construction of new information protection systems that increase the stability of these systems to similar influences. These are issues related to the problems of data transmission and processing of large volumes of unsystematized information, and related problems of sharply increasing volumes of unsystematized information. These are issues related to the problems of electric power safety and the stability of the functioning of technical systems of the digital economy.

The proposed project of fundamental scientific research is based on the hypothesis that methods based on the theory of disasters and nonlinear dynamics will increase the predictability of the behavior of both individual elements of the digital economy, as well as its behavior and functioning as a whole. The study of catastrophe models in the tasks of predicting the behavior of elements or complex systems of the digital economy will allow us to «struggle» with the instability of such systems, as well as predict unstable behavior and, in the event such systems transition to such a state, quickly localize the consequences caused by such phenomena.

To achieve the goal of basic research, the following tasks are solved in the work:

IOP Conf. Series: Materials Science and Engineering **873** (2020) 012009 doi:10.1088/1757-899X/873/1/012009

- identify the main conceptual approaches to the study of the applicability of catastrophe theory and nonlinear dynamics to predict the behaviour of interconnected systems of the digital economy;
- determine the functional content of the application of catastrophe theory in economics and software and hardware systems;
- formulate the basic laws of the emergence of disasters and the predictability of their occurrences in digital economy systems;
- analyze the qualitative and quantitative certainty of the sustainability of the functioning and development of the digital economy and its elements;
- study the conditions and factors of increasing the sustainability levels of the digital economy and its system elements;
- conduct a comprehensive study of disaster models in the applicability of this theory in economic systems, in particular in digital economy systems;
- develop a methodology for interpreting disaster flags, which are indirect signs by which one can judge the possibility or presence of a disaster in the system;
- develop recommendations on the application of forecasting methods based on the theory of disasters and nonlinear dynamics;
- to ensure the stability of encryption algorithms and the security of information processes in the digital economy system, justify the use of the system of residual classes;
- study of the effectiveness of the proposed methods using simulation processes.

The solution of the tasks set in the work will make it possible in the future to obtain a scientific and technical reserve for the further implementation of the application of the theory of disasters and nonlinear dynamics in the tasks of predicting the behavior of systems in the digital economy.

The research being carried out is part of the direction of the Strategy for Scientific and Technological Development of the Russian Federation, namely, the transition to advanced digital, intelligent manufacturing technologies, robotic systems, new materials and design methods, the creation of systems for processing large amounts of data, machine learning and artificial intelligence.

3. Analysis of the current state of research in this field

The application of catastrophe theory and nonlinear dynamics in forecasting problems has not been practically studied, since for the economic and production systems the application of these theories is difficult for several reasons. One of these reasons is the difficult formalizability of factors affecting the behavior of such systems and the lack of sophistication in the theory of disasters, the definition of disaster flags.

In the 80s. there has been a leap in the development of catastrophe theory. Many different works on its study and application to various systems are published. An example is the work of such authors as A.V. Gaponova-Grekhova and M.I. Rabinovich «Nonlinear waves. Structures and bifurcations» (Gaponov-Grekhov, 1987), «Nonlinear waves. Dynamics and evolution»; G. Zaslavsky and R. Sagdeev «Introduction to nonlinear physics. From the pendulum to turbulence and chaos» (Zaslavsky, Sagdeev, 1998); R. Gilmore «Applied Theory of Disasters» (Gilmore, 1988).

In modern foreign science, economic security is studied in the context of two independent trends: catastrophe theory (I. Ziman, T. Oliva, E. Laszlo, G. Kolata, I. Casetti, etc.) and risk theory (R. Dembo, R. Keske , J. Clark, C. Arrow and others). However, there is a small number of works in which the authors directly examine the problem of interest to us (A. Westing, V. Kable, W. Albrecht) [4–7].

A fundamentally different approach to the study of the nature and specificity of the stability of economic systems, in our opinion, is possible on the methodological achievements of the natural and technical sciences in the field of self-organization (I. Prigogine, G. Haken) and the general theory of systems (R. Akoff, JI. Bertalanfi, I Blauberg, J. Clough). In this aspect, we note the evolution of methodology in economic research, the transition to interdisciplinary and system analysis in the works of domestic scientists - N. Moiseev, I. Shurgalina, V. Vasilkova, L. Evstigneeva and others.

IOP Conf. Series: Materials Science and Engineering **873** (2020) 012009 doi:10.1088/1757-899X/873/1/012009

It must be emphasized that questions on the topic under study, domestic scientists began to pay attention only in the early 90's in the face of a radical transformation of the socio-economic foundations of the country's life.

Despite the presence of significant research and fundamental approaches to the disclosure of the scientific problems posed in the draft, there is currently no single methodology for applying the theory of disasters and nonlinear dynamics in forecasting economic systems.

Several works can be distinguished from the works that are presented in the Web of Science database on the topic of research. For example, the following [8,9]. However, these works show one of the approaches, as a rule, highly specialized is applicable to one of the economic sectors of the financial market.

An analysis of the work of other authors, both Russian and foreign, allows us to conclude that there is no single integrated methodology for using the theory of disasters in economics. The practical application of catastrophe theory is usually of a private nature and it is not possible to apply the results of research for other economic or technical systems [10,11]. In this connection, the choice of the topic of our fundamental research was justified.

4. Scientific novelty of the studied subject

A fundamental scientific problem is the creation of a system reaction model (digital economy) and the prediction of systemic risk using the methods of catastrophe theory and nonlinear systems [12]. The applicability of such systems is possible in such components of the digital economy as the development of models of interaction in the Internet of things, in decision-making tasks in the smart city or smart home system. Scientific novelty consists in the development of a comprehensive methodology for applying the theory of disasters and nonlinear dynamics for the fundamental problems indicated above.

Among the sources of the fundamental problems of the digital economy are the following:

1. «Hacking» of systems (unauthorized access to data, or specialized attacks that provoke the system to exit the established (stable) mode of operation;

2. Poorly protected IoT devices can become entry points for cyber attacks, giving attackers the opportunity to reprogram the device or fail, devices that do not work correctly also create security vulnerabilities;

3. Problems of collecting unsystematized data, inadequate processing of such Big Data;

4. Control or regulation systems built 20, 30, or even 40 years ago transmit data in different formats, and no one will change the infrastructure of industrial control systems, including sensors, as this entails high capital costs;

5. The intensive exchange and use of large data streams reduce the confidentiality of the information used and contribute to the creation of a number of digital threats.

To conduct basic scientific research, the following research methods will be applied in the work:

1. Mathematical modeling.

2. The study of models of catastrophe theory and nonlinear dynamics in the formulated problems

3. The study of the possibility of increasing the security of the system using the system of residual classes.

4. Based on the results of simulation modeling, it is supposed to verify the adequacy of the proposed models and methods and, if necessary, to adjust them.

The main task of the economic system is optimization, i.e. maximizing the profit function or minimizing the cost function [13,14]. However, as our studies have shown, for managing the components of the digital economy, in particular technical or software, standard management methods can be used, but based on neural network management and the use of residual class systems.

The key expected results that are of practical importance include the following:

1. Development of a comprehensive methodology for the application of catastrophe theory in the tasks of predicting the behavior of digital economy systems.

2. To increase the security of these systems and eliminate factors that can lead to "catastrophic" consequences, a system of residual classes will be applied.

3. Systematization of the methods obtained as a result of research for their further application in science.

4. Identification of the main conceptual approaches to the study of the applicability of catastrophe theory and nonlinear dynamics for predicting the behavior of interconnected systems of the digital economy.

5. Determination of the functional content of the application of catastrophe theory in economics and software and hardware systems [15].

6. The formulation of the basic laws of occurrence of accidents and the predictability of their appearances in the digital economy systems.

7. Analysis of the qualitative and quantitative certainty of the sustainability of the functioning and development of the digital economy and its elements.

8. The study of the conditions and factors for increasing the sustainability levels of the digital economy and its system elements [16].

5. Conclusion

In any system, including the economic one, on which various factors act, not only smooth, but also abrupt, spasmodic changes occur. For the digital economy or technology of the Internet of things, any critical impact that could lead to a sharp change in all system parameters is not acceptable. The key to sustainable development of any system, and even more so having social significance, is the control of the behaviour of such systems. Therefore, forecasting and preventing sudden changes in such systems, which can lead to uncontrolled «folding» of such systems, is an extremely urgent task. As noted above, one of the approaches to effectively forecast and predict the future behaviour of such systems is the application of catastrophe theory.

To solve the tasks set, it is necessary to perform a number of key actions or steps. To identify the main conceptual approaches to the study of the applicability of catastrophe theory and nonlinear dynamics for predicting the behavior of interconnected systems of the digital economy. Determine the functional content of the application of catastrophe theory in economics and software and hardware systems. It is necessary to clearly formulate the basic laws of occurrence of accidents and the predictability of their appearances in the digital economy systems [17].

It is necessary to analyze the qualitative and quantitative certainty of the sustainability of the functioning and development of the digital economy and its elements. Including studying the conditions and factors for increasing the sustainability levels of the digital economy and its system elements.

It is necessary to conduct a comprehensive study of models of the theory of disasters and nonlinear dynamics to predict the systemic risks of managing the digital economy. Based on the analysis, to develop methods for interpreting disaster flags, which are indirect signs by which to judge the possibility or presence of a disaster in the system. To develop recommendations on the application of forecasting methods based on the theory of disasters and nonlinear dynamics.

A completely new approach to ensure the stability of encryption algorithms and the security of information processes in the digital economy system is proposed to justify the use of the system of residual classes. Conduct a study of the effectiveness of the proposed methods using simulation processes.

The solution of the tasks set in the study will allow in the future to obtain scientific and technical groundwork for further implementation of the application of the theory of catastrophes and nonlinear dynamics in the problems of predicting the behavior of systems in the digital economy.

References

[1] Răzvan Șerbu 2019 *Several Contemporary Economy Features, Consequences of Internet Expansion and I.C.T. Innovations in the World* Stud. Bus. Econ. 14 pp 175–81.

[2] Scott R C and Sattler E L 1983 *Catastrophe Theory in Economics* J. Econ. Educ. 14 pp 48.

[3] Tikhonov E E, Chebanov K A and Burlyaeva V A 2019 *Improvement of Information Protection and Data Transmission Methods in the Power Industry Using Neural Networks and a System of Residual Classes* 2019 International Multi-Conference on Industrial Engineering and Modern Technologies (FarEastCon) 2019 International Multi-Conference on Industrial Engineering and Modern Technologies (FarEastCon) (Vladivostok, Russia: IEEE) pp 1–5.

[4] Zahler R S and Sussmann H J 1977 *Claims and accomplishments of applied catastrophe theory* Nature 269 pp 759–63.

[5] O'Donovan N 2020 From *Knowledge Economy to Automation Anxiety: A Growth Regime in Crisis?* New Polit. Econ. 25 pp 248–66.

[6] Žikeš F, Baruník J and Shenai N 2017 *Modeling and forecasting persistent financial durations* Econom. Rev. 36 pp 1081–110.

[7] Rosser J B 2007 *The rise and fall of catastrophe theory applications in economics: Was the baby thrown out with the bathwater?* J. Econ. Dyn. Control 31 pp 3255–80.

[8] Pázsit I, Dykin V, Konno H and Kozlowski T 2020 *A possible application of catastrophe theory to boiling water reactor instability* Prog. Nucl. Energy 118 pp 103054.

[9] Lin X, Song S, Zhai H, Yuan P and Chen M 2020 Using catastrophe theory to analyze subway fire accidents Int. J. Syst. Assur. Eng. Manag. 11 pp 223–35.

[10] Cobb L and Zacks S 1985 *Applications of Catastrophe Theory for Statistical Modeling in the Biosciences* J. Am. Stat. Assoc. 80 pp 793–802.

[11] Cazzolli A, Misseroni D and Dal Corso F 2020 *Elastica catastrophe machine: theory, design and experiments* J. Mech. Phys. Solids 136 pp 103735.

[12] Oliva T A, Desarbo W S, Day D L and Jedidi K 1987 *Gemcat: A general multivariate methodology for estimating catastrophe models* Behav. Sci. 32 pp 121–37.

[13] Vošvrda M and Baruník J 2008 *Stock market crashes modeling: stochastic cusp catastrophe application* Polit. Ekon. 56 pp 759–71.

[14] Wagenmakers E-J, Molenaar P C M, Grasman R P P P, Hartelman P A I and van der Maas H L J 2005 *Transformation invariant stochastic catastrophe theory* Phys. Nonlinear Phenom. 211 pp 263–76.

[15] Cobb L 1981 *Parameter estimation for the cusp catastrophe model* Behav. Sci. 26 pp 75–8.

[16] El Hilali W, El Manouar A and Janati Idrissi M A 2020 *Reaching sustainability during a digital transformation: a PLS approach* Int. J. Innov. Sci. 12 pp 52–79.

[17] Kristoufek L and Vosvrda M 2014 *Measuring capital market efficiency: long-term memory, fractal dimension and approximate entropy* Eur. Phys. J. B 87 pp 162.

Elaboration of architecture of the enterprise of dairy animal breeding

O V Kochetkova, E V Shiryaeva, D P Arkov, A S Matveyev

Federal State Budgetary Educational Institution of Higher Education "Volgograd State Agricultural University", Faculty of Ecology and Reclamation, Universitetskiy Avenue, 26,400002, Russia

E-mail ovk555@bk.ru

Abstract. The actual task of dairy enterprises is to increase the efficiency of resource use, optimize business processes, reduce water and energy consumption, deploy a reliable traceability system that improves transparency, safety and product quality, providing detailed information about the product for the consumer. The specific task for dairy farms is monitoring the behavior of cows and their maintenance in order to improve their well-being and prevent morbidity. In turn, it will lead to better enterprise management, increasing of productivity and quality of production. It is possible to solve the set of tasks only with the help of digital technologies, which are vital for achieving the competitive advantage of dairy farms. It requires the elaboration of enterprise architecture for optimizing many disparate processes in an integrated environment, which will achieve the right balance between IT efficiency and business innovation. Enterprise architecture provides a strategic context for the development of IT systems in response to the ever-changing needs of the dairy industry environment. The research presents the “As Is” model of the architecture of typical dairy farms, developed using the ArchiMate language. As a result of its analysis, the discrepancy between business and IT requirements was found. An insufficiently high level of application of digital technologies was noted, allowing only to identify animals, but not giving the opportunity to determine their condition, to signal about its change and the need for urgent measures. The insufficient level of application of digital technologies does not provide support for decision-making by employees and managing specialists, and also does not allow to monitor in real time the parameters of the conditions of the animals, the quality of the products, the preparation of balanced feed and their individual dosing, automatic milking, storage, sorting and packaging products. The conducted studies are the basis for the creation of the standard architecture “To be” of dairy farms using modern digital technology and sensor systems for providing high quality and efficiency of production of dairy products.

Keywords: enterprise architecture, dairy farming, digital technology

1. Introduction

Nowadays the activity of dairy livestock enterprises is characterized by ever-increasing complexity and multidimensionality, because their sustainable activity requires improving the efficiency of resource use, safety and quality of products, compliance with environmental requirements, and reducing resource consumption.

The recently observed growth trend in the size of dairy farms [1] and an increase in the number of large farms present new, higher requirements to management [2]. In this case, an important management tool [3] is the introduction of reference business processes and best management practices supported by digital technologies [4].

The dairy farming enterprise is a complex system, the feature of which is the need to track and maintain the optimal state and development of biological objects and the environment. For this purpose, not only the state of the biological object and the environment, but also the nutrition, energy and technological impacts actually received by the animals and their environment are constantly measured, recorded and monitored. It allows you to identify patterns of development of animals and the formation of a product of production depending on the composition and amount of nutrients and energy, environmental parameters and the state of the cow. Important processes such as monitoring, planning and control can be supported by a wide range of digital technologies, hardware and software systems [5]. In addition, digital technology is an important factor in reducing labor costs, increasing animal productivity and reducing the likelihood of disease [6].

Creating a model and regimes for the formation of conditions conducive to the best development of animals and optimizing the objective function of production of maximizing the product or profit, taking into account real labor and technological resources and food and energy resources, is also impossible without the use of digital technologies. Therefore, there is an increasing role of IT in ensuring the efficiency and effectiveness of business processes on dairy farms, and in connection with it there is a constant increase in the requirements for the resource characteristics of information systems.

For the effective management of a dairy farming enterprise, which is a complex, multi-component system, it is very difficult, almost impossible to coordinate its parts and various aspects of the work; to keep in memory both the activities of people, and the work on managing the regime of keeping animals, and software, and the work of hard ware with communication and system soft. That is why, some authors of research in the dairy industry note that they did not find a real obvious economic benefit after the introduction of digital technologies [7], or it took a long time to find out how to use the advantages of digital technologies with the greatest return [8, 9].

To ensure the main characteristics of the enterprise and its activities, a common vision of the structure and functions of business systems and IT systems throughout the organization and a single strategic development of IT in accordance with business requirements, it is advisable to develop an architecture model. This is a system model that allows you to control the inherent complexity of systems and manage it in a changing environment.

The architecture of the enterprise allows you to make strategic decisions based on a holistic view at predefined goals, to change management, avoid duplicate costs of IT infrastructure, and also manage project risks [10]. The architecture allows you to formalize the problems and requirements to IT on the part of the business; shows how specific tasks are solved and how business requirements are realized in IT systems. Therefore, the development of the architecture of a dairy farming enterprise is an actual and important task, since it provides a stable and flexible environment for the enterprise.

The implementation of enterprise architecture requires two main components: the Architecture Framework, the results of which are architecture artifacts (models, diagrams, documents) and the methodology for its implementation [11, 12].

To demonstrate how the various components of an enterprise, such as organizational units, business processes, employees, and information systems, are interconnected and work in general to achieve goals [13], first the enterprise architecture "As Is" is developed. The second stage of this process is the assessment of the created architecture according to the criteria of efficiency, effectiveness, maturity, quality of results and functionality [14]. Based on the results of the assessment, the desired architecture (To be) is elaborated and the plan is presented for the transition from the existing to the desired enterprise architecture.

The objective of this study is to develop the "As Is" architecture of a typical Russian dairy farming enterprise.

2. Materials and methods

The study is based on a systematic approach, since the architecture of the enterprise is a complex system that includes IP as a fundamental component, as well as people, processes, information and technology of the enterprise, their relationship with each other and with the external environment. In addition, we used functional and logical analysis, as well as a synthesis of domestic and foreign theory and practice of creating enterprise architecture and features of the dairy industry of livestock.

To create the architecture of the enterprise, the study used the TOGAF technique, which is a tool for creating a wide range of diverse architectures and describes the process of creating architecture artifacts. Its utilized description format provides the management and the non-specialized target group with an easy-to-understand method for the development of all layers of architecture, adaptation to new business conditions and the possibility of implementing new plans in the company.

For the projecting of enterprise architecture, the ArchiMate language was used [15]. This language provides the ability to abstractly construct multi-layer enterprise architecture and the flexibility to work with various models and subject areas and describe the architecture from the point of view that is necessary for solving specific problems.

3. Results and Discussion

Based on the study and generalization of the activities of Russian dairy enterprises, the architecture of a typical enterprise was established, which includes several levels (Fig. 1). The business level describes the business elements and their relationships in the field of business architecture. The application level supports the business level and describes the elements of the architecture of the software application and their relationship. The infrastructure level describes the elements of an IT architecture domain and the relationship between software and hard ware. In a service-oriented architecture, a service is a unit of functionality that a system provides to an external environment. To obtain a list of services, the activity of business users of a typical dairy farming enterprise was analyzed for operations performed during work and access to certain sources of information.

After obtaining general information, the data was grouped by functional attribute into the following list of services: artificial insemination, management of animal welfare, veterinary control, production of dairy products, cooling dairy products, primary processing of products, sales of products.

These services provide the functionality necessary for operations in the activities reflected in the name of the service. In addition to services, the main external and internal business roles of the enterprise, representing the abstract user of business services were identified.

The analysis revealed the following internal business roles: artificial insemination specialist, livestock breeding specialist, veterinarian, technical services, workers, milkmaids. External business roles are: a supplier of insemination material, a supplier of equipment, a supplier of consumables, state regulatory authorities, buyers of dairy products, and buyers of calf meat. After the description of the main business roles and business services, each service and business role that is implemented by the business functionality were matched.

The next step in filling out the business layer was to highlight the levels of business functions and business processes. It was preceded by a lot of work on the study of the activities of dairy farms, formalization and analysis of the business processes taking place in them [16]. As a result of the work done, the main business functions of the enterprise were identified: herd reproduction, organization of the dairy herd, production of livestock products, management of the regime of keeping animals. Business processes that implement the herd reproduction business function: insemination, calving, rearing.

Business processes that implement the business function "Organization of the content of the dairy herd": placing the herd, feeding the herd, caring for the herd, veterinary control.

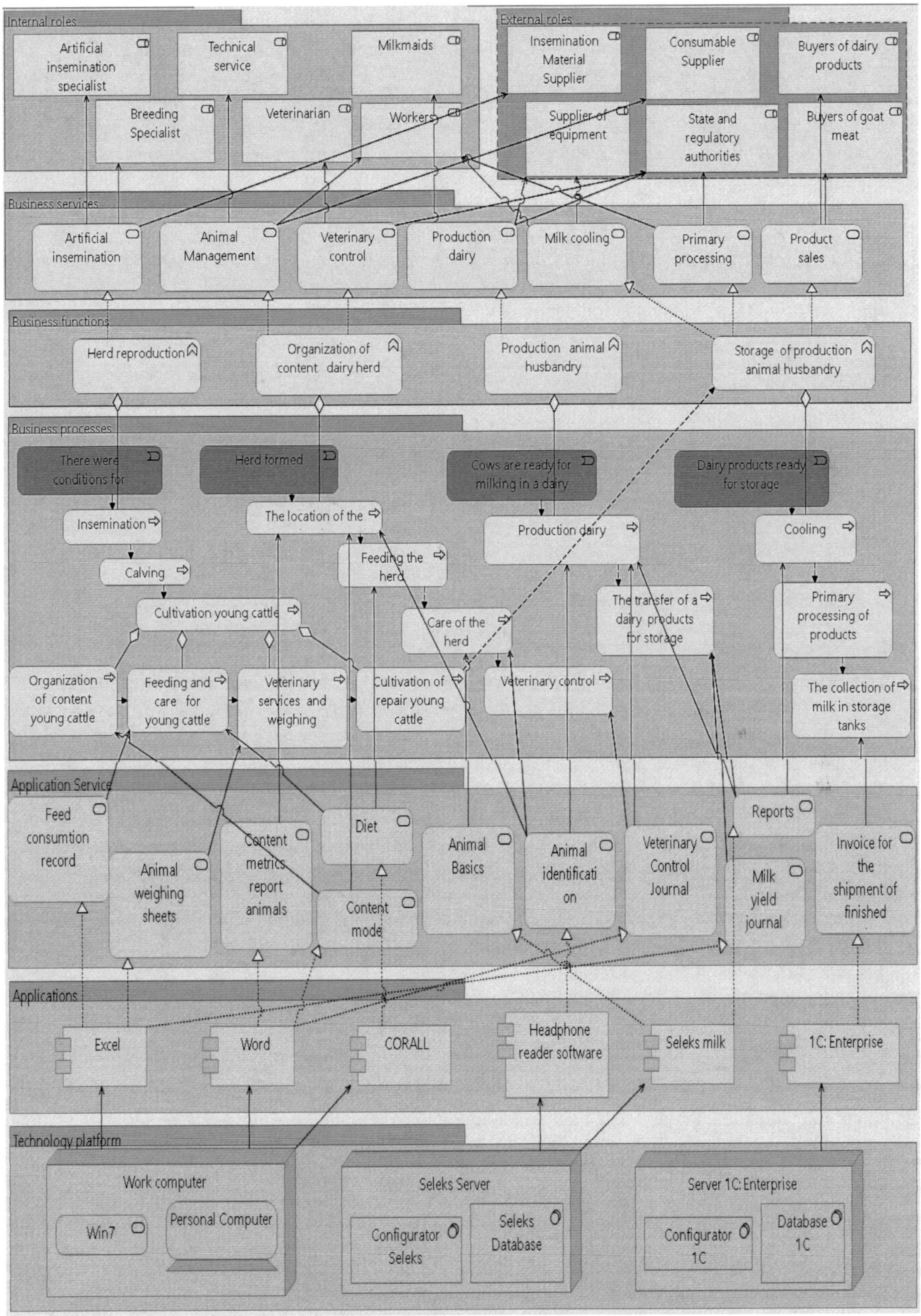

Figure1. Architecture model of a dairy farm

Business processes that realize the business function “Production of animal breeding”: dairy production and transfer of dairy products. Business processes that implement the business function "Organization of the content of the dairy herd": placing the herd, feeding the herd, caring for the herd, veterinary control. Business processes that implement the livestock production business function: dairy production and transfer of dairy products. Business functions “Management of the sale of livestock products is decomposed into a business process for cooling, primary processing and collecting milk in a storage tank.

The formation of the business layer of architecture is being completed by establishing links between business processes and services that are implemented using this business process.

During the description of application level, the service level and the level of application components were highlighted in it. Application services expose application functionality to users. During the analysis, the following application services were identified: statement of feed consumption, statement of animal weighing, report on indicators of keeping, diet, basic information about animals, keeping regime, identification of animals, journal of veterinary control, various reports, waybill for shipment of products.

The next step after studying application components is to describe the level of application components and to map the components and their services. The main applications used at the enterprise are: MS Excel, MS Word, software for reading information from the ear tags of cows, the information system "1C: Enterprise", a software product for compiling feeding diets "CORALL" and the information system "Seleks Milk".

The elaboration of enterprise architecture ends with the formation of a technological layer. The enterprise architecture at the technological platform level is represented by hardware in the form of a working computer, servers running Windows Server and network equipment.

During the elaboration of the architecture for provision the required performance and scalability, it is necessary to simulate equipment and system software carefully and thoroughly. The elaboration of enterprise architecture makes consider the availability of its modules and layers carefully, to establish relationships showing how the activities of the enterprise serve customers, the software services the work of those responsible and the work of other programs, the hardware services the work of the programs and the work of another hard ware. As a result, on the resulting architecture models, the mismatch of activity and the lack of proper software support becomes clearly visible. Therefore, there is a need to develop new architectural solutions "To be" taking into account the existing "As is". The evaluation of the developed model of architecture of a typical dairy farm in terms of efficiency allows us to help the company achieve its goals [17], identify shortcomings and make practical decisions to improve performance.

The analysis of the developed architecture model of typical Russian dairy farms made it possible to establish that they do not fully utilize advanced production tools, high-tech, digital technologies, and effective management methods based on decision support for specialists, that reduces the possibility of achieving high production results and makes it difficult to achieve strategic enterprise goals.

4. Conclusion

The elaboration and analysis of the architecture of typical dairy farming farms showed a low level of efficiency and maturity of the enterprise architecture due to insufficient support for production processes and regimes of keeping animals using digital technologies. At dairy farms the following are automated: animal identification, individual metering of milk production, calculation of deviations of production parameters from the reference ones and automation of accounting. Till the present date, a single information space has not been formed in the field of livestock breeding at various levels, ensuring the unification of pedigree accounting data and the adaptation of the legal framework to the real conditions of livestock breeding. There are no decision support systems that can be used to inform farmers of the best infrastructural and managerial changes to improve return on investment.

The use of digital systems is required in the implementation of processes that provide specified parameters for the execution of technological processes and living conditions in livestock buildings, the selection of machine operating modes, including operations such as preparing balanced feed mixtures, individual dosing of feeds, automatic milking, deep cooling of milk, automatic regulation of climate parameters indoors, sorting and packaging of finished products.

Therefore the developed "As is" architecture model of typical dairy farms allowed us to reveal insufficient support and compliance with goals of business and IT. There is a need to create a "To be" model that will help to radically modernize the infrastructure, develop compromise proposals for interested parties, manage changes in the company and achieve the goals set by the company.

References

[1] Barkema H W, von Keyserlingk M A, Kastelic J P, Lam T J, Luby C, Roy J P, LeBlanc S J, Keefe G P and Kelton D F 2015 *Invited review: Changes in the dairy industry affecting dairy cattle health and welfare* J. Dairy Sci. **98** pp 7426–7445

[2] Winsten J R, Kerchner, C D, Richardson A, Lichau A and Hyman J M 2010 *Trends in the Northeast dairy industry: Large-scale modern confinement feeding and management-intensive grazing* J. Dairy Sci. **93** pp1759–1769

[3] Barragan A A, Workman J D, Bas S, Proudfoot K L and Schuenemann G M 2016 *Assessment of an application for touchscreen devices to record calving-related events in dairy herds and monitor personnel performance* J. Dairy Sci. **99** pp 5662–5670

[4] Cummins C, Berry D P, Sayers R, Lorenz I and Kennedy E 2016. *Ques-tionnaire identifying management practices surrounding calving on spring-calving dairy farms and their associations with herd size and herd expansion.* Animal **10** pp 868–877.

[5] Kruize J W, Wolfert J, Scholten H, Verdouw C N, Kassahun A, Beulens A J M 2016 *A reference architecture for Farm Software Ecosystems* Computers and Electronics in Agriculture **125** pp 12-28

[6] Yazdanbakhsh O, Zhou Y, Dick S 2017 *An intelligent system for livestock disease surveillance* Information Sciences **378** pp 26-47

[7] Steeneveld W, Hogeveen H, Oude Lansink A G J M 2015 *Economic consequences of investing in sensor systems on dairy farms* Comput. Electron. Agri. **119** pp 33-39

[8] Ferrández-Pastor F J, Garcia-Chamizo J M, Nieto-Hidalgo M, Mora-Pascual J, Mora-Martinez J 2016 *Developing ubiquitous sensor network platform using internet of things: application in precision agriculture*/Sensors (Basel), **16** (7) 10.3390/s16071141

[9] Bewley J, Russell R A, Dolecheck K A, Borchers M R, Stone A, Wadsworth B A, Tsai I-C 2015 *Precision Dairy Monitoring Opportunities, Limitations, and Considerations* Paper presented at Western Dairy Management Conference, Reno, USA.

[10] Plataniotis G, de Kinderen S, Proper H A 2013 *Capturing decision making strategies in enterprise architecture–a viewpoint* Enterprise, business-process and information systems modeling Springer **147** pp 339-353

[11] Iacob M, Meertens L 2014 *From enterprise architecture to business models and back* Software & Systems Modeling **13.3** pp 1059-1083

[12] Simon D, Fischbach K, Schoder D 2014 *Enterprise architecture management and its role in corporate strategic management.* Information Systems and e-Business Management **12** pp 5-42

[13] Tamm T, Seddon P 2012 *How does enterprise architecture add value to organizations.* Communications of the Association for Information Systems **28**(1) pp. 141-168

[14] Giachetti R E 2012 *A flexible approach to realize an enterprise architecture* Procedia Computer Science **8** pp 147-152,

[15] Asif Qumer Gill 2015 *Agile enterprise architecture modelling: Evaluating the applicabilityand integration of six modelling standards* Information and Software Technologies **67** pp 196-206

[16] Kochetkova O V, Khoroshavin V S 2019 *Research of business processes of dairy animal breeding with purpose of their following automation* Lower Volga Agro University Comp. **4**(56) pp 332-342.

[17] Fatemeh Nikpay, RodinaAhmad, ChiamYin Kia 2017 *A hybrid method for evaluating enterprise architecture implementation* Evaluation and Program Planning **60** pp 1-16

Modification of entropy-based algorithm for determining words boundaries in noisy conditions

A S Karpov, V I Drozdova, G V Shagrova and M G Romanenko

Department of Information Systems and Technologies, North-Caucasus Federal University, 1, Pushkin Street, Stavropol 355017, Russia

E-mail: victoria_drozdova@rambler.ru

Abstract. A modification of the algorithm for finding word boundaries in continuous speech based on an analysis of the entropy of a speech signal is described. At the preprocessing stage, the signal is divided into frames; the entropy value is calculated for each frame and compared with the threshold value of the entropy γ. The algorithm is carried out in two stages, first a rough and then more precise definition of the word boundaries is made. At each stage, its own minimum distance between words is used. A modification is proposed, which consists in the fact that the entropy threshold γ is determined on the basis of the calculated SNR coefficient taking into account the experimental selection of signal parameters at the stage of precise determination of word boundaries in a noisy speech signal. Checking the functionality of the modified algorithm on noisy signals has shown that taking into account the effect of SNR on the selection of signal parameters increases the accuracy of word boundary determination.

1. Introduction

Automation of speech recognition systems is relevant both for traditional applications and for new breakthrough technologies. Automatic speech recognition began at the end of the 60s on the basis of the dynamic programming method for individual commands [1], followed by Markov chains, Fourier transform and wavelet transform, Format analysis and signal decomposition into empirical modes (EM) [2,3], cepstral analysis [3], correlation analysis, randomized algorithms of stochastic approximation [4], algorithms based on the calculation of speech signal entropy [5,6] and neural networks [7,8]. New and rapidly developing areas of application of speech recognition systems include: telephony: automation of processing of incoming and outgoing calls by creating voice self-service systems; Voice control system interface "Smart Home"; voice control of household appliances and robots; voice input in computer games and applications; voice control in the car; medical diagnostics of borderline human mental disorders [9] and a method for detecting psychoemotional human disorders based on the decomposition of speech signals into empirical modes and their format analysis [2,3]; systems that automatically generate dictated text [1].

Continuous speech consists of discrete components from phoneme to utterance, which are analyzed at various stages of signal processing. In particular, at the preprocessing stage, after filtering and noise suppression, informative sections of speech signals are extracted. The definition of word boundaries in the speech stream is an inherent part of the speech recognition process. It is at this stage that fragments that do not carry speech information are excluded from the signal, which increases the speed of the speech recognition system. The implementation of algorithms for determining word boundaries based on statistical processing of signals [4], the signal analysis in time domain, where the envelope of the signal is calculated and compared with previous created pattern stored in memory [10], entropy calculation [11] does not require the use of large computational resources, which can be useful for servicing mobile devices [6], recognition of keywords and individual commands, as well as the use of modern intelligent methods at the preprocessing stage [8]. Methods based on the calculation of speech signal entropy retain their effectiveness even under conditions of sharp fluctuations in signal amplitude caused by noise or the acoustic characteristics of the pronunciation of certain sounds [6, 11]. The approach considered in this paper makes it possible to use the results of analysis of the entropy of a speech signal to search for word boundaries. The aim of the work is to improve the accuracy of determining the boundaries of words in a speech signal by modifying the algorithm described previously by the authors in [12].

2 Materials and methods

To determine the word boundaries, an algorithm based on the calculation of the entropy of the speech signal is used, according to which in the preprocessing step, the following is performed:

1) normalization of digitized speech signal at which amplitude values range from 1 to -1;

2) dividing the normalized signal into frames by 25 milliseconds, overlapping by 25-50%.

Then, for each frame of the analyzed speech signal, the entropy value is calculated according to the technique described in [12]. As a result, the histogram of values of entropy of a recognizable speech signal for each frame turns out ($1 \leq j \leq m$). This histogram is called the entropy profile ξ:

$$\xi = [H_1, H_2, \dots H_m]. \tag{1}$$

The obtained values of the entropy of the speech signal are compared with the value of the threshold of entropy γ, which is calculated by the formula:

$$\gamma = min(\xi) + (max(\xi) - min(\xi)) \cdot k \tag{2}$$

where k is the coefficient that was selected experimentally.

Any value of H_j equal to or greater than the threshold of entropy is considered a speech, and all that is less is silence or noise. However, as a result of the comparison, some frames containing speech may be discarded as non-speech data. To avoid the erroneous definition of word boundaries in a speech signal, the concepts of the minimum word length (h) and the minimum distance between words (δ) [5] are entered. Both of these values are characterized by the number of frames. With h and δ in mind, each recognized speech segment λ_i must have a certain minimum length, which is indicated as a constant.

That is, if $\lambda_i < h$ and the distance between adjacent segments $d_{ij} > \delta$, then this segment is discarded as a segment that does not contain speech information. Two segments of the analyzed speech signal, defined as speech, are combined into one if the distance between them (d_{ij}) is less than the specified number of frames. The described algorithm is carried out in 2 stages, first a rough and then more precise definition of word boundaries is performed. At each stage, its own minimum distance between words (δ) is used. The presented algorithm copes well with determining the word boundaries of continuous speech recorded under ideal conditions (without distortion and noise) [12]. In this work, this algorithm has been tested for noisy audio signals, which are a recording of continuous speech, as well as modification, which allows to more accurately determining the boundaries of words in the noisy speech signal.

To estimate the influence of noise on the speech signal, the SNR indicator was used, which is expressed in decibels (dB) and can be calculated.

$$SNR = \frac{P_{signal}}{P_{noise}} = \left(\frac{A_{signal}}{A_{noise}}\right)^2 \tag{3}$$

where P is the average signal strength (noise); A - RMS value of signal (noise) amplitude.

The higher the SNR, the less noise affects the characteristics of the speech signal [13, 14, 15].

$$SNR(dB) = 10\, log_{10}\left(\frac{P_{signal}}{P_{noise}}\right) = 20\, log_{10}\left(\frac{A_{signal}}{A_{noise}}\right) \tag{4}$$

In telephone systems, the SNR is about 30 dB [4], so it was assumed that the operation of the algorithm should determine the word boundaries in a speech signal for which the SNR is 30 dB or more.

Taking into account the estimate of signal power and noise depending on the SNR value of the signal under study, the entropy threshold is calculated as follows:

$$\gamma = \begin{cases} SNR(Db) > 30, min(\xi) + (max(\xi) - min(\xi))\cdot 0{,}9 \\ SNR(Db) < 30, min(\xi) + (max(\xi) - min(\xi))\cdot 0{,}95 \end{cases} \quad (5)$$

To test the algorithm's performance under noisy conditions, white noise of various powers was artificially superimposed on the recording under study. Checking the original algorithm on the studied sound signals showed that it does not allow determining the boundaries of all words in a noisy signal. Therefore, a modification of the algorithm described in [12] was proposed, which consists in the fact that the entropy threshold γ is determined at the stage of exact determination of word boundaries in a noisy speech signal based on the calculated value of the SNR coefficient taking into account the experimentally selected coefficient k.

Thus, a modified algorithm for determining word boundaries, as well as described previously in [12], is performed in two stages. The first stage is reduced to the following operations:

1. Pretreatment. Dividing the signal into frames.
2. Rough definition of word boundaries.

The second stage of the modified algorithm is characterized in that the exact definition of word boundaries is made taking into account SNR values and is reduced to the following operations:

1. Calculation of threshold values of entropy taking into account SNR according to (5).
2. Comparison of the values of entropy with the threshold of entropy

$$\xi = \begin{cases} H_j, & H_j \geq \gamma \\ 0, & H_j < \gamma \end{cases}.$$

3. Checking the criteria

$$\begin{cases} \lambda_i < h, \ d_{ij} > \delta \\ (\lambda_i \text{or}\, \lambda_j) > h, d_{ij} < \delta \end{cases}$$

4. Formation of groups of frames corresponding to individual words.

3. Results

The results of the original and modified algorithms are analyzed using the example of the phrase λ"Last spring was very warm" (in Russian). The figure 1 shows the result of the original algorithm obtained by determining the word boundaries of continuous speech recorded under ideal conditions (without distortion and noise). At the second stage of the algorithm, only those groups of frames that were formed after the first stage were considered. For them, the parameters were used: k = 0.75, and the minimum distance between words was 3 frames, since the minimum duration of sounding of phonemes is approximately 50 ms. As shown in Figure 1, the original algorithm identified the boundaries of all five words.

The experimental results obtained on sound signals of continuous speech under conditions of noise of different power are shown in Table 1. The table shows the number of selected frame groups corresponding to the number of words in the analyzed speech signal for different values of the SNR and k parameters. The table shows that for k = 0.9, the algorithm provides an accurate definition of word boundaries in the signal for which SNR = 30 dB or more. An increase in the coefficient (k) to 0.95 facilitated the accurate determination of word boundaries in a speech signal for which SNR = 22 dB, but negatively affected the operation of the algorithm with signals for which SNR = 35 dB or more. Thus, a change in SNR affects the operation of an algorithm based on determining the entropy of a speech signal.

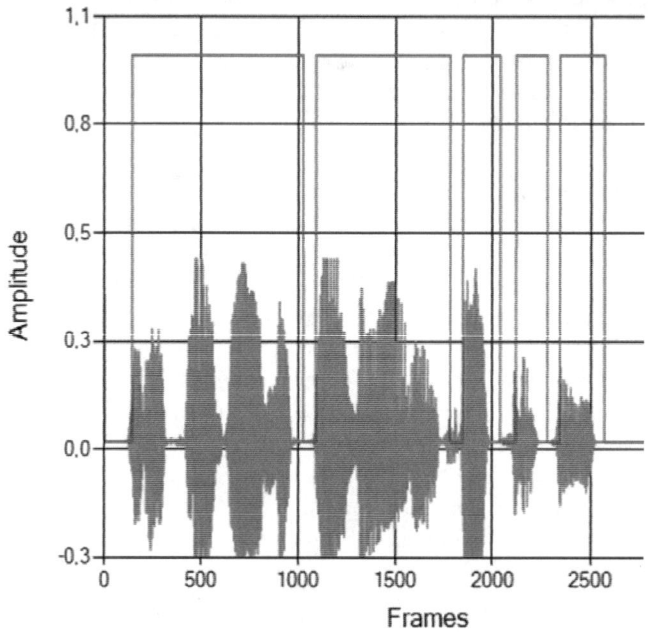

Figure 1 - The result of the algorithm for the phrase "Last spring was very warm"

Table 1. The number of selected groups of frames

	SNR =22 dB	SNR =30 dB	SNR =35 dB	SNR =40 dB	SNR =50 dB
$k = 0{,}75$	6	4	6	6	6
$k = 0{,}8$	6	6	3	4	3
$k = 0{,}85$	8	7	6	6	7
$k = 0{,}9$	4	5	5	5	5
$k = 0{,}95$	5	5	7	6	7

Figure 2a presents the results of the initial algorithm in which the entropy threshold at the second stage was calculated by formula (2), and a modified algorithm in which the entropy threshold at the second stage was determined according to (5). From the results presented in figure 2, it follows that the original algorithm did not cope with the task of determining word boundaries, highlighting four groups of frames. A modified algorithm accurately determined the boundaries of all five words. By calculating the SNR for the input speech signal, it is possible to vary the coefficient (k) depending on the degree of noise of the recognized speech stream. The modified algorithm allocated 5 groups of frames in the conditions of noise. Therefore, to more accurately determine the word boundaries in a noisy signal, it is proposed to use formula (5) to determine the entropy threshold.

4. Discussion

As a result of the analysis of the data of a computational experiment, it was found that the original algorithm allows you to accurately determine the word boundaries in continuous speech recorded under ideal conditions. When applying noise of different power to this speech signal, this algorithm does not allow determining the boundaries of all words in the analyzed message. Therefore, to more accurately determine the word boundaries in a noisy sound signal, it is advisable to use an approach in which the entropy threshold at the second stage of the algorithm is calculated based on the coefficient k, which must be selected taking into account the values of the SNR parameter. This approach provides a more accurate definition of the boundaries of speech signal words in noise conditions. However, for noisy signals it is advisable to check the values of the variable coefficient on a more representative sample.

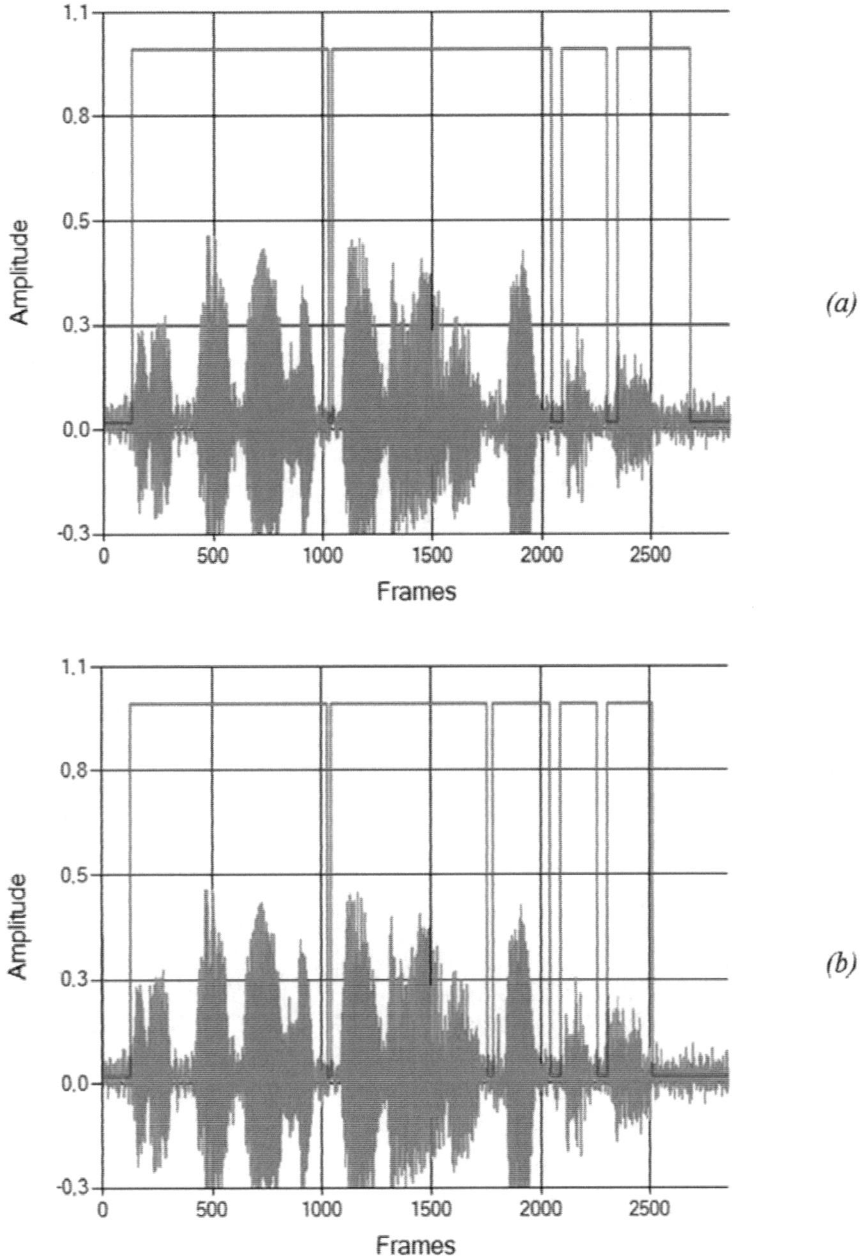

Figure 2.The result of the original (a) and modified (b) algorithms in a noisy environment

5. Conclusion

An algorithm is presented that allows you to select individual words in continuous speech based on the definition of the entropy of a speech signal. The relative simplicity and low cost of implementing the algorithm can find application at the stages of signal preprocessing. A modified algorithm for determining word boundaries in continuous speech based on the calculation of the entropy of a speech signal is presented. The process of determining word boundaries is carried out in two stages. At the first stage, a rough selection of large groups of frames containing voice information is performed. At the second stage, more detailed segmentation of speech fragments obtained at the first stage takes into account the SNR value.

References

[1] Tampel I B 2015 *Journal of Information Technologies, Mechanics and Optics* **15**:6 957 (In Russian) doi:10.17586/2226-1494-2015-15-6-957-968
[2] Alimuradov A K, Tychkov A Yu, Kuzmin A V, Churakov P P, Ageykin A V and Vishnevskaya G V 2017 *Proc.Int.Conf. Open Innovations Association FRUCT (Helsinki: Finland/ IEEE) p26*
[3] Alimuradov A K, Tychkov A Yu, Ageykin A V and Churakov P P 2018 *Proc.Int.Conf* on Biomedical Engineering, Radioelectronics and Information Technology (USBEREIT) (Yekaterinburg: Russia / *IEEE)* p 128
[4] Granichin O N, Shalymov D S , Avros R and Volkovich Z 2011 *Autom. Remote Control*, **72**:4 754
[5] Obin N and Liuni M 2012 *IEEE workshop on Spoken Language Technology*, United States
[6] Mustafa M K 2016 On-device Mobile Speech Recognition – *Dr.Ph Thesis Nottingham Trent University (*Nottingham:UK) p 149
[7] Romanenko A N 2018 *Scientific and Technical Journal of Information Technologies, Mechanics and Optics* **18** : 2, 350 (in Russian) doi: 10.17586/2226-1494-2018-18-2-350-352
[8] Nazarov A V , Yakimov V L and Avdeev V A 2015 *Information and Control Systems (http://www.i-us.ru)*: **75** :2 57 (In Russian) doi:10.15217/issn1684-8853.2015.2.57
[9] Alimuradov A K, Tychkov A Yu, Churakov P. P.and Ageykin A V 2017 University proceedings. Volga region Engineering sciences. Computer science, computer engineering and control **43**: 3 4 (in Russian) doi: 10.21685/2072-3059-2017-3-1
[10] Raczynski M 2018 *Proc.Int. Interdisciplinary PhD Workshop (IIPhDW)(Swinoujście:Poland)* p 27 doi: 10.1109/IIPHDW.2018.8388238
[11] Kochetkov A V and Fedotov P V 2015 *Internet journal "Naukovedenie"* (http://naukovedenie.ru)/, **7**(6) p1(in Russian) doi: 10.15862/72TVN615
[12] Karpov A S , Drozdova V I , Shagrova G V. and Shevchenko AV 2018 *CEUR Workshop Proceeding* **2254** 162
[13] Recommendation ITU-T G.160 2008 Objective measures for the characterization of the basic functioning of noise reduction algorithms (*Geneva: Switzerland,)* p 72
[14] Stolbov M B and Quan Trong 2020 *Journal of Information Technologies, Mechanics and Optics* **20**(1) 147 (in Russian). doi: 10.17586/2226-1494-2020-20-1-147-154.
[15] Kim C and Stern R M 2008 *Proc. Proc.Int.Conf. INTERSPEECH* (Brisbane: Australia) P 2598

Designing the algorithm of profiler's decision-making

A M Troshkov, S V Bogdanova, A N Ermakova, S N Dronov, AV Shuvaev

Stavropol State Agrarian University, Zootekhnichesky Lane, 12, Stavropol, 355017, Russia

E-mail a-troshkov@mail.ru

Abstract. In this paper, a package of information profile characteristics is proposed and formed based on the specified requirements, which are decomposed into three packages (kinect, thermal imaging, and visualization). They are processed on different hardware and software platforms. Designing a profiling algorithm only allows you to make a guess about possible involvement in an act of unlawful interference or an attempt to perform an unauthorized action. The actual threat can be identified only on the basis of an in-depth personal search of the alleged offender and the items transported and transported by them. Specialists in the use of face identification methods are called profilers, but there are not enough special information applications to improve the efficiency of such specialists ' work. Detection of suspicious fragments in human anthropometry in appearance and behavior occurs during psychological testing by a Profiler, which is understood here as a visual diagnosis of the internal state of a person through visible manifestations of characteristic signs that may indicate an upcoming illegal act, so the use of information technologies will allow more accurate and memorizing, as well as comparison increases the probability of a correct Profiler decision. An algorithm for generating decision-making Profiler theoretically increase the analytical work of specialists, and study of this issue was included in the study qualification of the employees of the linear police Department on transport, in particular at the airport and Stavropol train station as there is a conclusion of the chief of the police Department.

Keywords: WS-profiler, video profiling, Kinect profiling, commutation system, human mnemoanthropometric structure, multi-factor profiler characteristics, profiler, decision-making support system, thermal-imaging profiling, digital information.

1. Introduction

There currently exists a problem of constant worldwide fight against terrorism. However, this problem is still insufficiently resolved due to the deficiency of common global counter-terrorism information space and common counter-terrorism policy, politicization of recognized characteristics by experts, lack of clear and qualitative definition of terrorism, the concealment of terrorist practices and acts of terrorism in almost all countries of the world. At present fight against terrorism comes to the fore; moreover, there increased a number of unfair competitors, who penetrate into different spheres of operational dynamics of organizations with different forms of property. Based on the above profiling becomes a topical trend of research. The goal of research is implementing a sub-system of decision-making by an expert-profiler in order to improve the quality and increase the efficiency of executing the decision on the detection of suspicious persons.

First successful investigations were carried out in cooperation with transit police officers on the basis of anthropomorphic mnemosigns with the use of Kinect software platform and equipment. This enabled to expand the range of research activities and to practice the tactics of multi-factor profiling (adding thermal-imaging scanning and visual comparison). Profiling research enabled to perform observations even more carefully, to intentionally study the environment and the behavior of a person, as well as to search for new suspicious signs and to establish cause-and-effect relationship and to unconventionally address the problems, which arise in course of work execution. Present day international situation does not provide any encouraging forecasts that problems in the field of security will disappear in the 21st century.

Profiling is based on a concept that any person may appear to be a terrorist or an unfair competitor and any object may appear to be an explosive device, weapon or a device for information interception, etc. Therefore, any actions performed as a part of the technology under consideration are aimed to either prove or disprove this assumption. Consequently, profiling should be understood as the identification of potentially dangerous people and situations in course of profiling activities [1]. In order to ensure adequate safety and security it is reasonable to use classical methods of criminal intelligence operation, identification and technical devices for video monitoring in combination with up-to-date psychological methods, such as profiling, which is used to identify persons prone to illegal activities [1]. Concepts of oriental medicine and vast clinical experience of “face reading” have also been studied [2]. Based on the above profiler’s decision-making system has been suggested.

2. Materials and methods

The beginning of the research work is reflected in the program of primary personal identification by mnemoanthropometry of human skeleton [3]. Such investigations were facilitated by the discussion of information system for biometric identity authentication with experts.

Profiling is a new direction in psychological science that has appeared relatively recently and allows you to decipher nonverbal components used in interpersonal communication, but information technologies in this direction are not used enough. The analysis of scientific papers gives the right to say that profiling is an opportunity to get the most objective information about the potential danger of the subject and can be effectively used for preventive measures in the first echelon of inspection to prevent terrorist acts on transport and places of mass congestion of people. Profiling has already been used on the territory of the Russian Federation and in some countries on international flights at the airports of Pulkovo, Sheremetyevo, Domodedovo, etc. This technology is also used abroad: in the United States, Israel, and other countries in a number of airlines to ensure security at airports.

Research includes the following methods: system analysis, logical method, search and cognition methods, abstraction method, and other methods of scientific research. After designing the decision support system, the Profiler developed a block diagram of the multi-factor profiling software platform. The Profiler's decision support system is designed as a computer program. Taking into account the proposed three directions of automation of the Profiler's work with the arm, block diagrams of video profiling, thermal imaging profiling, and Kinect profiling were developed.

Simulation of profiling processes was performed in MP NPO “OOO IT-Video” laboratory; as a result, the possibility of further research activities has been confirmed.

3. Results

Processed information passes through the commutation system (commutator), which is divided into three streams (Kinect, thermal-imaging and video) with subsequent integration into a single common packet in accordance with result-based division (packets No. 1-3). In order to transfer the information of the common packet it is suggested to use data-transmission system, which converts special carrier-signals into the form that is suitable for processing and submission to WS-profiler. Security specialists take their final decision by means of decision-execution system based on the results and analysis of the received information (since humans are social by nature, i.e. humans are always judged by other humans – it is a long-lasting evolutionary mechanism [4]) (Figure 1).

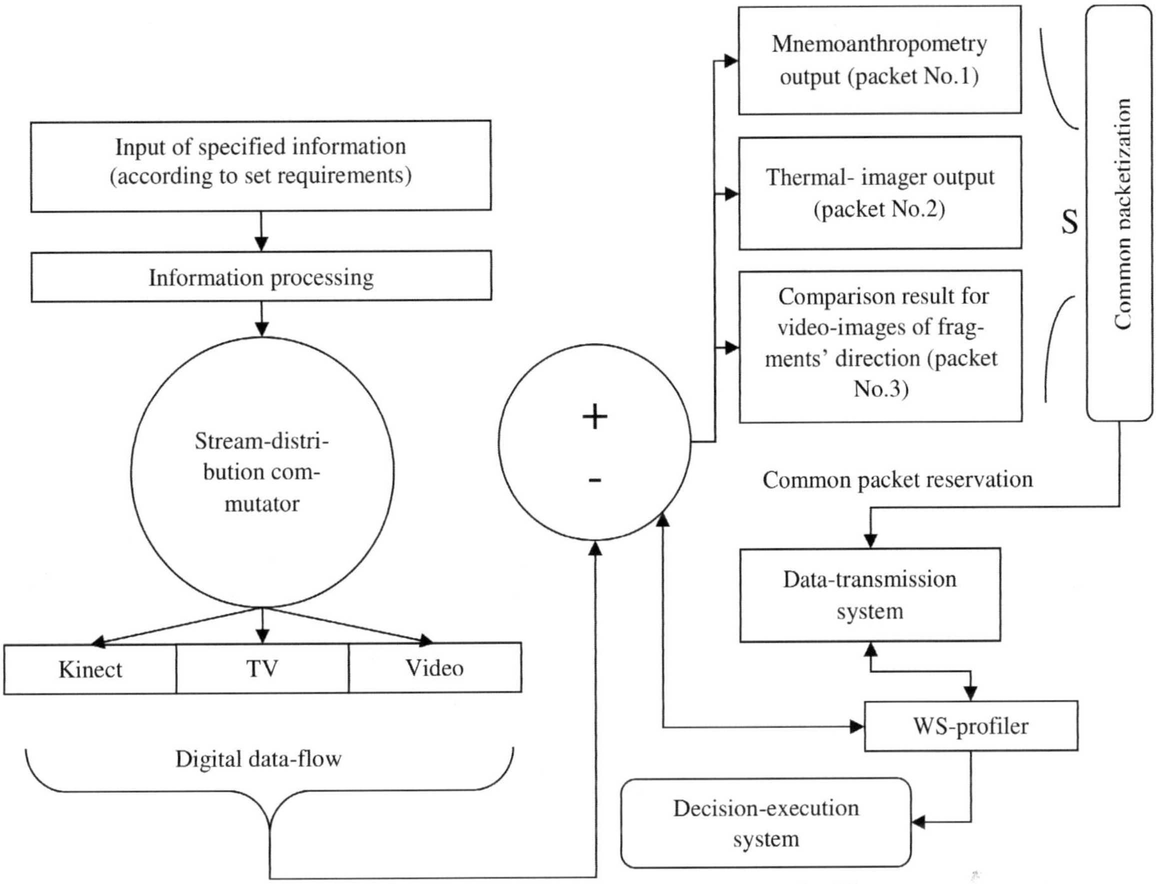

Figure 1. The formation of decision-making by a Profiler

The requirements, which are used for the operation of decision-making support system have been specified. Such requirements have been determined based on evaluation parameters of three components: Kinect, thermal- imager and video. Kinect parameters are based upon mnemoanthropometric differences and the calculation of deviations. Thermal- imager parameters are derived from the segments of human thermal radiation (red glow on pre-defined body areas). Video images enable to determine psychological facial expressions, which have been evaluated by psychologists. After that the information moves to processing system, where it is compared according to specified requirements and, further on – to the input of the commutator, which distributes streams by their identity: Kinect, thermal- imager and video. Digital information signals at the outputs of all devices are integrated into a single digital flow. Then digital information flow is divided into three packets П=1-3, which are unfolded into a common digital packet П1-3 by means of S multiplexer. Common digital packet П1-3 moves to profilers' workstation, where delivered information is processed and sent to the input of the decision-execution system by means of profiler's software platform (the request for program is sent in order to obtain the certificate).

Based on the results of designing profiler's decision-making support system there has been developed the flowchart for the software platform of multi-factor profiling [5] (Figure 2).

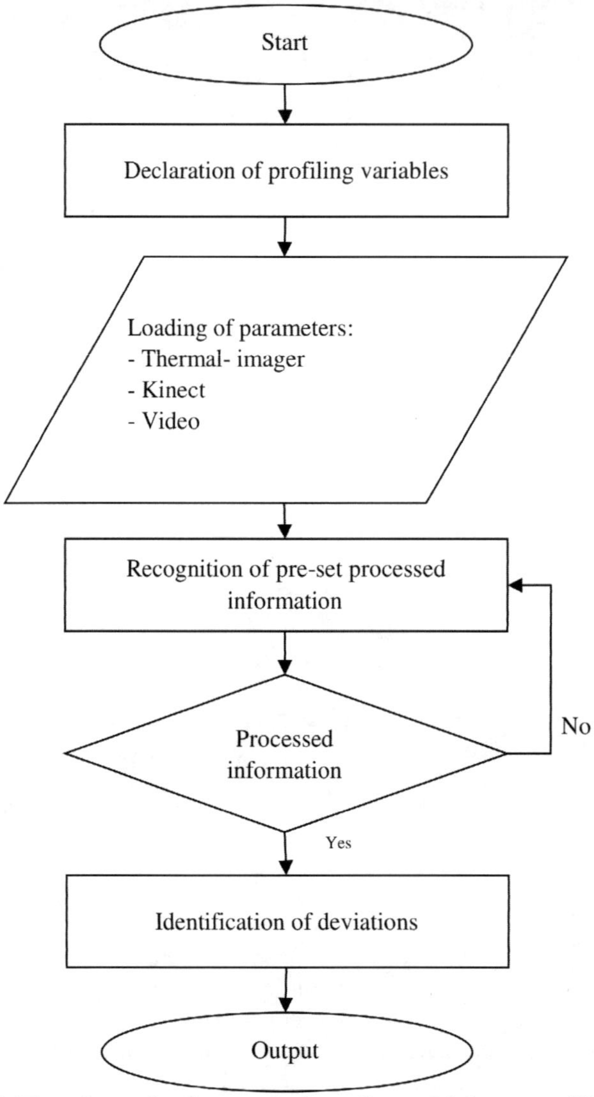

Figure 2. Flowchart of software platform for multi-factor profiling

The system of profiler's decision-making support is represented by a computer program. The program can be used on Windows operation systems together with installed MS Office 2007 programs (and later versions), with any browser. The system does not require pre-installation and is able to support both local and network operation, which is an advantage of this product. For network operation the user should preliminary provide access to information resources of a PC (server), where this software application will be located. Further on, it is required to use remote access or conventional network technologies in order to run a decision-support system executive file and to start system operation.

The designed profiler's decision-making support system is represented by a program, which opens a dialog window with the following active components:

1. Active hyperlink «Emotions».
2. Active hyperlink «Intonation».
3. Active hyperlink «Gestures».
4. Active hyperlink «Postures».
5. Active hyperlink «Facial gestures».
6. Active hyperlink «Thermal-imager».

IOP Conf. Series: Materials Science and Engineering **873** (2020) 012012 doi:10.1088/1757-899X/873/1/012012

Based on Figures 1 and 2 following algorithm of profiler's decision-making process has been suggested (Figure 3).

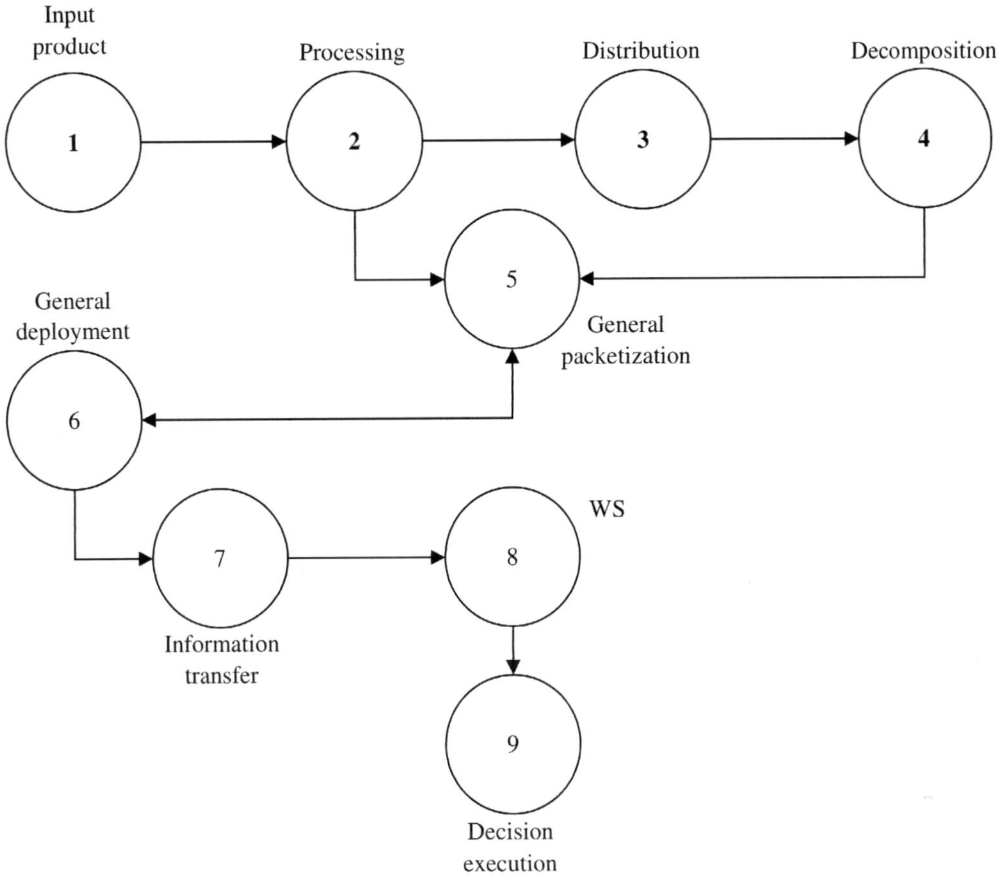

Figure 3. Algorithmization of the process of profiler's decision-making

Structural charts of video profiling, thermal-imaging profiling and Kinect profiling have been developed with reference to three suggested directions of automation of profiler's work with the WS.

The structural chart of video profiling is based upon the data of video monitoring and information-processing device. The parameters selected for decision-making are divided into two groups: facial gestures and pantomimicry, which are based upon human psychological profile, in particular upon the changes of facial areas and emotions. It is possible to perform the correction of facial expression and the selection of video simulation for the determination of operation modes (Figure 4).

IOP Conf. Series: Materials Science and Engineering **873** (2020) 012012 doi:10.1088/1757-899X/873/1/012012

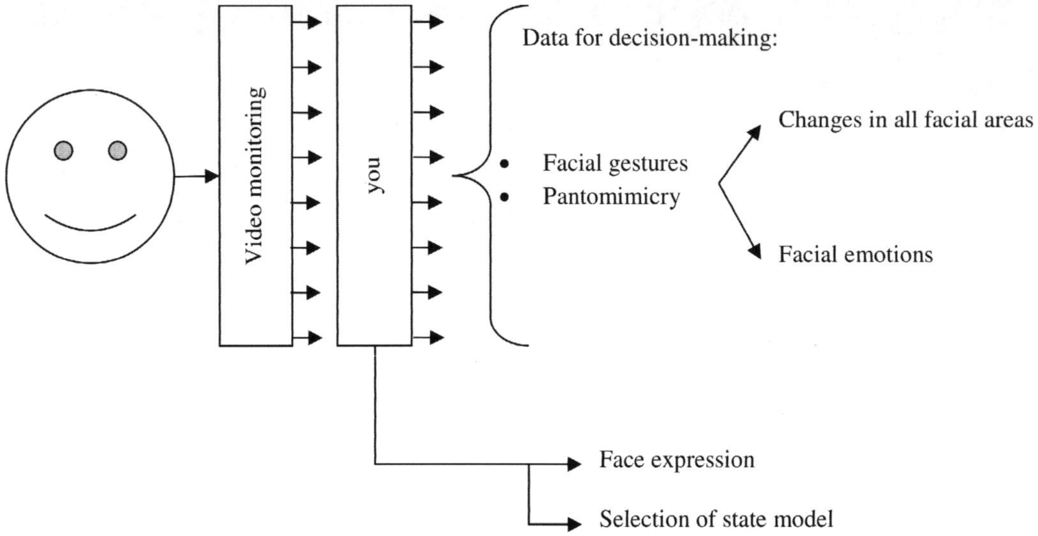

Figure 4. The structural chart of video profiling

The structural chart of thermal-imaging profiling operates on the basis of a thermal-imager, which identifies thermal areas of a human. Investigations show that thermal distribution within human body is not uniform under certain circumstances. Main control segments are distributed within the following areas: legs-and-chest, arms-and-chest, abdomen-and-chest. In order to receive more detailed information it is suggested to use gastroetiological, cardiological and spot measurement on human body (Figure 5).

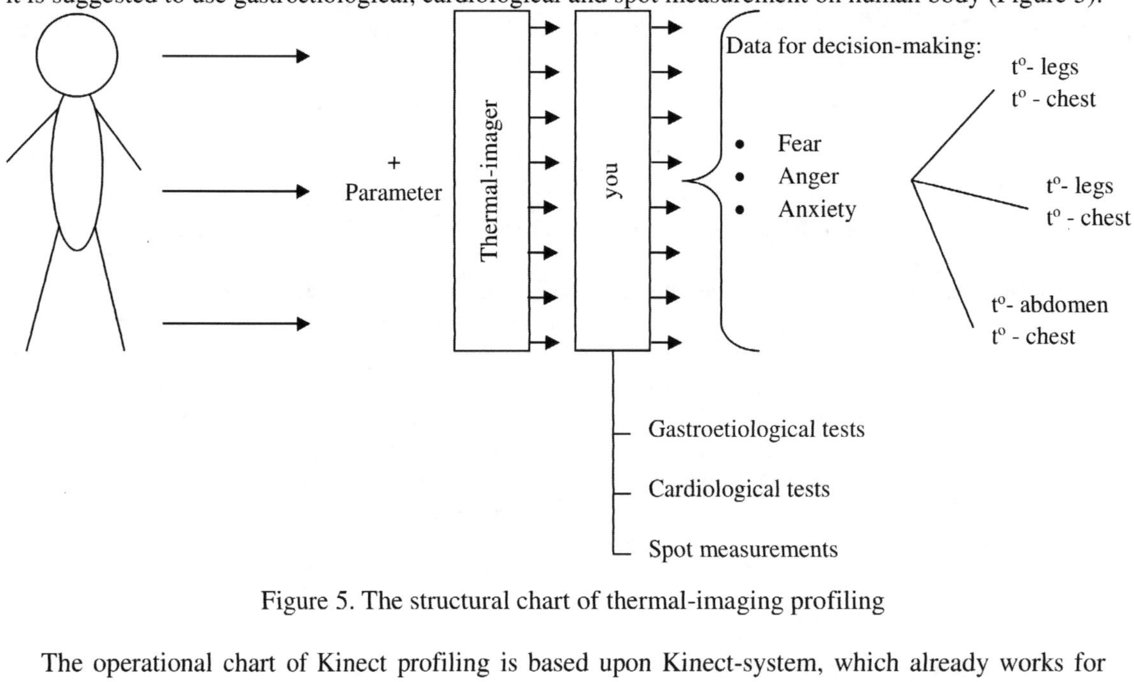

Figure 5. The structural chart of thermal-imaging profiling

The operational chart of Kinect profiling is based upon Kinect-system, which already works for mnemoanthropometry and investigated deviations from 0-axis into different directions in order to determine the behaviour of a human. Correction is performed based upon axis visualization and deviation points (Figure 6).

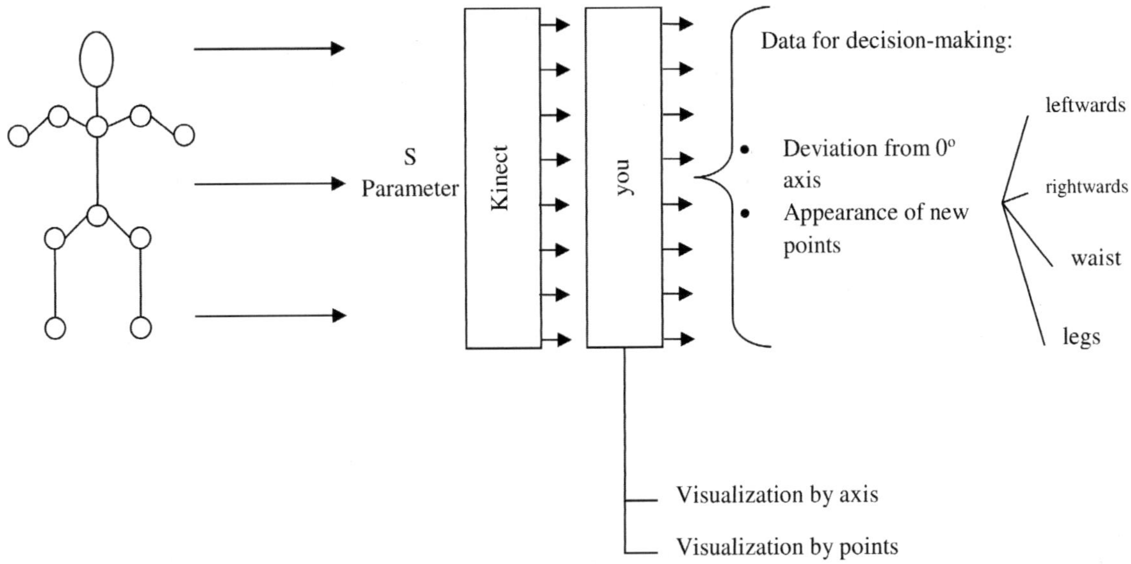

Figure 6. The structural chart of Kinect profiling

Based on charts in Figures 4, 5 and 6 following structural chart of multi-factor profiling is suggested (Figure 7).

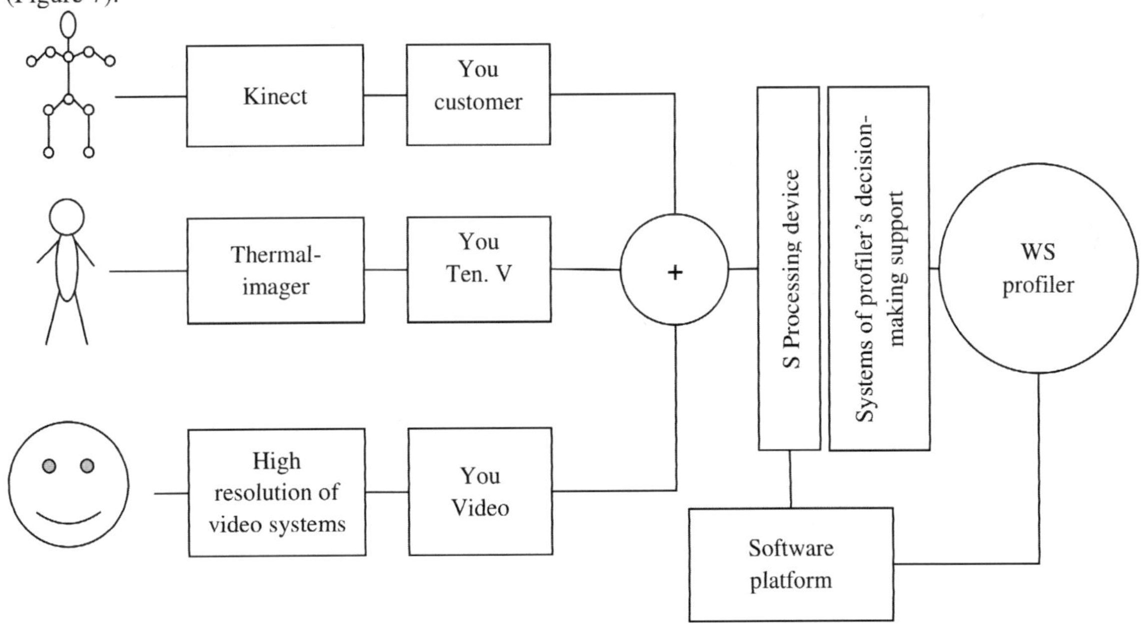

Figure 7. The structural chart of multi-factor profiling

The structural chart operates as follows. At the first step mnemoanthropometric structure of a human is investigated with the help of Kinect [3]. Kinect conversions are sent to adder unit, set within the first packet of digital information and coded, if necessary [6]. At the second step information from thermal-imaging device is examined [7]; it also moves to adder unit and is processed as the second packet of digital information. During the third step human facial expression (including facial gestures and pantomimicry [8]) is identified; this information is also sent to adder unit, is digitized and forms the third packet of information. Thus, a three-packet multi-factor code of digital information is composed [9]. S information processing device uses software platform of profiler's information and creates a special-purpose code [10-15], which is used by the system of profiler's decision-making support and is displayed in WS.

4. Conclusion

Due to constant development of information technologies and software platform of multi-factor profiling it is possible to improve the speed of decision-making and the reliability of personal identification.

The described results of the research paper are able to contribute to the solution of practical issues related to visual testing of human emotions, which are associated with simulation, suppression and concealment of expressive demonstrations, as well as the impact of individual-and-typical facial characteristics on facial gestures and some others. The accuracy of facial expression identification is inversely related with the field size of expressive meanings: the wider is the field size, the less is the accuracy of identification. The dynamic implementation of profiling information technologies into the operation of transportation security structure should provide significant benefit in this field. In fact, as laboratory tests show, even providing training on profiling with the use of multi-factor system and software platform for security officers by itself has positive effect on their subsequent attitude to their direct responsibilities, making them consciously change approach to their direct duties. It is significant that the implementation of profiler's decision-making support system as yet another potential aspect in security maintenance will qualitatively contribute to the overall situation of fight against terrorism. The more practical activities based on innovation technologies in this field are implemented, the more theoretical approaches and trends with their practical implementation of information technologies are involved, the higher is the security level. Thereby we provide indirect impact on the situation, reducing the risk of new terrorist attacks.

References

[1] Kulik A 2015 *The notes of a Profiler* (OOO LitRes) pp 6-7
[2] Gonikman E 2007 *The art of face diagnostics* (OOO Firma Veligor) pp 19-20
[3] Troshkov A M and Sapozhnikov V P 2016 *Program of primary identification of personality by the anthropometry of human skeleton* (Certificate of a computer program registration No 2016612872 of March 16)
[4] Spiritsa E 2015 *The psychology of deceit The way to expose a liar* (Saint Petersburg: Piter) pp 16-17
[5] Troshkov A M, Kosova T V, Dalvadyants V V and Fitisov D S 2011 The estimation of information content of sign systems by psychophysiological selection of experts in agribusiness industry *The collection of articles includes: Simulation of production processes and development of information systems (2nd International research and practice conference)* (Daugavpils University, Latvia, European Union; Belarusian State University of Economics, Belorussia; Kazakh National Technical University, Kazakhstan; North-Caucasus State Technical University, Russia; Stavropol State University, Russia; Stavropol State Agrarian University, Russia) pp 72-73
[6] Babkov V S and Sobolev E G 2012 The development of sub-system of interactive communication as a part of a training system with the use of Microsoft Kinect, IMS and KM 2012 platform *The proceedings of III All-Ukrainian scientific and technical conference of students, post-graduate students and young scientists* (Donetsk: Don NTU) pp 685-689
[7] Troshkov A M, Avilova M K and Bulanova T N 2013 Thermal-imaging visual diagnostics of the life of specimen within the colony of bees *The collection of articles includes: information systems and technologies as a growth factor of regional economics (II International research and practice conference)* pp 148-150
[8] Volkov V 2013 *Age psychology (From school childhood to late adulthood)* (Moscow: Vlados) pp 432-433
[9] Troshkov M A 2003 Increasing the quality of USB-range radio lines with the use of block pulse-code modulation and codes on the basis of the group of orthogonal signals *Collection of scientific*

articles of Povolzhskiy State University of Telecommunications and Informatics (SF PGATI) pp 11-12

[10] Zhuk A P 2003 The estimation of relative efficiency of DOS groups bandwidth by the value of the first lateral peak of the correlation function *Physics of wave processes and radio technical systems* Vol 6 (Samara) chapter 5 pp 28-29

[11] *Single window (Electronic Materials)* Access mode: http://window.edu.ru/ (access date 27.05.2019)

[12] Firulin A M 2015 The creation of Web interface for the expert system *Challenges and opportunities of technical sciences* pp 208-212

[13] Abrukov V S, Nikolaeva Ya G, Makarov D N, Sergeev A A and Karlovich E V 2008 The application of data mining tools (Data Mining) for the study of non-completely defined systems *Vestnik ChGU* vol 2 pp 233-241

[14] Grigorash A S, Kureychik V M and Kureychik V V 2017 Clustering software solution *Software products and systems* Vol 2 pp 261-269

[15] Brovko S A and Dikunova M S 2014 Planning and its basic principles *The collection of articles includes* (Contemporary practice conferences) 14 parts pp 14-16

IOP Conf. Series: Materials Science and Engineering **873** (2020) 012013 doi:10.1088/1757-899X/873/1/012013

Transaction Costs in the New Age of Digital Economy

O V Shumakova

Omsk State Agrarian University, Department of Economics, Accounting and Financial Control, 1 Institutskaya Square, Omsk, 644008, Russia

E-mail shumakovaov@mail.ru

Abstract. The present article is devoted to peculiarities in the formation of transaction costs in a period of increasing influence of digital technologies on economy and transaction practices. The integrated concept of transaction costs themselves is reported, while their classification is updated. The present article is meant for specialists in the fields of economy, education, science and technology transfer.

Keywords: transaction costs, digital economy, transformation, technology, institutions

1. Introduction

The digitalization of economy has become pertinent as the result of major breakthrough occurring in society. Current processes testify a new age for the management of the production and commercialization of goods, works and services, as well as in the main processes due to the end-to-end application of innovative information technologies, ranging from artificial intelligence to technologies of the industrial Internet of things. The interaction models based on the connection of electronic platforms lead to the implementation of new economic technologies. New means and methods of data elaboration for a task-oriented creation, transmission, conservation and presentation of content (data, ideas, knowledge) influence the volume of transaction costs in the interaction of economic agents.

In the digital segment of the economy of developed countries, the transaction sector accounts for more than 70% of the national GDP. This sector includes: public administration, information services, consulting, wholesale and retail trade, provision of different social and public services, finances. In those spheres which depend on the Internet, such as transport, commerce, logistics and others, the share of the transactions sector roughly accounts for about 10% of the GDP.

The improvement of the classification system of transaction costs, which should guarantee a proper approach to their management, is a crucial task for science, which has to provide consultative support of current economic processes, as well as of processes related to the conversion to digital economy.

The aim of the present research consists in the search for specific mechanisms aimed at simplifying the economy conversion process towards the introduction of innovative technologies and the possibility to reduce transaction costs in the new digital institutional environment.

2. Research methodology

The systematization of knowledge was obtained through a system of regulatory principles, procedures and methods through which an objective view can be obtained as regards the study of the influence of the spread rate of digital technologies on the availability of methods and possibility to reduce the volume of transaction costs.

Three main methods of scientific and educational activity were applied while carrying out the present research. Among the so-called universal methods of enquiry, the following ones were applied: analysis – in the characterisation of certain elements of the digitalisation process; synthesis – in the classification of transaction costs; generalisation – in the determination of the general properties and characteristics of transaction costs; induction and deduction – in the elaboration of a structure for this category of costs for a specific sector of economy, which can be recommended for application in other sectors as well.

Empirical methods were used in order to obtain practical knowledge while studying the process of the activity of economic subjects in digital economy based on target-oriented observation and experiments within the institutions.

All the applied methods are consistent with the following underlying principles: the principle of objectiveness (estrangement of the researcher from the object at study); the principle of continuity (research was carried out from 2005 to 2020); the principle of replicability (all the steps and phases of the research process can be replicated and verified by other researchers).

In order to determine the conceptual framework of the research the main paradigms of transaction costs were studied: I. Barzel, J. Commons, R. Coase, C. Ménard, R. Nelson, P. Milgrom, D. North, K. Polanyi, R. Richter, J. Roberts, O. Williamson, J. Wallis, E. Furubotn, T. Eggertson; methodology for the assessment of transaction costs at the level of economy in general: D. North and J. Wallis; the quantitative assessment of transaction costs on specific market: H. Demsetz and T. Loeb; issues related to the interconnection of information and transaction costs in economic systems: A.A. Auzan, S.B. Abdasheva, V.G. Belolipetsky, B.L. Inozemtsev, O.V. Inshakov, R.I. Kapelyushnikov, S.G. Kirdina, G.B. Kleiner, V.E. Kokorev, V.A. Kundius, D.S. Lvov, A.K. Lyasko, S.V. Malakhov, A.A. Migranyana, A.D. Nekipelov, A.N. Oleinik, V.M. Polterovich, V.V. Radaev, I.A. Strelets, V.F. Stukach, V.L. Tambovtsev, A.E. Shastitko, A.A. Yakovlev and others.

Among the paradigms of scientific knowledge, the concept of positive dialectical cognitive theory can be considered the most fitting and appliable in the present research, being the study of system, which consists of interconnected elements and opposed aspects. This system should be described as logically and consistently as possible, by creating behavioral models along the binary opposition between objectivism and humanism, by facing and integrating the aims of society development and competence development in economic subjects.

The theoretical and methodological foundation of the research is composed of scientific works and conceptions elaborated by domestic and foreign scientists, in which issues related to the development and transformation of economic sectors and on the consequences of mass digitalization on social life are taken into consideration.

3. Transaction costs in a neo-institutional environment

Occurring transactions are represented by the exchange of data and their interpretation, which determines the character of subsequent interactions; on its turn, this leads to the formation of relations among participants, the elaboration of behavioral rules, the change of behavioral motives, the transformation of the set of values, including behavioral rules, in the new circumstances of spreading of the results of information and communication technologies (hereafter, ICT).

The concept of transaction costs emerged within the framework of neoinstitutionalism, which developed as the negation of neoclassical models. Its principles are defined in the works of scholars such as Coase [1], Arrow [2], Kornai [3], Davis, Word. Problems related to transaction costs first appeared in the 1960-70s in the USA and Western Europe in works by scholars such as Demsetz [4], Jay, Paily, Olson [5] and others. This approach was more comprehensively elaborated by Williamson [6]. The empirical implementation of the theory of transaction costs was presented in works by the American scholars J. Wallis [7] and D. North [8]. In Russian scientific literature, such problems were faced while attempting to interpret costs which configure themselves as transaction costs in terms of interests, behavioral models, cultural traditions, previous social experience. Therefore, the works were dedicated to problems of business ethics, bureacratization and corruptibility of the authorities, formation of a Russian business elite, social sources for the reinforcement of entrepreneur groups. Attempts to introduce the term "transaction costs" into use are present solely in the development of scientific categories in economic theory. Rare exceptions are represented by the works of scholars such as V. Kokorev, R. Kapelyushnikov [9], V. Kurchenkov [10] and V. Radaev [11]. The digitalization of economic processes finds broad support in scientific circles, in particular among supporters of institutionalism.

4. Results

Institutional theory, applying information theory with the help of the synthetic method, analyzes transaction costs and enables a search for ideal economic relations. Russian economy, if compared to other countries, is characterized by a high level, in particular in the field of the statement and protection of ownership rights, which is quite often discussed in scientific literature [12].

Transaction costs in a narrow sense include expenses for the collection of information and its elaboration, the conduct of negotiations, the monitoring of the compliance of contract relations. However, due to the major breakthroughs occurring in society, production and commercialization of goods, works and services related to the spread of the application of innovative end-to-end and information technologies, this requires a transformation in the conceptual and categoric system, in particular – changes in the key definition. In the digital age, *transaction costs*[1] in a broad sense should be understood as the economic assessment of losses emerging as a result of the implementation of market transactions between economic agents, in particular based on digital partnership networks, the efficiency of which is guaranteed by a set of fitting institutions which carry out transactions in traditional and remote ways as well.

The complexity of the assessment of transaction costs is determined by the fact that the minimum level of transaction costs cannot be established (for example, as a result of the impossibility in different cases to draw a boundary between transformation, organization and transaction costs) and can change in time [13]. One of the most popular interpretations of transaction costs within the framework of institutionalism is part of the theory of P. Milgrom and J. Robertson [14]. They consider the shortcomings of coordination between economic agents as their source (for example, the incompleteness of information). This shortcoming can be exponentially decreased by introducing digital certificates and blockchain technologies. Such hybridization of services in transactions and the rejection of the uncoordinated provision of goods are of crucial interest in the new economy based on electronic platforms, banks and digital partnership networks. New institutional structures in the corporate ecosystem, by integrating the informational, organizational and material resources of economic subjects and blending the distinct boundaries between markets, create favorable conditions for the minimization of transaction costs for the transaction parties [15].

The advantages of ICT can be acknowledge with a good command of computer work, in particular using the cyberspace, and developing a specific mindset which can be called "open-mindedness". The rejection of the traditional approach to the management of the activity of any economic subject from house management to major corporations shows the tangible positive effects of the application of ICT – the reduction and simplification of time procedures for the preparation, arrangement, registration and tracking of transactions, the mitigation of risks in the conclusion of a transaction, which reduces transaction costs.

Institutional units which are well-informed, "advanced" in a communicational sense, mobile and striving to improve their position on the market, having an enhanced access to data content, can autonomously switch from one sector of the market to another and at the same time improve the structure of the market itself and the composition of transaction costs. However, in the Russian economy the amount of economic subjects with the stated characteristics is unsufficient and unable to have a material impact on the market. The causes can be identified in the following phenomena: in the inertness (conservatorism) of economic behaviour; in the lack of receptivity to the instruments of digital economy which allow for rational organization in the solution of any issues (due to the absence of opportunities

[1] As proposed by the author.

or skills for their application); in the aversion to spend time for the acquisition of new products; in the low degree of trust in ICT (risk of fraud, disclosure of confidential information); in the unsufficient familiarization with such services. The fact that the consumer determines the competitive position of the production in conditions of information asymmetry complicates the forecasting process [16].

The detractors of overall digitalization point at the following issue: the existing state information resources are not always completely available for all the subjects of the market, which can be explained by the absence of an integrated regulatory basis that might control their creation at federal, regional and local levels. This way, the amount of transaction costs increases in the exchange of information, for example, at interinstitutional level due to the low updating operativity, the weak relevancy and accuracy of the information provided settle the information asymmetry while making any particular managerial decision.

Numerous istances of empirical and practical research have proven that interaction on the present-dat market, meant as the sum of all the participants and the relations among them, is accompanied by a growth in transaction costs. In order to broaden the markets of merchandising and reduction of expenses, a system for the management of transaction costs for the participants of the corporate ecosystem consisting in the sequential realization of stages (the first of which is the necessity to classify the category of costs) is needed, as schematically shown in Figure 1.

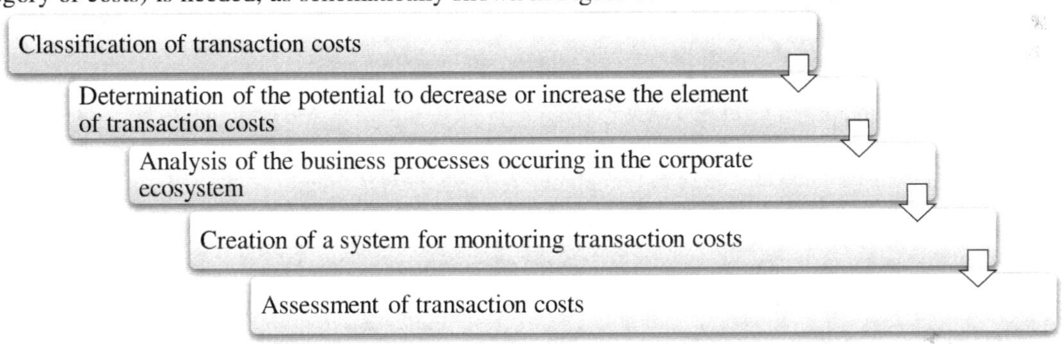

Figure 1. System for the management of transaction costs for the participants of the corporate environment

The whole classification set of transaction costs as evidenced through long-term research requires a systematic updating in the condition of digital economy against the background of the development and cheapening of personal and industrial electronics (computers, laptops, tablets, mobile phones, servers, telecommunication devices) as well as development and spreading of online services. As part of the research, the peculiarities of the performance conditions of economic subjects have been determined, while the composition and form of each type of transaction costs have been specified. The results are shown in Table 1.

Such structure of transaction costs includes all the possible expenses related to the settlement of a transaction within the corporate ecosystem in the conditions of digital economy. The organization of all the field of activity in the process of interaction with market subjects (institutions) accounts for transaction costs. The diversification in the form, essence and sources of transaction costs must be taken into account while introducing a mechanism for their management.

Only the systematic collection of information for subsequent analysis gives the possibility to determine the measures necessary to reduce the costs of each type. Transaction costs allow for a significant reduction in the scale of activity. The development of an infrastructural settlement as a consequence of the efficient management of transaction costs in "digital circumstances" minimizes their impact, increasing the benefits of the interactions among participants of the market and corporate ecosystem.

Table 1. Classification and composition of transaction costs in the conditions of digital economy

Type	Peculiarities in the new age	Elements / components
Information retrieval costs	Simplification in the collection and elaboration of information Correction of information distortion through mechanisms of consumer assessment and feedback Online search services Target-oriented advertising Marketplaces	Expenses in terms of time and resources for the formation and management of databases
		Expenses for information sourcing (specialized publications, traditional mass media, Internet, online services)
		Expenses in terms of time for competitive intelligence, in particular using paid marketing services
		Expenses for IT outsourcing
		Expenses for advertising and promotion
		Losses due to the commercialization of products, works and service at prices against self-interest
		Expenses for the participation to exhibitions, trade fairs, expositions, conferences, congresses, forums, sales pitches and briefings
Negotiation costs	Conduct of negotiations using the World Wide Web from any place on earth by means of audio and video conferences Reduction of transport expenses Online trade systems allowing the automatization of communication E-commerce in the consumer sector Chat bots for client communication Threat of competitive IT "wars" among providers (anonymous spread of misinformation)	Expenses for the allowance of contract negotiation managers
		Expenses for digital services and facilities: access to the Internet, mobile communication, local and long-distance telephony
		Legal outsourcing, including consulting services
		Representational expenses
Measurement costs	The volume of transaction increases due to a range of specific features Bots as software simulators and wastage of advertising budgets of advertisers when using "one-click payment" pricing models Accumulation of measurement methods form big volumes of information to recognize and integrate end-to-end big data technologies	Reimbursable financial loss as a result of qualitative factor mismatch
		Extra costs to increase the level of customer orientation in the production of goods, execution of works and provision of services
		Expenses for warranty repair and maintenance
		Expenses for the compulsory compilation of goods quality certificates
		Expenses for the acquisition of the documentation for quality control systems
		Expenses for the acquisition of measuring instruments
		Expenses for the allowance of the administrative subdivision responsible for quality control
		Non-material damage due to the low quality of production which has consequences on the entreprise reputation
		Substitution of faulty production

		Payment for the services of specialized institutions (standardization, metrologic and other centers)
Contracting costs	Possibility to draft standard contracts Acceleration of the contract signing process using e-signatures	Expenses for the formation and maintenance of intermediate institutions for project implementation
		Expenses for the external design of final documentation on e-commerce platforms
		Expenses for the preparation of contract documentation, including outsourcing
		Expenses for project lobbying (expenses for the creation of social awareness, gifts)
		Brokage
		Exchange rate fluctuations
		Payment for legal and notary services
		Losses due to incorrect legal compilation of the arrangement – complaints, fees and others
		Loan interests
		Missed profit due to unconcluded contracts, overspending, failure to meet project deadlines, failure to obtain support for grant applications
		Participation to tenders
Ownership right statement and protection (coordination) costs	Preponderance of informal institutions in the digital environment From the legal point of view, specific digital objects are not objects of ownership rights Without formal institution in the redistribution of ownership rights, transaction costs increase	Reimbursment of complaints to third parties
		Government fees
		Expenses for the registration of a juridical person in audit and oversight authorities, for license acquisition, for banking account opening
		Expenses for the allowance of outsourced security providers
		Insurance expenses
		Production labelling with conformity marks which are counterfeit-proof
		Losses in the physical volume of production (beside natural loss rates) in the commercialization process
Opportunist behavior (motivation, protection from third parties, "politicization") costs	Risks related to risk and trust in technologies on part of the society, state authorities and other institutions Development of the field of IT and cyber security Lock out of IP adresses, domain names, protocols, messengers, social networks and other services as the end of the age of free Internet	Expenses for the correction of compliance risks
		Expenses for the allowance of specialists of the (accounting and legal) subdivision responsible for the exaction of debts or accounting and legal outsourcing
		Expenses for the allowance of the legal department
		Expenses for the exaction of debts – legal expenses, wages of negotiators, payment for debt recovery services
		Losses due to workers' negligence
		Charges of paid fees, penalties, as well as interests provided by contracts in case of non-compliance with their conditions
		Corruption-ridden expenses in business processes
		Expenses for double-entry bookkeeping

5. Conclusion

The present article has solved the problem of generalization of behavioral features of transaction costs in the age of the increasing impact of digital technologies on economy and transactional practices. As a result of research, the integrated concept of transaction costs themselves has been reported and their classification has been updated, the conceptual and categoric system for the execution of procedures for the control of the assessment of the transaction costs of economic subjects has been transformed. This proposition should be taken into account in the decision-making process at an intermediate management layer. The efficiency of the subject functioning is influenced both by the rational use of resources and by the establishment of efficient economic connections with external contractors (producers and consumers), with the monetary and financial system using end-to-end technologies and ICT.

There is a reason to assert that not only the level of transaction costs, but also their structure and distribution among the participants of an economic exchange are relevant, since all this reflects the specific configuration of the institutions, therefore the task before state authorities consists in the creation and support of a functioning new digital infrastructure, in the provision of digital economy with acting regulatory acts, guaranteeing the statement and protection of ownership rights with minimal expenses, as well as in the development of digital skills in the population.

An important trend in the development of the corporate ecosystem is represented by the introduction of the principles of digital economy, which enables the formation of an institutional environment in compliance with the current requirements and circumstances, as well as a decrease in transaction costs and a general increase in economic efficiency.

As a result of the research, the propositions of the scientific bases of the mechanism of the formation of transaction costs in the structure of interaction among economic subjects in the conditions of the general digitalization of the categories of institutional economy have been refined.

References

[1] Coase R 1993 *Firma, rynok i pravo* transl. into Russian by B. Pinsker (Moscow, Delo LTD with the participation of Catallaxy publishing) 192 pp

[2] Arrow K 1995 *Informatsiya i yekonomicheskoe povedenie: voprosy teorii* Voprosy yekonomiki 5 p 98-107

[3] Kornai J 1990 *Put k svobodnoi yekonomike* Moscow: Yekonomika 149 pp

[4] Shalneva V, Blazhevich O 2019 *Osnovnye metodologicheskie podkhody k opredeleniyu transaktsionnykh zatrat* Nauchnyi vestnik: finansy, banki, investitsii vol 7 URL: https://cyberleninka.ru/article/n/osnovnye-metodologicheskie-podhody-k-opredeleniyu-transaktsionnyh-zatrat

[5] Sukharev O *2017 Institutsionalnaya teoriya M. Oslona: "Printsip rasshireniya"* JIS vol 2 URL: https://cyberleninka.ru/article/n/institutsionalnaya-teoriya-m-olsona-printsip-rasshireniya

[6] Williamson O 1996 *Yekonomicheskie instituty kapitalizma. Firmy, rynki, "otnoshencheskaya" kontraktatsiya* transl. into Russian byY Blagov, V Katkalo at al Saint Petersburg: Lenizdat; CEV Press 702 pp

[7] Denisova A 2019 *Metody otsenki transaktsionnylh izderzhek* Academy vol 40 URL: https://cyberleninka.ru/article/n/metody-otsenki-transaktsionnyh-izderzhek

[8] North D 1997 *Instituty, institutsionalnye izmeneniya i funktsionirovanie yekonomiki* transl. into Russian by A Nesterenko Moscow: Fond yekon. knigi "Nachala" 180 pp

[9] Kapelyushnikov R *Gde nachalo togo kontsa?* Voprosy yekonomiki vol 1 p 138-156

[10] Tashmatov R 2016 *Transaktsiya kharazhatlarini koshqarish kontsepstsiyasi* Yekonomika i finansy vol 10 URL: https://cyberleninka.ru/article/n/transaktsiya-harazhatlarini-bosh-arish-kontseptsiyasi

[11] Radaev V 2002 *Rossiiskii biznes: na puti k legalizatsii?* Voprosy y ekonomiki 1 p 68-87

[12] Litvintseva G, Gakhova N 2016 *Dinamika transaktsionnogo sektora yekonomiki Rossii: kak uchil D. Nort* Journal of Institutional Studies 2 p 38-50

[13] Kotlyarov I 2017 *Transaktsionnye izderzhki i funktsionirovanie khozyaistvuyushchikh subektov* Journal of Institutional Studies 1 p 69-76
[14] Syrovatkina T, Fedorova O 2018 *Tsifrovaya yekonomika kak institutsionalnaya sreda transaktsionnykh izderzhek uchastnikov rossiiskogo rynka zhilya* KYe 2 URL: https://cyberleninka.ru/article/n/tsifrovaya-ekonomika-kak-institutsionalnaya-sreda-transaktsionnyh-izderzhek-uchastnikov-rossiyskogo-rynka-zhilya
[15] Yakubanis N 2011 *Razvitie integrirovannykh biznes-grupp kak faktor modernizatsii promyshlennogo kompleksa strany* SRRM 5 URL: https://cyberleninka.ru/article/n/razvitie-integrirovannyh-biznes-grupp-kak-faktor-modernizatsii-promyshlennogo-kompleksa-strany
[16] Averyanova V 2017 *Marketingovaya politika v novykh yekonomicheskikh realiyakh* Izvestiya SPbGAU vol 47 URL: https://cyberleninka.ru/article/n/marketingovaya-politika-v-novyh-ekonomicheskih-realiyah

Model of phase synchronization system operating under conditions of violent ionospheric disturbances

V A Tsimbal, S N Shimanov, D V Mokrinskiy, V E Toiskin

MGEI Institute of Engineering Physics, Udarniy Bolshoi per., 1-a, Serpukhov town, 142210, Moscow region, Russia

E-mail dmitrimokrinski1991@mail.ru

Abstract. A model of the phase synchronization system based on the circuit of digital phase-locked-loop frequency control, adaptive equalizer, and a system of frequency-matched filters is proposed. This model provides the required value of the bit error probability at the output of the digital demodulator under the influence of strong ionospheric disturbances in the decameter communication channel.

Keywords: synchronization, phase-locked loop frequency, adaptive equalizer, transfer function, loop filter, frequency decay.

1. Introduction

It is known that the decametric channel (DCW) is characterized by a complex of negative effects that significantly complicate the qualitative detection of the received signal. One of the factors (along with additive and multiplicative noise [1,2]) that significantly affect the quality of reception is the presence of a low frequency in the communication channel. Frequency decay can be associated with the movement of the transmitter and receiver relative to each other as well as with the uniform rise or fall of the reflecting region of the ionosphere for a limited time interval. In the second case, there will be a classic Doppler effect, i.e. a change in the frequency of the received signal. According to P. Grin and M. P. Dolukhanov [3,4], a value of the maximum decay of the DCW frequency of the communication line during strong ionospheric disturbances can reach 100 Hz. Currently, in order to compensate for frequency loss in the digital communication systems the adapted classical methods are proposed based on the implementation of digital (software-implemented) circuit of the phase-locked loop (PLL), adaptive equalizers (AE), and a set of frequency-matched filters [5-9]. However, the problem of frequency loss compensation in the presence of fading in the communication channel is not sufficiently studied. One of the main problems of such systems is the presence of circuit synchronization failures with significant frequency deviations [10,11]. The long process of restoring synchronism leads to irremediable reception errors, which negatively affects the noise immunity of the communication as a whole. Thus, an urgent task is to develop a model of the received signal phase synchronization system that can meet the specified requirements for noise immunity under the influence of strong ionospheric disturbances.

2. Formal statement of the problem

Source data:

1. The structure of the DCW communication protocol, a feature of which is the sequential alternation of information sequences and sequences for learning AE:

$$F_r\left(fr_n, fr_m\right), \tag{1}$$

where fr_n is the length of information sequence; fr_m is the length of AE learning sequence.

2. Parameters of the adaptive filtering subsystem:

$$F_a(F_{a\_type}, F_{a\_n}, F_{a\_m}), \tag{2}$$

where F_{a_type} is the type of adaptive filtering algorithm; F_{a_n}, F_{a_m} is the number of forward and reverse branches of the AE (determine the order of the AE)

3. The PLL circuit parameters:

$$F_{PLL}\left(K_d, F_{Filt\_Order}, G_1, G_2, K_{NCO}\right), \tag{3}$$

where K_d is the gain of the phase detector (FD); F_{Filt_Order} is the order of the loop filter; G_1, G_2 is the proportional and integral gain of the loop filter; K_{NCO} is the gain of the voltage - controlled generator (VCO).

4. Type of signal modulation: F_m.

5. Parameters of communication channel:

$$F_C = \left\{F_{C\_AWGN}, F_{C\_Fad\_type}, I, W_{ci}, T_{ci}, \theta, \Omega T, \Omega^2 T, f_c\right\} i = \overline{1, I}, \tag{4}$$

where F_{C_AWGN} is the parameter of additive noise (value of the ration signal-noise at the demodulator input; $F_{C\_Fad\_type}$ is a fading type; I is the number of rays distributions; W_{ci}, T_{ci} are beam power and delay relative to the principal one; θ is a jump phase value; ΩT is a frequency jump value, $\Omega^2 T$ is a value of frequency drift rate; f_c is a value of maximum frequency decay.

6. Sample rate of the received signal: f_D.

7. Number of frequency-matched filters: N.

It is necessary to justify the structure of the phase synchronization system (F_{PLL}, N) that provides the required noise immunity of communication, with the specified F_r, F_a, F_m, F_C, f_D.

The indicator of noise immunity as a manifestation intensity of this property, we assume the probability of a bit error at the output of the demodulator (BER) of the radio receiving digital device E_b.

3. Mathematical model

To analyze the functioning and stability features of the system under study, the mathematical apparatus of z-transformations was used. A linearized model of a typical PLL circuit in the approximation of small phase angles is shown in figure 1. This system is described by a transfer function (TF) of the following type [12,13]:

$$H_c(z) = \frac{\phi(z)}{\psi(z)} = \frac{\sqrt{2P}K_d N(z)F(z)}{1+\sqrt{2P}K_d N(z)F(z)}, \tag{5}$$

where $N(z)$ is the VCO TF; $F(z)$ is the loop filter TF; $2P$ is a square of the amplitude of the received signal; $\phi(z)$ and $\psi(z)$ are the z-images $\phi(n)$ и $\psi(n)$ (of input readouts and ones from VCO output).

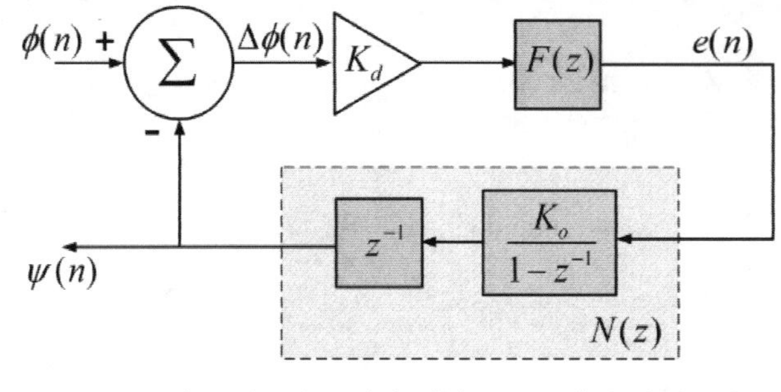

Figure 1. Linearized model of the PLL digital circuit

In the course of the researches carried out to substantiate the structure of the phase synchronization system (F_{PLL}) an option was offered based on the PLL and AE included in the synchronization circuit (figure 2). Taking into account that the AE is a linear system at each fixed time, the AE was represented as a TF linear adder [14]:

$$H_{AE}(z) = \frac{1+\omega_0 z^{-1}}{1-\omega_1 z^{-1}} = \frac{z+\omega_0}{z-\omega_1}, \tag{6}$$

where ω_0, ω_1 are the weight coefficients of the linear adder.

Considering the known expressions [13] for describing the VCO TF (7), the TF of the loop filter of the first (F_1) and second (F_2) order (8) and the expression for the TF by mistake (9):

$$N(z) = \frac{K_{NCO} z^{-1}}{1-z^{-1}} = \frac{K_{NCO}}{z-1}, \tag{7}$$

$$F_1(z) = G_1, \quad F_2(z) = G_1 + \frac{G_2}{1-z^{-1}}, \tag{8}$$

$$H_e(z) = 1 - H_c(z); \tag{9}$$

the following expressions were obtained:

$$H_1 = \frac{z-1}{z-1+K_1\left(\frac{z+\omega_0}{z-\omega_1}\right)G_1}, \quad H_2\frac{(z-1)^2}{(z-1)^2+K_1\left(\frac{z+\omega_0}{z-\omega_1}\right)(G_1(z-1)+G_2 z)}, \tag{10}$$

where H_1 is a TF of the first-order circuit; H_2 is aTF of the second-order circuit; $K_1 = \sqrt{2P}K_d K_{NCO}$.

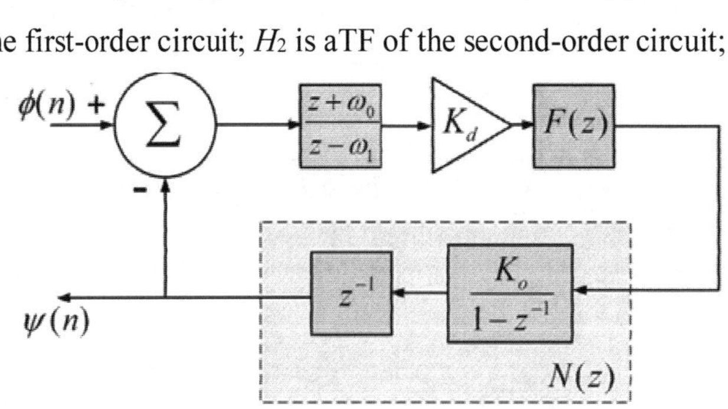

Figure 2. Linearized model of phase synchronization system based on PLL and AE

The pulse characteristics of TF (10) under the influence of typical disturbances (in the form of a phase jump H_{e1}, a frequency jump H_{e2} and a linear frequency decay H_{e3} [13])

$$H_{e1} = \left(\frac{\theta z}{z-1}\right),\ H_{e2} = \frac{\Omega T z}{(z-1)^2},\ H_{e3} = \frac{\Omega^2 T z(z+1)}{2(z-1)^3}. \tag{11}$$

are shown in figure 3. Thus: for H_1, the qualitative characteristics of a typical first-order PLL are preserved (first-order astatism); for H_2, the circuit is unstable when exposed to disturbances in the form of a frequency jump and linear frequency decay. Variation of initial parameters gives stable solutions only at abnormally low values of G_2 ($G_2 < 0.01$). Based on this the choice of the loop filter order is justified: $F_{Filt\_Order} = 1$

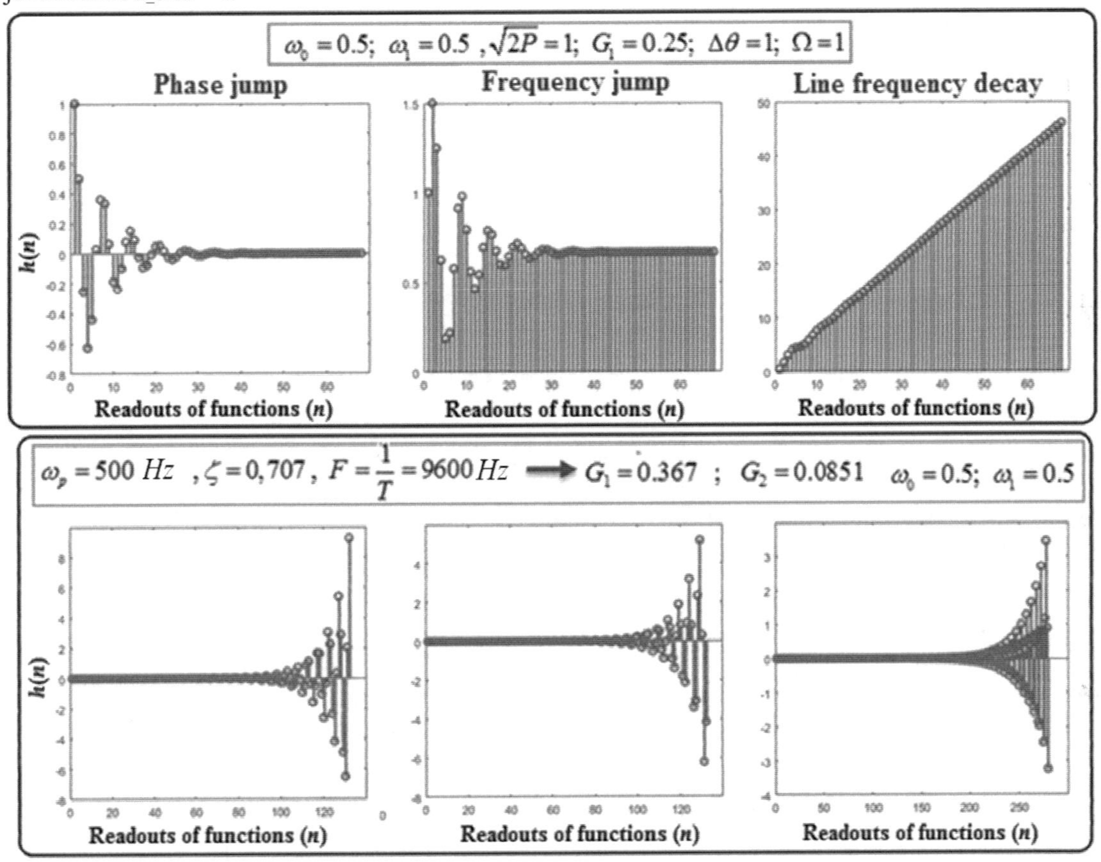

Figure 3. The pulse characteristics of the first and second order circuits

Consider the behavior of the system with a fixed frequency shift when the overall gain of the loop displaces $\left(K = K_1 G_1 = \sqrt{2P} K_d K_{NCO} G_1.\right)$. The system response for frequency jumps of 1 and 10 Hz is shown in figure 4.

The analysis shows that for each frequency shift value, there is a quasi-optimal value of the total loop gain (K), which minimizes the phase error $\Delta\varphi$ at the output of the system. Based on this, it is proposed to minimize the phase error by adjusting the overall gain of the loop. It is physically feasible [12,13] to use G_1 as a control parameter.

Let the communication channel has a linear frequency departure from 0 to f_c. Divide the entire frequency decay range into N discrete intervals. In each interval

$$f_{(m-1)c} \le f_m < f_{mc};\ 1 \le m \le N \tag{12}$$

the system operates with G_{1m} selected in such a way that the total phase error of the selected interval $\Delta\varphi$ does not exceed the required value $\Delta\varphi < \Delta\varphi_{max}$. In this case the system of equations describing the

IOP Conf. Series: Materials Science and Engineering **873** (2020) 012014 doi:10.1088/1757-899X/873/1/012014

functioning of this system has the form:

$$\begin{cases} 0 \le f_1 < f_{1c} \to H_{e1} = \left(\dfrac{\Omega_1 Tz}{(z-1)^2}\right)\left(\dfrac{z-1}{z-1+K_1\left(\dfrac{z+\omega_0}{z-\omega_1}\right)G_{11}}\right) & \Delta\varphi < \Delta\varphi_{\max}; \\ f_{1c} \le f_2 < f_{2c} \to H_{e2} = \left(\dfrac{\Omega_2 Tz}{(z-1)^2}\right)\left(\dfrac{z-1}{z-1+K_1\left(\dfrac{z+\omega_0}{z-\omega_1}\right)G_{12}}\right) & \Delta\varphi < \Delta\varphi_{\max}; \\ \ldots \\ f_{(N-1)c} \le f_N < f_c \to H_{eN} = \left(\dfrac{\Omega_N Tz}{(z-1)^2}\right)\left(\dfrac{z-1}{z-1+K_1\left(\dfrac{z+\omega_0}{z-\omega_1}\right)G_{1N}}\right) & \Delta\varphi < \Delta\varphi_{\max}. \end{cases} \quad (13)$$

By adjusting G_1, you can ensure that the phase error in the steady-state mode does not exceed the required value. As a result, we get a system that has a non-zero phase error at the output. The effect of the residual phase error on the final probability of the E_b bit error ($\Delta\varphi$) for different types of modulation is well studied and described [15,16].

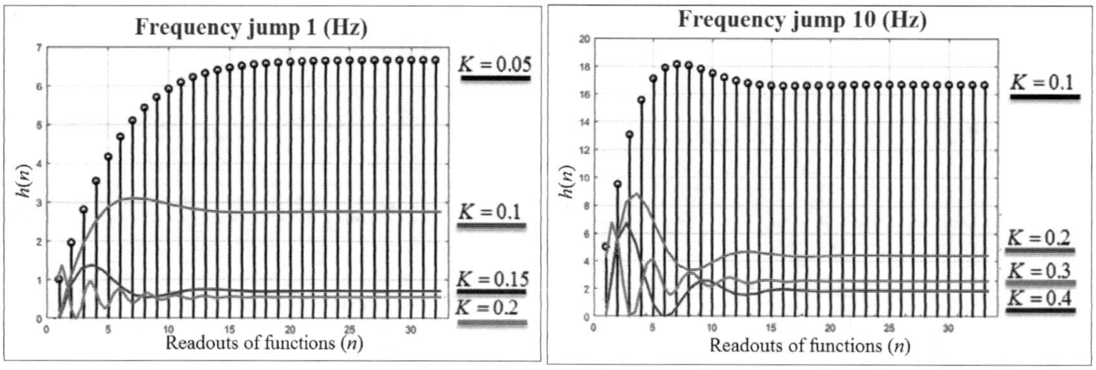

Figure 4. System response to frequency jump at different values of the overall circuit gain

The resulting mathematical model has the following limitations:

Limit 1: the parameters of additive and multiplicative noise in the communication channel are not taken into account: Fc_{AWGN}, Fc_{Fad_type}, I, W_{ci}, T_{ci}.

Limit 2: the type of modulation of the received signal is not taken into account: F_m.

Limit 3: the quantitative characteristics of the AE are not taken into account: F_a.

Limit 4: the specifics of the Protocol used are not taken into account: F_r.

A special feature of the model is the need to adjust G_1 depending on the instantaneous frequency decay in the f_{Nc} communication channel, in this case, the f_{Nc} search is performed using a set of matched filter blocks, each of which is consistent with the expected signal with a certain carrier frequency shift [1], while reducing the frequency uncertainty is limited by the number of matched filters in the system

(N).

Thus, according to the formal statement of the problem, the mathematical model justifies the structure of the phase synchronization system (F_{PLL}). At the same time, in order to justify the number of agreed filters (N), avoid the specified restrictions, and in order to move from a qualitative analysis of the system's functioning to a quantitative assessment, it is necessary to use simulation tools.

4. Simulation model

Based on the mathematical model of the phase synchronization system (12), a simulation model was built in the visual-oriented programming environment Simulink (MatLab package), its general structure is shown in figure 5.

Figure 5. General structure of the simulation model

The transmitting side contains the following basic structural elements: random binary sequence generator, AE training sequence generator (ASG), package structure generator (F_r is taken into account.), converter of binary sequence to numeric one, digital modulator (F_m is taken into account.).

The communication line contains: the block of frequency-phase mismatch of the channel (implements the effect of typical disturbances θ, ΩT, $\Omega^2 T$, f_c), the block of multiplicative noises ($F_{C\_Fad\_type}$, I, W_{ci}, T_{ci} are considered.), the block of additive noise generation (F_{C_AWGN} is taken into account).

The receiving side consists of a phase synchronization system, training sequence generator, demodulator, redundancy elimination unit and message receiver.

The simplified block diagram of the phase synchronization system is shown in figure 6 and contains the following basic structural elements: AЭ (considers F_a), CBPE – a calculation block of phase error (is a part of PD (considers K_d)), BPF - bandpass filter (considers F_{Filt_Order}, G_1, G_2), NCO – numerically controlled oscillator – digital analog of VCO (considers K_{NCO}).

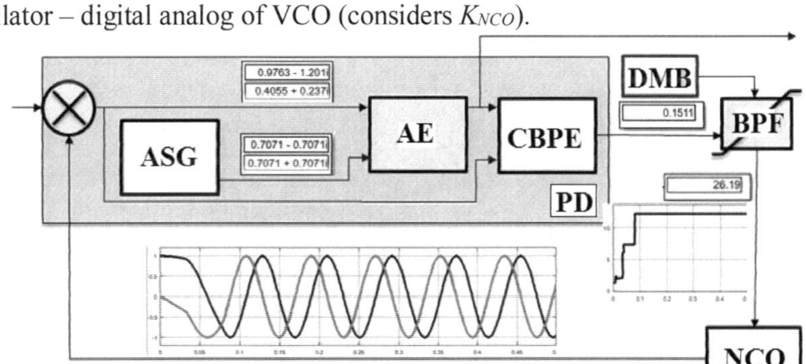

Figure 6. Simplified diagram of the synchronization system simulation model

The G_1 parameter of the bandpass filter is adjusted based on the f_{Nc} value using a set of matched filter blocks, the estimation is based on the highest value of the output signal (figure 7). The main requirement for this system is to design a set of filter blocks that cover the entire range of possible frequency shifts, as well as a decision-making block that implements control of the synchronization system parameters depending on the maximum output of one of the filters.

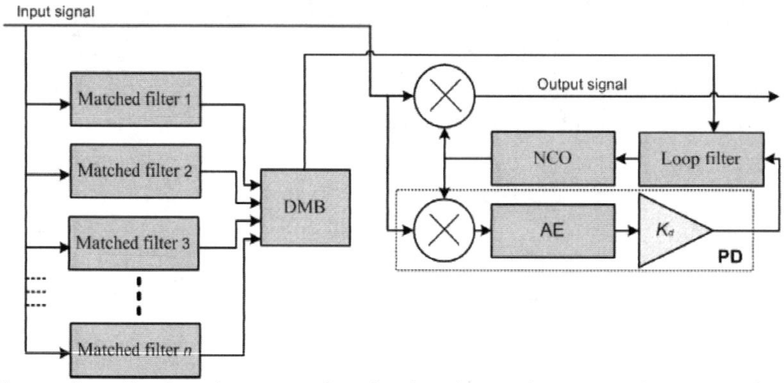

Figure 7. Scheme for adjusting the proportional gain when using a set of matched filter blocks

5. Simulation results

Figure 8 shows the results of simulation under the following initial conditions: structure of data transmission packet (fr_n=*20* bits, fr_m=*20* bits); F_{a_type} - RLS (recurrent calculation of weight coefficients based on the criterion of minimum sum of error signal squares); $F_{a\_n}=20$; $F_{a\_m}=20$; K_d = 1; G_1 - *adjustable parameter*; G_2 = 0; K_{NCO} = 1; F_m: - QPSK; T = 20 (c) - simulation time.

Characteristics of the communication channel:

Option 1 (blue curve on the diagram): I = 1; $Fc_{\_AWGN}$ – signal / noise ratio (SNR) (1-30) (dB); $Fc_{\_Fad\_type}$ - amplitude distribution of the incoming ray according to the Rayleigh rating; disturbance parameters: f_c = 72 (Hz), Ω^2T - 4.4(Hz/s).

Option 2 (red curve on the diagram): I = 2; $Fc_{\_AWGN}$ - (1-30) (dB); W_{c2} = - 3 (dB); T_{c2}= 0.003 (s); $Fc_{\_Fad\_type}$ - amplitude distribution of the incoming rays according to the Rayleigh rating; disturbance parameters: f_c = 72 (Hz), Ω^2T - 4.4(Hz/s).

The simulation was performed for the following configurations of frequency matched filters: N = 4 (corresponds to the maximum value of the phase error $\Delta\varphi$ = 14,7^0), N = 7 (corresponds to the value of the phase error $\Delta\varphi$ = 8,4^0), N = 14 (corresponds to the value of the phase error $\Delta\varphi$ = 4,1^0).

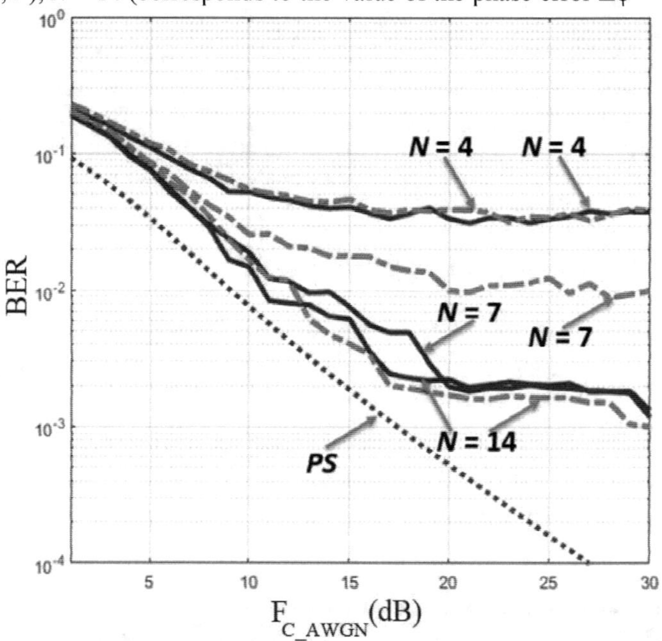

Figure 8. Dependence of the bit error probability on the signal-to-noise ratio for different communication channels (PS - case of perfect synchronization)

Based on the results of the simulation, the following conclusions can be drawn:

For option 1 - in the presence of additive and multiplicative interference (taking into account a single propagation beam), increasing the number of frequency-matched filters gives a significant gain in noise immunity. The difference in reception characteristics between $N = 4$ and $N = 7$, can range from 15% (SNR 10 (dB)) to 40% (SNR 30 (dB)). The further increase in N – does not lead to a significant improvement in the noise immunity index. The difference between $N = 7$ and $N = 14$ is on the average 7%.

For option 2 - in the presence of additive and multiplicative noise (taking into account two propagation beams), an increase in the number of matched filters (N) has a stronger effect on the quality of detection of the received signal. Thus, the difference in reception characteristics between $N = 4$ and $N = 9$, can range from 15% (SNR 10 (dB)) to 25% (SNR 30 (dB)). The further increase in N - leads to a significant improvement in the noise immunity index (in contrast with option 1). Thus, the difference in reception characteristics between $N = 7$ and $N = 14$, can range from 7% (SNR 10 (dB)) to 35% (SNR 30 (dB)).

6. Conclusion

Based on the developed mathematical model, a phase synchronization system (F_{PLL}) based on PLL and AE was proposed. The analysis of the obtained TF in the z-region revealed that the system under study has the first order of astaticism, which makes it necessary to adjust the proportional gain of the loop filter depending on the values of the instantaneous frequency decay in the communication channel. Using a set of frequency-matched filters as a control mechanism, it is possible to control the residual value of the phase error at the output of the synchronization system. Increase of the number of frequency-matched filters (N) reduces the residual phase error $\Delta\varphi$, which allows you to achieve the desired level of noise immunity. The most demanding to N are the communication channels, in which the effect of multipath propagation of the radio signal is more pronounced. The simulation results show that to ensure the required level of bit error probability for a radio channel model with two propagation beams, twice as many frequency-matched filters are required as compared to the single-propagation channel model.

References

[1] Yoshii K, Saitou M, Liu J and Shimamoto S 2019 *Proc. IEEE Int. Conf. on Communications Workshops* (Shanghai, China) p 875

[2] Bianchi C, Baskaradas J.A, Pezzopane M, Pietrella M, Sciacca U and Zuccheretti E 2013 *Advances in Space Research* **52** 403

[3] Dolukhanov M P 1971 *Fluctuation processes in the propagation of radio waves* (Moscow: Svyaz Press) p.187

[4] Jiang Y, Liu Y and Ying W 2013 *Dianbo Kexue Xuebao* **28** 775

[5] Shu Z, Shen D, Wang G, Tian X, Chen G, Pham K and Blasch E 2017 *Proc. IEEE Aerospace Conf.* (USA, Montana) p 794

[6] Mario S C Dolecek G 2010 *53rd IEEE Int. Midwest Symp. on Circuits and Systems* (USA, Seatle) p.656

[7] Blaickner A, Nagy O and Grunbacher H 2000 *Lecture Notes in Computer Science* **1896** 322

[8] Wasenmüller U, Gimmler C and Wehn N 2010 *Advances in Radio Science* **8** 123.

[9] Kinjo Sh Ochi H 2007 *IEICE Transactions on Fundamentals of Electronics, Communications and Computer Sciences* **8** 1601

[10] J Wu P Fan A 2016 *IEEE Access* **4** 450

[11] Nikolić B, Dimitrijević B, Milošević N and Dordević G T 2012 Radioengineering **21** 831

[12] Kudrewicz J Wasowicz S 2007 *Equations of phase-locked loops* (London: Word Scientific Publishing Co.Pte.Ltd.) p 235

IOP Conf. Series: Materials Science and Engineering **873** (2020) 012014 doi:10.1088/1757-899X/873/1/012014

[13] Stephens R 2002 *Phase-locked loops for wireless Communications. digital, analog, and optical Implementations 2nd Edition* (New York: Kluwer Academic Publishers) p. 434
[14] Tsymbal V A, Mokrinskiy D V and Stefan V I 2019 *Theory and technique of radio communication* **4** 108
[15] Artemenko A A, Maltsev A A and Rubtsov A E 2007 *Radiophysics* **2** 81
[16] Sklar B 2001 *Digital Communications: Fundamentals and Applications* 2nd edition (USA, Prentice-Hall PTR) p.1079

The use of intelligent technologies of Business Intelligence Platforms and Data Science and Machine Learning Platforms for monitoring the socio-economic indicators of the administrative districts of Moscow

P A Sakhnyuk[1], T I Sakhnyuk[2], M V Korshikova[3]

[1]National Research University Higher School of Economics, 20, Myasnitskaya str., Moscow, 101000

[2]Moscow City University, 4, 2-nd Selskohozoyastvenny street, Moscow, 129226

[3]Stavropol State Agrarian University, Zootekhnichesky Lane, 12, Stavropol, 355017, Russia

E-mail pav-sahnyuk@yandex.ru

Abstract. The article discusses the possibilities of studying the state of the social sphere according to the repository of the Moscow Government open data portal by administrative districts and city districts using Business Intelligence Platforms and Data Science and Machine Learning Platforms intellectual technologies. Opportunities are presented for using machine learning technologies for business analytics platforms to identify hidden patterns in order to make informed management decisions.

Keywords: data science and machine learning platforms, business intelligence platforms, machine learning, artificial intelligence, augmented analytics technology.

1. Introduction

Currently, when collecting and analyzing data obtained from databases of state statistics services, portals of open data and other sources, there is a problem of obtaining high-quality and visual information. As a result, the information available to executive authorities and individual citizens is largely expert and heuristic in nature. Therefore, there is an objective need to use intelligent business intelligence (BI)-platform technologies for monitoring socio-economic indicators of administrative districts of Moscow [1, 2].

Augmented analytics technology has become the main development driver of BI platforms and platforms for data processing and machine learning. Augmented analytics technology, based on machine learning, provides insight into huge amounts of data. This technology also includes natural language processing as a way to query data and create narratives to explain drivers and graphics. Nowadays, up to half of all analytical queries in BI platforms are generated using search, natural language processing or generated automatically. Natural language processing and conversational analytics are increasing the spread of business intelligence among civilian professionals, including new classes of users, especially employees of front offices and city municipalities.

IOP Conf. Series: Materials Science and Engineering **873** (2020) 012015 doi:10.1088/1757-899X/873/1/012015

A data set of families receiving subsidies in Moscow in various sections and data on citizens ' requests to the open data portal were selected for intellectual analysis [3].

2. Visual data analysis using Business Intelligence Platforms

Figure 1 shows a collection of different visualizations created in the Power BI service.

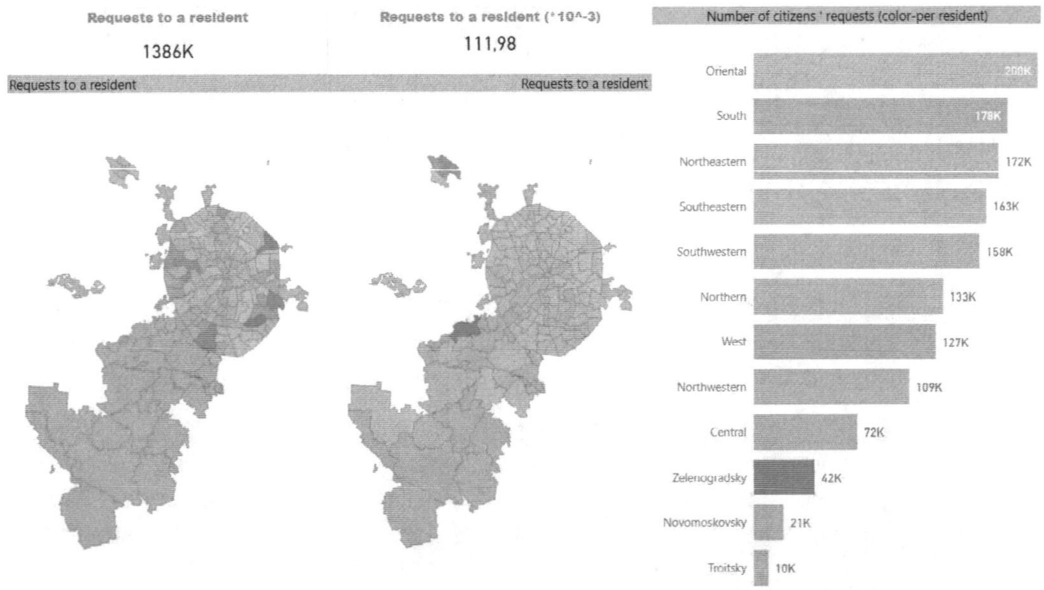

Figure 1. Dashboard number of citizens requests to the open data portal (in absolute units and per resident of administrative divisions and districts of the city)

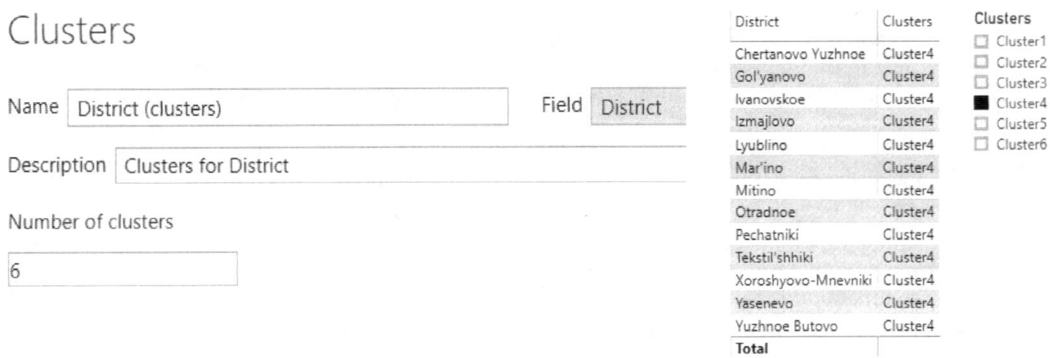

Figure 2. Results of clusterization of Moscow districts using the k-means Power BI algorithm

Let us find "similar" objects among 146 districts of Moscow. Searching for similar objects is one of the most common tasks in data analysis. We determine the similarity based on the characteristics of the data set "the number of families receiving subsidies in Moscow in various sections": the total number of families, low-income families, multi-child families, single-parent families, retired families, students' families, unemployed families and the number of requests to open data portal. In Power BI clustering support provides powerful analytical capabilities and Power BI supports the k-means algorithm, the most popular clustering method. One of the most difficult tasks in clustering is to determine the number of clusters (Figure 2). To facilitate this task, Power BI provides both automatic and manual control options.

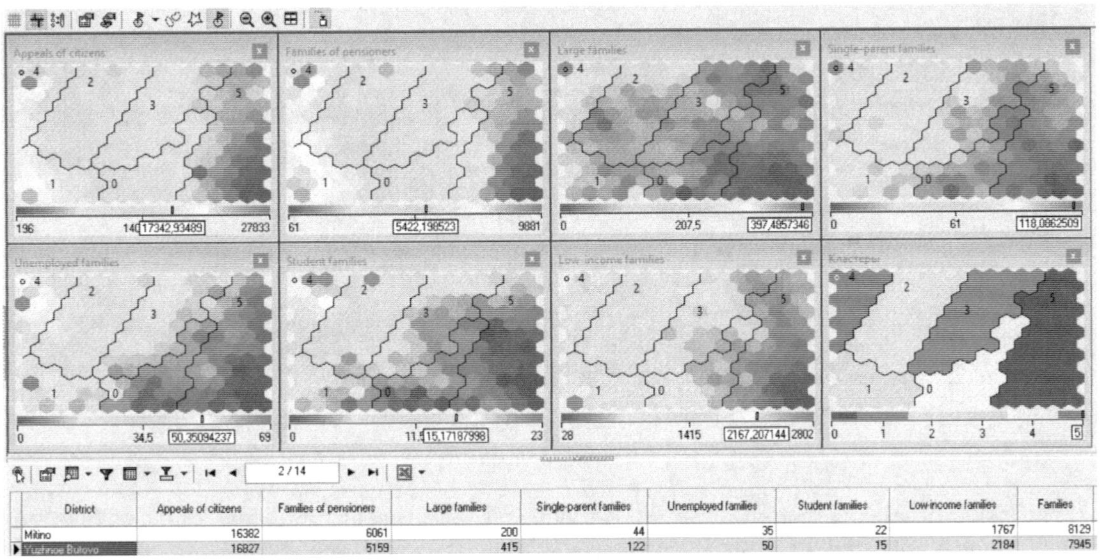

Figure 3. Kohonen maps for the set "Number of families receiving subsidies in Moscow in various sections" (cluster 4 shows the values for the Southern Butovo district)

To check the results, we will perform clusterization of Moscow districts using other clusterization algorithms. As a tool for cluster analysis, we will use the Kohonen neural network, implemented in Deductor Studio analytical platform, the advantage of which in comparison with other algorithms is the ability to visually analyze multidimensional data: similar objects fall into neighboring map cells (Figure 3).

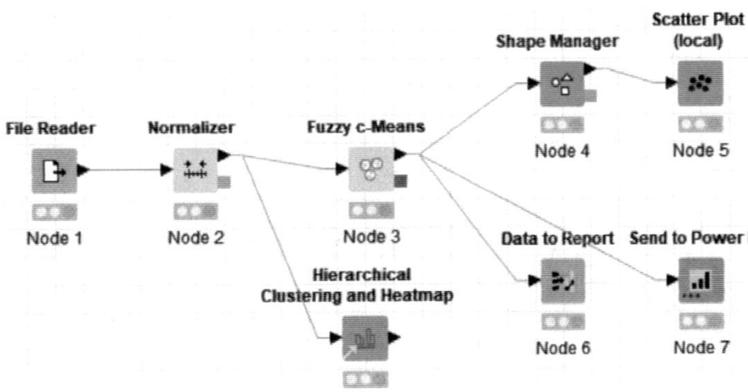

Figure4. Workflow FCM кластеризации in the KNIME Analytics Platform

Verification of clustering results using Kohonen maps and k-means can be performed in the KNIME Analytics Platform (Figure 4), where it is easy to perform fuzzy c-means (FCM) clustering, the feature of which is that each data point is assigned to a cluster with a membership function that varies in the range from 0 to 1 [4-8]. Clusters are represented as fuzzy sets and in addition, the boundaries between clusters are also fuzzy. The degree of belonging is determined by the distance from the object to the corresponding centroids.

IOP Conf. Series: Materials Science and Engineering **873** (2020) 012015 doi:10.1088/1757-899X/873/1/012015

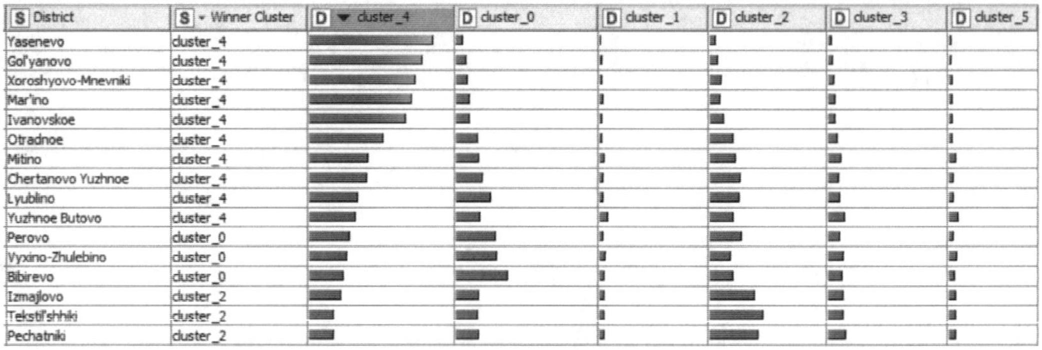

Figure 5. Results of FCM clustering of Moscow districts in the KNIME Analytics Platform

The results of fuzzy clustering are shown in Figure 5. Comparing them with the results of clustering using Kohonen maps built in Deductor Studio and k-means Power BI (Figure 3, 4), we make sure that they are objective. The difference in results is observed only for the districts of Perovo, Izmailovo, Textilschiki and Pechatniki. FCM algorithm assigned them to the second cluster with a slight difference in the values of the membership function.

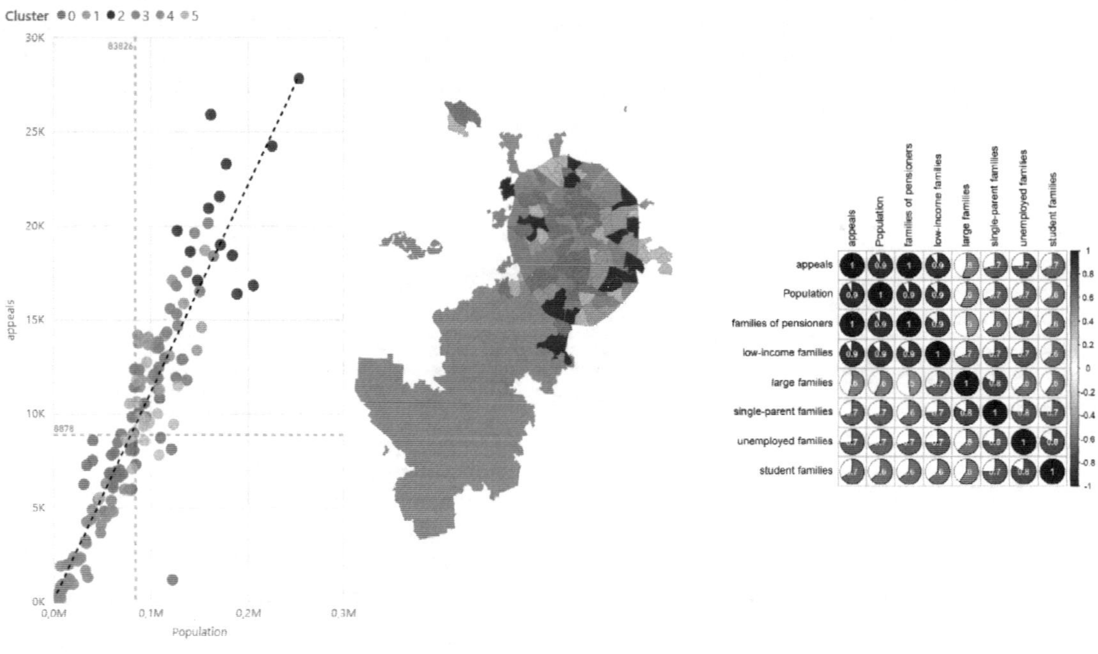

Figure 6. Dashboard Power BI by Moscow districts in the context of found clusters

Corelation between the characteristics of our data set reflecting the social status of residents of city districts are investigated using Power BI dashboards. They include a correlation matrix of the studied characteristics, a dot diagram of the population of the cluster districts and the number of requests from citizens to the open data portal and a map of the city districts (Figure 6).

For more in-depth data research, we use the Qlik Sense platform, which combines free-form associative research, provides context-sensitive suggestions and automatically constructes visualizations based on artificial intelligence. The platform uses the Qlik Cognitive Engine to provide automated analysis suggestions that help you look at data in a new way by automatically generating and prioritizing analytics based on the overall data set and user search criteria. The augmented intelligence platform offers insight and automation of visual analytics by understanding questions and responding in natural language.

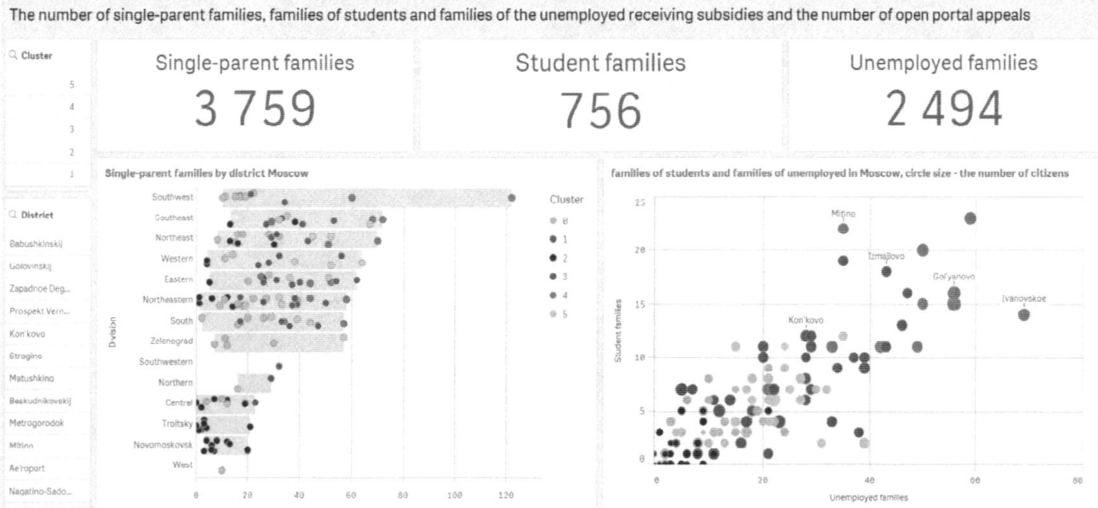

Figure 7. Dashboard Qlik Sense: distribution of Moscow districts depending on correlation between the number of student families, unemployed families and single-parent families receiving subsidies and the number of requests to the open data portal in the context of found clusters

An analysis of the dashboard (Figure 7) shows that as the number of families of unemployed, single-parent families and families of students receiving subsidies increases, so does the number of requests from citizens to the open data portal (typical for the districts of the second cluster).

To identify key factors of influence, we use the appropriate visualizer in Power BI that will help you to understand what factors affect the metric under study (in our case, the cluster number that the Moscow region belongs to). Using machine learning in Power BI helps to explore data, allowing to perform deep analysis to automatically search for patterns, interpret results in a clear way and predict results. Using machine learning (a linear regression model), the factors of influence are ranked from the most to the least significant and a probability indicator and text description are provided to explain the influence. For example, for districts in cluster 4 (Figure 8), the most significant factor is the number of families of students receiving the grant.

For the districts of cluster 0, the most significant factor is the number of unemployed families receiving subsidies. For districts of cluster 1, the most significant factor is the number of single-parent families receiving the subsidy. For cluster 3 districts, the most significant factor is the number of families of students receiving the subsidy. For the districts of cluster 4, the most significant factor is the number of families receiving the subsidy. For the districts of cluster 5, the most significant factor is the number of unemployed families receiving subsidies.

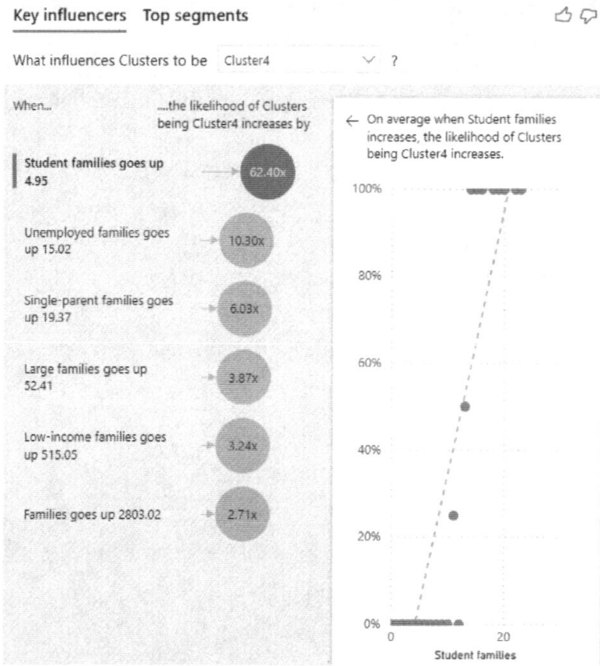

Figure 8. Visualizer of searching for key power BI influence factors in the districts of cluster 4

One of the fastest ways to get an answer from data is to ask a question in natural language. This feature is provided by the Q&A function in Power BI. Q&A is interactive option, in which often one question leads to another, because visualization opens up interesting ways to achieve the goal. In the Power BI service, the dashboard contains sheets pinned to one or more datasets, so you can ask questions about any data contained in the dataset. The Q&A tool recognizes the entered words and finds out where (in which data set) to find the answer. Q&A will also help to form a question with autofill, recalculation and other text and visual suggestions. The answer to the question is displayed as an interactive visualization and updated as the question changes.

3. Application of machine learning technologies

There is a close correlation between the number of families receiving subsidies and the number of requests from citizens to the open data portal (Figure 9). To identify hidden patterns in data and automate the main tasks of machine learning, we use the customizable artificial intelligence platform H2O Driverless AI. The platform provides automatic engineering of functions, checking and configuring models, selecting and deploying models, interpreting machine learning, creating custom scenarios in model construction, processing time series and texts, and automatically generating lines for model scoring.

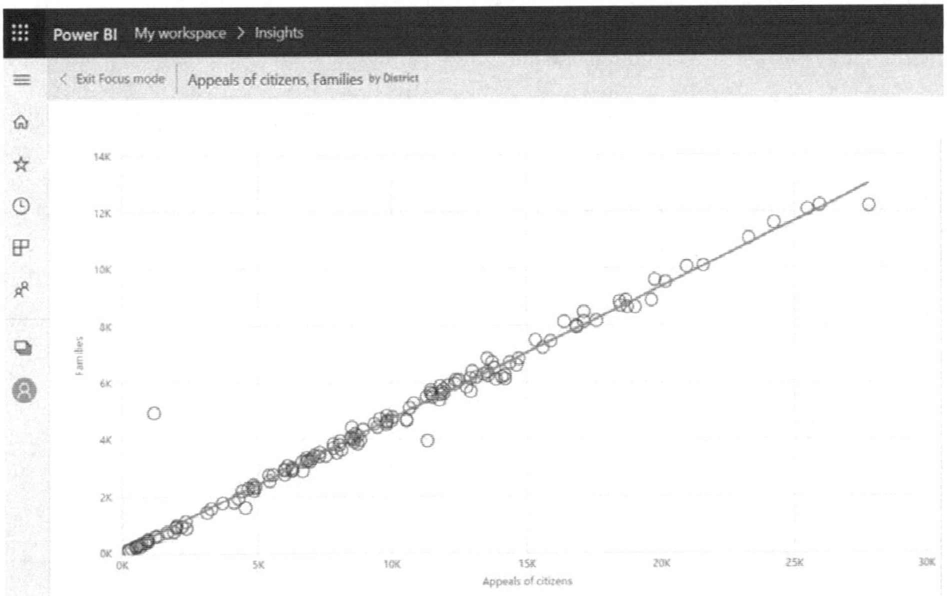

Figure 9. Brief analytics visualizer for finding trends in a data set: scatter plot demonstrating the relationship between the number of families receiving a subsidy and the number of citizens applying to an open data portal

H2O Driverless AI performs the function of designing input features and output variables to determine their optimal representation for building a machine learning model. Various stages of function transformations appear throughout all iterations. You can view them by hovering over iterative data points in the Variable Importance (Figure 10) section. Conversion to H2O Driverless AI is applied to columns of a dataset. Numerical, categorical, time and text converters design functions that support machine learning algorithms. Function design is a fundamental complex and expensive process in machine learning.

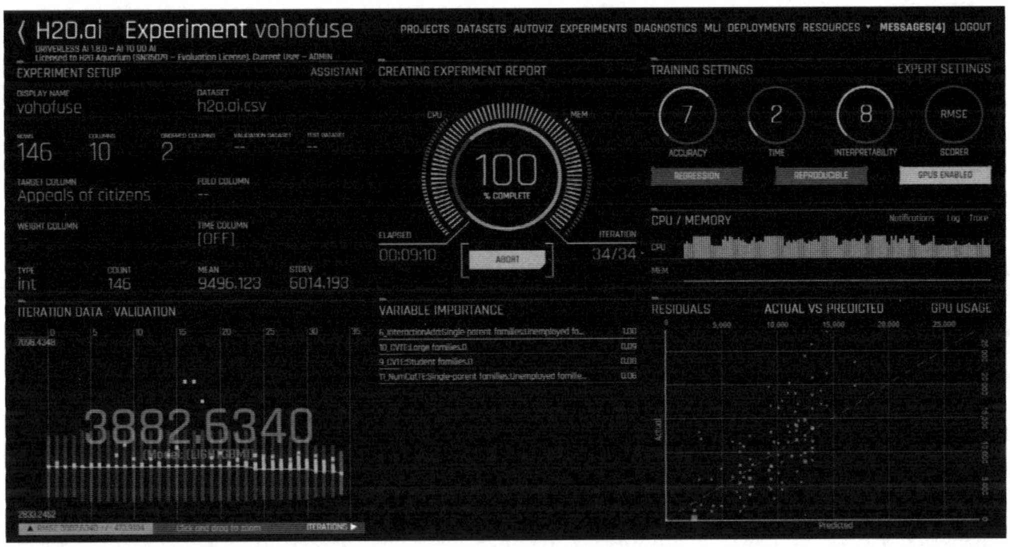

Figure 10. Experiment dashboard

The best model in our experiment is LightGBM, a gradient boosting framework developed by Microsoft that uses decision tree-based learning algorithms. It was specifically designed to reduce memory usage, increase learning speed and improve efficiency. Like XGBoost, it is one of the best gradient boosting implementations available [9-13]. It is also used to fit Random Forest models inside Driverless AI. The results are found in the experiment summary at the bottom right of the experiment page. Once the best predictive model is found, we can interpret it. MLI Dashboard presents various types of explanations concerning the model and its results [14-16]. All graphics on the dashboard are interactive. The number of single-parent families, families of the unemployed, families of students and large families receiving subsidies are of the greatest importance for forming a model for predicting the number of requests from citizens to the open data portal. Moreover, the contribution of these features to the model is different. Due to the high correlation with the output variable, the following indicators are: the number of families, retired families and low-income families receiving subsidies do not participate in the formation of the model. Additionally, H2O Driverless AI allows to download automatically generated documents, such as" Download experiment report "and" MLI Report", at the touch of a button. The developed model can be used to forecast the number of requests from citizens to the portal of the Moscow Government.

4. Conclusion

The results confirm the hypothesis that there is a positive correlation between the number of families receiving subsidies in Moscow in various sections and the number of requests from citizens to the open portal of the Moscow Government. This indicates that it is possible to influence the mechanism of participation of active Moscow citizens in the life of the city. The obtained objective information can be used to develop a strategy and make decisions on the development of the city.

5. Acknowledgment

The work is partially supported by the RFBR grant # 18-413-770006.

References

[1] Frolov Yu V, Sakhnyuk P A, Sakhnyuk T I 2018 Use of geoanalytics tools for training bachelors and masters in the direction of "business Informatics" *Bulletin Of the Russian University of peoples friendship. Series: Informatization of education* **15** 2 151-164 doi: 10.22363/2312-8631-2018-15-2-151-164

[2] Mamaev I I, Sakhnyuk T I, Sakhnyuk P A 2016 Analysis of the main socio-economic indicators of the districts of the Stavropol territory for 2011-2015 by means of business Analytics systems *Scientific journal of Kubgau* **123** 09

[3] Open data portal of the Government of Moscow (data.mos.ru) https://data.mos.ru/opendata/7701236617-kolichestvo-semey-poluchayushchih-subsidiyu-v-gorode-moskve

[4] T Siddiqui, A Ausaf 2018 Data mining tools and techniques for mining software repositories: A systematic review *Big Data Analytics. Springer, Singapore* 717-726.

[5] Verma K, Bhardwaj S, Arya R, UL Islam M S, Bhushan M, Kumar A, Samant P. 2019 Latest Tools for Data Mining and Machine Learning *International Journal of Innovative Technology and Exploring Engineering (IJITEE)* **8** 9S doi: 10.35940/ijitee.I1003.0789S19

[6] R Alcalá, MJ Gacto, J Alcalá-Fdez 2018 Evolutionary data mining and applications: A revision on the most cited papers from the last 10 years (2007–2017) *Wiley Interdisciplinary Reviews: Data Mining and Knowledge Discovery* **8** 1239

[7] Jean N, Burke M, Xie M, Davis WM, Lobell DB, Ermon S 2016 Combining satellite imagery and machine learning to predict poverty *Science 353(6301)* 790–794

[8] Froelich W. 2017 Towards improving the efficiency of the fuzzy cognitive map classifier. *Neurocomputing* 232 83-93

[9] Mitchell R 2017 Gradient boosting, decision trees and XGBoost with

CUDA. https://devblogs.nvidia.com/parallelforall/gradient-boosting-decision-trees-xgboost-cuda/. Accessed 25 Feb 2018

[10] Mitchell R, Frank E. 2017 Accelerating the XGBoost algorithm using GPU computing *PeerJ Computer Science* 3:e127

[11] Shi S, Wang Q, Xu P, Chu X 2016 Benchmarking state-of-the-art deep learning software tools 2016 7th international conference on cloud computing and big data (CCBD) *IEEE* 99–104.

[12] Tianqi Chen and Carlos Guestrin 2016 Xgboost: A scalable tree boosting system *Proceedings of the 22Nd ACM SIGKDD International Conference on Knowledge Discovery and Data Mining* 785–794

[13] Nguyen G, Dlugolinsky S, Bobák M 2019 *et al.* Machine Learning and Deep Learning frameworks and libraries for large-scale data mining: a survey. *Artif Intell Rev* 52 77–124

[14] Patrick H 2019 On the Art and Science of Machine Learning Explanations *KDD '19 XAI Workshop Proceedings*

[15] Patrick H, Navdeep G, Nicholas S 2019 Proposed Guidelines for the Responsible Use of Explainable Machine Learning. arXiv:1906.03533v3 [stat.ML]

[16] Sorelle A F, Chitradeep D R, Carlos S and Dylan S 2019 Assessing the Local Interpretability of Machine Learning Models. arXiv preprint arXiv:1902.03501,

Intellectual Technologies in Digital Transformation

P A Sakhnyuk[1], T I Sakhnyuk[2]

[1]National Research University Higher School of Economics, 20, Myasnitskaya str., Moscow, 101000

[2]Moscow City University, 4, 2-nd Selskohozoyastvenny street, Moscow, 129226

E-mail pav-sahnyuk@yandex.ru

Abstract. Artificial intelligence and machine learning helps to improve the quality of customer service and change the methods of companies' activities. For this reason, enterprises should consider integrating these technologies into digital transformation plans to remain competitive. Low-code machine learning platforms allow companies and business professionals with minimal coding experience to create applications and fill in the gaps of the personnel in their organization. Automated machine leaning (AutoML) technology represents the next step in the evolution of machine learning, providing non-technical companies with the ability to create machine learning applications quickly and cheaply.

Keywords: Digital transformation, data science and machine learning platforms, automated machine learning, artificial intelligence.

1. Introduction

Information is a key resource in the contemporary world. Humanity generates huge amounts of digital data, which not only occupy a place in cloud storage and data lakes but also help companies have done business. Collection, storage, structuring and analyzing data are necessary to take full advantage of the available information. The variety, complexity and speed of this data as well as its many sources changes the methods of companies' activities. Perhaps, the biggest factor in data growth is advances in technology that have made data collection and data exchange easier and more accessible. Modern users have also become more technically competent and use digital technology to create and share more data and applications [1-2].

Digital transformation is a strategic transformation aimed at a business can operate in a rapidly changing environment, identify key goals and develop the trajectories necessary to achieve these goals. Digital transformation allows integrate digital technology in all business areas, so business processes can change for the better. Digital transformation achieves its goals due to advanced technologies such as big data (BD) or artificial intelligence (AI). These technologies are aimed at processing information flows, on the basis of which it is possible to make decisions, adapt offers to specific clients and predict their behavior. AI and machine learning (ML) process and analyze data quickly and accurately. These technologies are a key part of revealing the value of company and customer data, allowing make decisions faster and clearly.

2. Informal statement of the problem

AI is one of the main areas of activity and the main engine of digital transformation. It has become an excellent tool for companies undergoing digital transformation. Data grows exponentially, and with it grows the potential for extracting valuable business information from them, understanding problems and automating routine operations using ML.

The development of AI is pushing digital enterprises to transform into intelligent enterprises. This is mainly due to advances in ML combined with technology based on data science. AI technologies are already revolutionizing not only how we perceive and conduct business, but also the business environment and its overall IT landscape. Many companies invest in the development of AI technologies, but they are hampered by the lack of skills in the practical application of the developed models.

Unfortunately, ML models are often expensive to develop and require a lot of data to build and train algorithms. Traditional enterprise-level AI / ML implementation methods are complex and typically require expensive, highly trained Data Science professionals. The widening gap in the qualifications of business users and Data Science professionals is also creating problems for organizations. Smart technology projects can include months of data preparation, followed by the design, development, testing, tuning, and deployment of ML models.

It is also important that many developed solutions based on AI / ML are difficult to trust. They consist either of complex non-linear models and ensembles of models that are difficult or almost impossible to interpret by people, or of pre-built models that do not give an idea of how the predictions are made. When using the "black box" model, the input data are known and receive the output data, but it is not known what happens between them. This model may not be acceptable because it is important that the business trust your results. Understanding the drivers of certain behaviors is what allows you to make the right decisions and have a real impact on the business. That is why it is important to work in cross-functional teams with industry experts from the business environment. As AI and ML become an integral part of any part of the business, it is important that all employees have access to the tools and solutions that ML offers. It is clear that experts in the field of business, who have knowledge in business to identify the main problems and opportunities, need access to these tools. However, they do not have technical expertise. In other words, they cannot write code.

Thus, today there are three big problems in the application of intelligent technologies in digital transformation: lack of data and analytics experts; a long and expensive period of development and implementation of machine learning models; trust in the results obtained with their help. All of these together prevent organizations from implementing AI/ML for digital transformation at the enterprise level.

3. The ability to use the data science and ML platforms and automation-focused ML

Nonetheless, modern solutions use advanced ML capabilities even for companies with minimal experience in data processing. The accelerated implementation of AI and ML, combined with the availability of open-source software, allows you to create and train far more models than ever. Open-source software allows IT departments to access infrastructure, data sets, workflows and trained models (transfer training). This approach reduces the overall cost of implementing intelligent technologies by providing developers with tools for developing models and using the experience of other Data Science specialists. This is why companies are interested in Citizen Data Scientists. These new data processing specialists are closing the skills gap on the part of the business using automated and other powerful data processing technologies [3-5].

The term "open source software" refers to a tool with source code available over the Internet for free. The proprietary (closed) program code is private and distributed through licensing rights. For a business that is just starting its ML initiative, using the open-source tools Knime (Figure 1), H2O.ai, RapidMiner can be a great way to practice data science for free before choosing enterprise-level tools such as machine service Amazon machine learning et al.

IOP Conf. Series: Materials Science and Engineering **873** (2020) 012016 doi:10.1088/1757-899X/873/1/012016

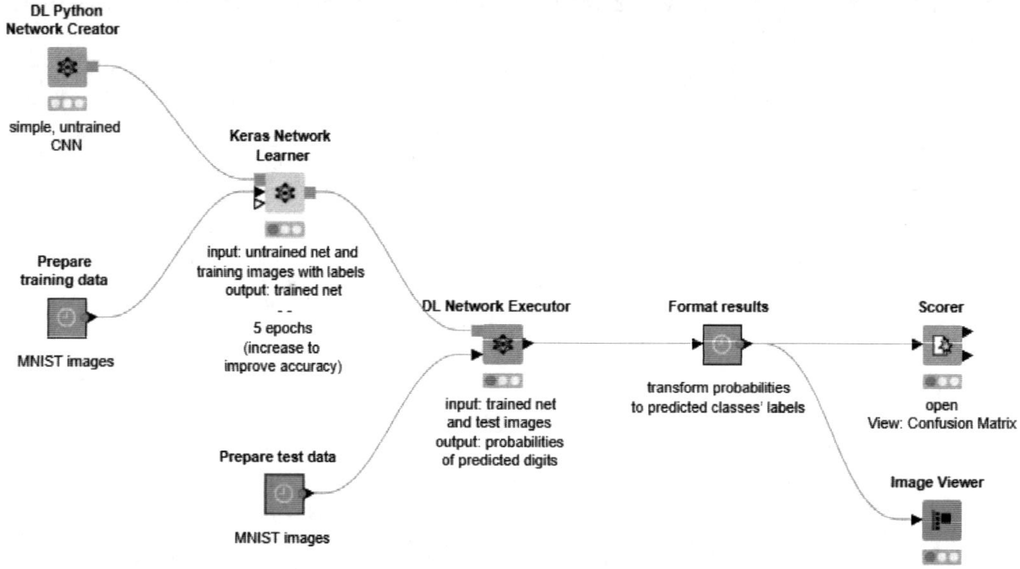

Figure 1. Workflow Knime trains a simple convolutional neural network on the MNIST dataset through Keras

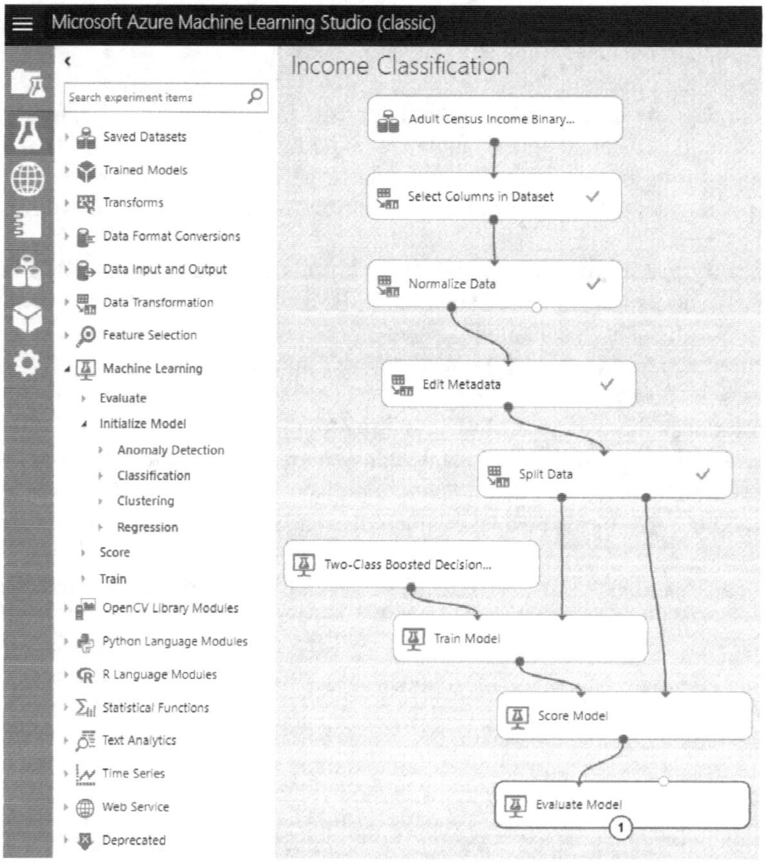

Figure 2. Web interface of Microsoft Azure Machine Learning Studio

The benefits of using open source tools are not limited to their availability. Typically, open-source projects have a huge community of data engineers and scientists seeking to share data sets and pre-trained models.

Data Science and Machine Learning Platforms tools use ML algorithms to scale Data Science workflows. Data Science project teams using platforms that combine data preparation, ML, and model deployment are more productive. Automated platforms scale the capabilities and resources of the data scientists team by providing enhanced functionality for data visualization, feature development, model interpretation, and low-latency deployment.

Among the Data Science and Machine Learning Platforms, platforms with Visual workflow designer, for example, Microsoft Azure Machine Learning Studio (Figure 2), are of great interest. Visual Workflow Designer allows increasing the productivity of the entire data scientists team, from analysts to experts, to speed up and automate the creation of predictive models in the drag and drop visual interface. To fine-tuning the models of the platform, the R & Python scripting nodes written in the Jupyter Notebook, specialized and advanced ML libraries are combined in one workflow, providing a developed solution for the rest of the team.

Low-code is a visual approach to application development. Low-Code Machine Learning does most of the development work using drag-and-drop components and model-driven logic through a graphic user interface. Low-Code ML frees non-technical developers from having to write code while supporting professional developers while abstracting from the tedious infrastructure tasks required when developing applications. Working together Citizen Data Scientists and of data and analytics experts create, train and deploy models in an order of magnitude less time than traditional methods. Low-Code Machine Learning creates a full range of applications for solving problems of classification, regression, clustering, anomaly search and recommendation systems [6-10].

Tools such as Visual workflow designer in Azure Machine Learning and automatic ML capabilities are changing this paradigm. ML used to be a black box or was perceived as unattainable, very complex and unusable by anyone who could not code.

Previously, deploying a trained model in a production environment required the experience of a data specialist, knowledge of coding, model management, container maintenance, and web service testing. Using the constructor, the user can drag and create an experiment and then run it, which means that any business user can deploy a trained model with just a few clicks. This provides the real value of AI / ML projects for business.

4. The ability to use the Automation-Focused Machine Learning

Much of the time in a typical data science project is devoted to building a model or choosing a model. This is a complex task requiring expensive calculations and time consuming data scientists.

Using representative structured data (flat file tables), a machine learning algorithm can build a model of what it expects from the data. Then it compares the results of the model with the reference data, finds the error as the difference between the target and actual values, and iteratively modifies its parameters to minimize the error. When choosing the right settings, taking into account and evaluating all possible features and basic trainees of the ensemble model, an optimal conveyor is obtained that provides the highest possible accuracy [11, 12].

The complex and iterative nature of the feature development makes qualified data scientists relevant, making it difficult to find them. At present, ML as technology has reached a point where the development of features and other important stages of model training can be performed automatically with a high degree of accuracy. This allows organizations to complete this important ML workflow faster by focusing on data scientists on the preparation of the presentation sample to be trained.

If the target or data changes (for example, new attributes are added), the process must be repeated. AutoML can help data scientists save time searching for the best model and its hyperparameters by focusing on a better understanding of the business problem being solved and the quality of the data.

An input for AutoML is data and a task (classification, regression, recommendations, etc.), an output is a model ready for production, capable of qualitatively solving forecast problems with new data. Hyperparameters of a model are parameters whose values are set before the model starts training and does not change during the training process. Each solution in a data-driven pipeline is a hyperparameter. The task of AutoML is to find such hyperparameters that could give a good result in a reasonable time though:

- to choose a strategy for data preprocessing: how to work with unbalanced data; how to fill in the missing values; remove, replace or save outliers and anomalies; how to encode the values of a categorical variable; how to avoid data leakage from input characteristics to the target variable, etc.
- to generate new feature and choosing the most significant;
- to choose the best model (generalized linear models, XGBoost, Random Forest, deep neural networks, etc.);
- to search for hyperparameters of the selected model (for example, the number of trees and parts of the dataset for models or architecture based on trees, learning speeds and the number of eras for neural networks);
- to create an ensemble of models to increase the accuracy of the forecast, if possible.

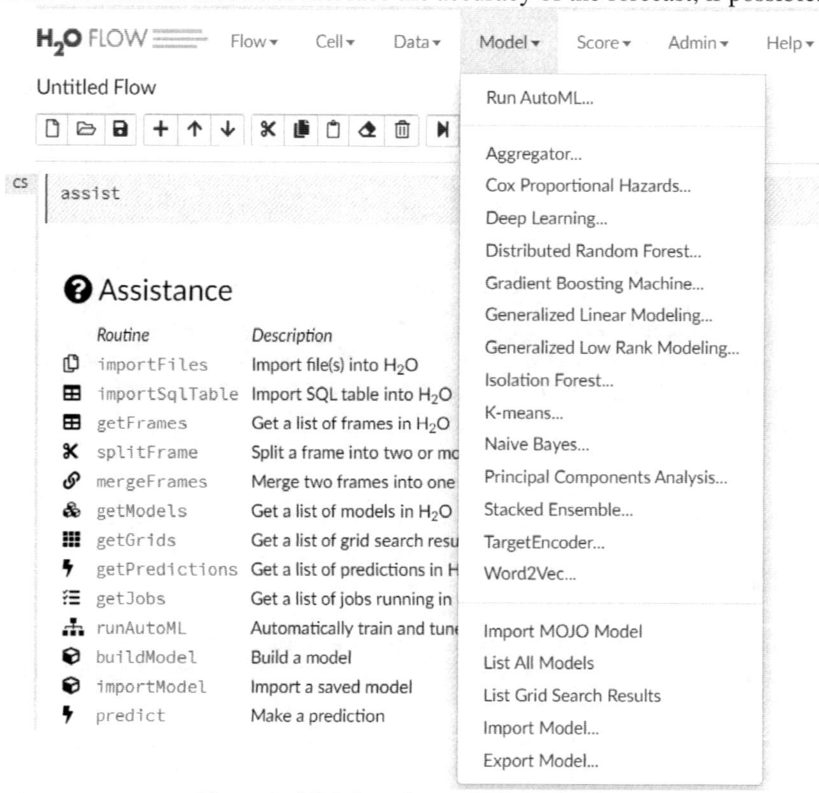

Figure 3. Web interface of H2O FLOW

The only free library for industrial use with automatic machine learning is H2O AutoML. The library provides an easy-to-use interface that automates data preprocessing, training and tuning a large number of candidate models (including models with several stacks for excellent model accuracy and performance). The result of AutoML is the “Leaderboard” of H2O models, which can be easily exported for use in production. AutoML is available in all H2O interfaces: R, Python, Scala, the web interface (Figure 3) and due to the distributed type of the H2O platform, it can be scaled across clusters to process big data. The open-source ML components H2O.ai are an industry standard that integrates many other platforms (for example, Alteryx, Dataiku, Domino, IBM, KNIME, RapidMiner, and TIBCO Software) [13-15]. The library also supports multi-core processor architectures and offers a software layer for significantly accelerating the GPU.

AutoML H2O.ai key features are:

- automatic selection of algorithms and tuning automates the test execution process for several algorithms and hyperparameter configurations. AutoML checks the results for accuracy and confirms that the optimal model and configuration has been selected for use. This significantly saves the time of data processing and analysis specialists and, more importantly, allows each of them to get the same results as the most experienced specialists;

- automatic selection of predictive features simplifies the creation and selection of features by automatically identifying key predictive features from large data sets;
- model assessment generates a complete set of assessment metrics and related visualizations to measure model characteristics with new data. It allows ranking models over time to ensure optimal behavior of the working version. Assessing a model is beyond the scope of a direct assessment of performance. In order to fully take into account the various effects of errors of the first and second kind (false positive and false negative), the expected basic behavior is taken into account and the cost model is used;
- model explanation: automatically provide explanations of the relative weight and importance of factors affecting the formation of the forecast.

5. Conclusion

Data Science and Machine Learning Platforms and Automated ML simplifies the process of developing the ML model and gives business users, regardless of their experience in the field of data science, the opportunity to define an end-to-end machine learning pipeline for solving Data Mining problems. Automated ML makes it easy for Citizen Data Scientist to create machine learning models without writing code. Automated ML includes the best machine learning practices from leading experts in the field of Data Science, and modern open-source libraries to make machine learning more accessible to the entire organization.

Designing a model using a traditional process is extremely time-consuming, repetitive and tedious. The AutoML application automatically performs model building tasks that typically require qualified Data Scientists. Instead of taking weeks or months, the automated ML system is fast and usually requires business users / analysts to have these days to create hundreds of models, predict and analyze. Automated ML for data analysts enables organizations to do more in less time.

AutoML allows enterprises in industries such as healthcare, FinTech, banking, and others to use advanced machine learning and artificial intelligence technologies that were previously limited to large business companies. By automating most of the tasks of training analytical models, AutoML allows business users and data analysts to easily integrate into business processes and pay more attention to solving complex business problems.

Data Scientist and business users from various industries can use automatic machine learning with a view:

1. to implement ML-based solutions without in-depth programming knowledge.
2. to save time and resources.
3. to use best practices in data science.

By providing accurate forecasts and recommendations, applications based on Low-Code Machine Learning and AutoML can improve customer service, improve product quality and reduce costs. For these reasons, enterprises should consider integrating these technologies into digital transformation plans in order to remain competitive.

6. Acknowledgment

The work is partially supported by the RFBR grant # 18-413-770006.

References

[1] Frolov Y V, Sakhnyuk P A and Sakhnyuk T I 2018 Use of tools of geoanalytics for training of bachelors and masters in the business informatics direction *Rudn J. Informatiz. Educ.* **15** 151–64

[2] Mamaev I, Sakhnyuk T and Sakhnyuk P 2016 Analysis of the main socio-economic indexes of districts in the Stavropol region in 2011-2015 by means of business analytics platforms *Sci. J. KubSAU* **9** 1–14

[3] Siddiqui T and Ahmad A 2018 Data mining tools and techniques for mining software repositories: A systematic review *Advances in Intelligent Systems and Computing* vol 654 (Springer Verlag) pp 717–26

[4] Kanupriya V, Sahil B, Resham A, Mir S, Ul I, Megha B, Ashok K and Samant P 2019 Latest

Tools for Data Mining and Machine Learning *Int. J. Innov. Technol. Explor. Eng.* **8** 18–23
[5] Alcalá R, Gacto M J and Alcalá-Fdez J 2018 Evolutionary data mining and applications: A revision on the most cited papers from the last 10 years (2007–2017) *Wiley Interdiscip. Rev. Data Min. Knowl. Discov.*
[6] Jean N, Burke M, Xie M, Davis W M, Lobell D B and Ermon S 2016 Combining satellite imagery and machine learning to predict poverty *Science (80-.).*
[7] Froelich W 2017 Towards improving the efficiency of the fuzzy cognitive map classifier *Neurocomputing*
[8] Mitchell R Gradient Boosting, Decision Trees and XGBoost with CUDA | NVIDIA Developer Blog
[9] Mitchell R and Frank E 2017 Accelerating the XGBoost algorithm using GPU computing *PeerJ Comput. Sci.*
[10] Shi S, Wang Q, Xu P and Chu X 2017 Benchmarking state-of-the-art deep learning software tools *Proceedings - 2016 7th International Conference on Cloud Computing and Big Data, CCBD 2016*
[11] Chen T and Guestrin C 2016 XGBoost: A scalable tree boosting system *Proceedings of the ACM SIGKDD International Conference on Knowledge Discovery and Data Mining*
[12] Nguyen G, Dlugolinsky S, Bobák M, Tran V, López García Á, Heredia I, Malík P and Hluchý L 2019 Machine Learning and Deep Learning frameworks and libraries for large-scale data mining: a survey *Artif. Intell. Rev.*
[13] Hall P 2018 On the Art and Science of Machine Learning Explanations *KDD '19 XAI Workshop, August 04–08, 2019, Anchorage, AK*
[14] Hall P, Gill N and Schmidt N 2019 Proposed Guidelines for the Responsible Use of Explainable Machine Learning *33rd Conference on Neural Information Processing Systems (NeurIPS 2019), Vancouver, Canada*
[15] Slack D, Friedler S A, Scheidegger C and Roy C D 2019 Assessing the Local Interpretability of Machine Learning Models *arXiv Prepr. arXiv1902.03501*

The assessment of gait features according to the data of a portable acceleration sensor in an intelligent monitoring system

N V Dorofeev[1], A V Grecheneva[1]

[1]Vladimir State University, Faculty of Radio-electronics and Computer Systems, 23 Orlovskaya str., Murom, Vladimir Region, 602264, Russia

E-mail dorofeevnv@yandex.ru

Abstract. The article is devoted to assessing the possibility of identifying individual characteristics of a person's gait according to the data of one accelerometric sensor, as well as the possibility of assessing impaired functioning of the musculoskeletal system with subsequent use in intelligent expert systems based on neural network algorithms. The article notes the prospects of using dynamic measurements to improve the effectiveness of orthopedic diagnostics. The prospects of using accelerometry along with complex medical motion capture systems are noted. The results of distinguishing characteristic gait indicators among a group of subjects according to one accelerometer sensor, as well as an assessment of the distribution structure of the articular angle when walking, are described. As subjects, people with impaired functioning of the musculoskeletal system and people without disorders are involved. The results obtained are in good agreement with the data of goniometric control systems built on the basis of several sensors. Thus, the results of the research show the possibility of obtaining a detailed picture of gait. The results obtained will allow us to form a preliminary structure of the neural network and evaluate its coefficients with the further development of automated control and monitoring systems based on existing portable wearable devices (phone, watch).

Keywords: registration, monitoring, geotechnical system, technical and economic analysis

1. Introduction

In the field of orthopedics modern scientists and researchers have found that disorders of the musculoskeletal system are best manifested in the process of performing motor actions [1-3]. So, for example, when performing multiple locomotion, the maximum value of the articular angle can be significantly larger than with a single bending of the joint. This is explained not only by the physicomechanical properties of biological tissues that make up the joint compound, but also by the stimulation of biochemical processes in the body (blood flow, volume dynamics of synovial fluid, etc.) [4-6]. However, most diagnostic techniques rely on data obtained by examining a person in a static position [7, 8]. This leads to a decrease in the reliability of the diagnosis of the real clinical picture of the disease, errors in the diagnosis and the wrong choice of rehabilitation procedures.

In a dynamic state diagnosis of diseases of the musculoskeletal system of a person is based on the allocation of motion parameters and is a difficult task in medical orthopedics, requiring the use of specialized measuring instruments and algorithms for processing recorded data. This task is complicated by factors, the main of which are:

- movement parameters of each person are individual$
- movement consists of a set of monosyllabic locomotion, in the process of which dozens of articular joints can be involved;
- when registering movement parameters, it is necessary to compensate for the biomechanics of muscle tissue, skin and fat;
- the need to isolate and classify movements in complex waveforms;
- dynamic time range characterizing the performance of the same movement;
- the need to compensate for the result of free mounting of sensors and their possible movement during movements.

Currently, motion capture systems are used to measure the dynamics of motion parameters in medical practice. Such systems use an optical channel for recording motion parameters and are characterized by the high complexity of processing the data provided and are quite expensive for mass implementation in medical institutions.

In the field of manufacturing microelectronic wearable devices the development of technologies allows us to create systems that can eliminate shortcomings and expand the functionality of traditional motion capture systems. The use of accelerometric sensors in micromechanical design has special prospects in solving the problem of recording motion parameters. These devices are technological and allow high-precision fixation of the acceleration of biokinematic links.

In the works of a number of scientists [9–13] the trends are traced in the use of accelerometers in solving problems of recording motion parameters, in which the effectiveness of this approach was confirmed. However, the effectiveness of the diagnosis of violations is largely determined by the data processing algorithms using the accelerometric control method [14-17], methods for identifying and classifying motion parameters [18, 19], and the number of attributes taken into account [20]. Currently, in portable wearable devices (telephones, watches, smart tags, etc.) the presence of accelerometric sensors provides an opportunity to assess individual characteristics of a person, his identification. However, researches have not been conducted enough to identify the features of movements based on a wearable acceleration sensor and the application of this approach in intelligent systems of medical monitoring, tracking and identification.

Thus, the aim of the research is to assess the possibility of identifying individual characteristics of a person's gait according to one accelerometer sensor, as well as the possibility of assessing impaired functioning of the musculoskeletal system.

2. The method and approaches

When evaluating gait parameters, studies were conducted involving 10 students of 3 and 4 courses at the age of 19 to 21 years. When forming the group, the similarity of the anthropometric profile was taken into account - height, weight, length of the lower extremities. Two out of 10 students had impaired motor function of the left leg due to recent knee surgery. During the week, gait was recorded and analyzed based on accelerometer data.

When walking measurement of the angle of deviation of the thigh was carried out on the basis of an external accelerometer and an accelerometer of a mobile phone. Both devices were in the pocket of the researcher closely fitting to the body. Data obtained from both devices correlate with each other.

The data processing algorithm includes: preliminary low-pass filtering at a frequency of 50 Hz; determination of medium-term and inter-quartile differences to assess the differences in the angle of movement. The data were processed both for each committed movement and for the daily interval.

3. Results

The objective of the research was to highlight the characteristic indicators of gait among a group of subjects, as well as to assess the distribution structure of the articular angle when walking. Initially, an assessment of the manifestation of a knee joint injury in gait parameters was made. figures 1-3 show the data for one step of the student with surgery on the knee joint (right leg - figure 1a) and the step of the student without violations (left leg - figure 1b).

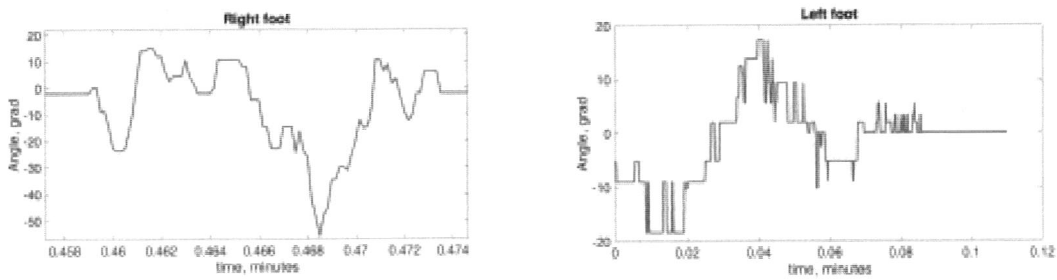

Figure 1. The change in the parameter of the articular angle over time in a test subject with a knee joint injury (left). The change in the articular angle parameter over time is normal (right)

In the course of studies of the step parameter of a healthy person, three phases of movement can be distinguished:

- The phase of the beginning of movement - a period of time with a sharp increase in the parameters of the articular angle lasts from 50 to 60% of the total step time. During this period, an increase in the speed of limb movement to the maximum value at the end of the period is recorded.
- The mid-motion phase is the time interval between the legs being brought together and the moment the tibia moves to the vertical position, begins immediately after the first phase and lasts within 10–13% of the total step execution time. There is a decrease in the speed of movement until a minimum peak is reached at the end of the period.
- The phase of the end of the movement is the time interval between the moment of the vertical leg of the leg and contact with the surface of the earth, begins immediately after the phase of mid-movement and lasts from 27 to 30% of the total step time. An increase in the speed of movement is recorded until the second maximum peak is reached at the end of the step.

In the case of analysis of the step of a person who suffered a knee joint injury, three phases of movement can also be distinguished. However, the steps of a person with a knee joint injury will differ in duration, as well as in the parameters of speed and amplitude of movement, which is reflected in the graphs (Figure 1 a and b).

To increase the reliability of experimental studies, daily observations were made of the dynamics of the articular angle of the subjects. Deviation of the angle of the hip joint during walking per day for the left (Figure 2) and right (Figure 3) legs of the subject from operations on the knee joint have pronounced surges and fluctuations, which is not observed in the gait of a healthy person (Figure 4). It should be noted about deviations in the movement of the injured leg, which is also seen from the presented diagrams.

For completeness of the analysis, Figures 5, 6 present diagrams showing the structure of the distribution of the angle when walking for 24 seconds for several days in a student who underwent surgery on the knee joint.

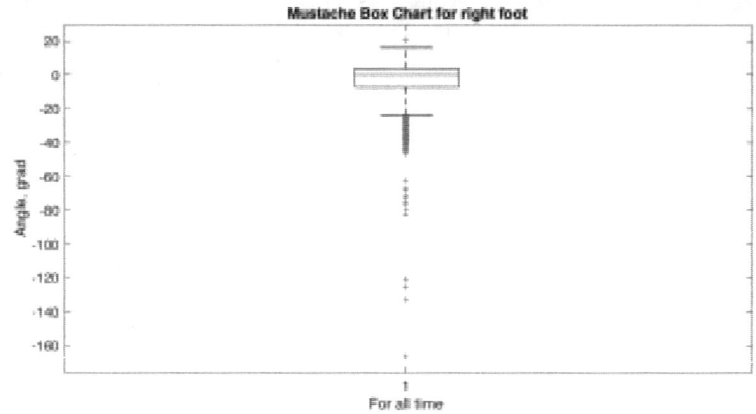

Figure 2. Daily deviation of the angle of the hip joint of the left leg when walking per day of the subject who underwent surgery on the knee joint

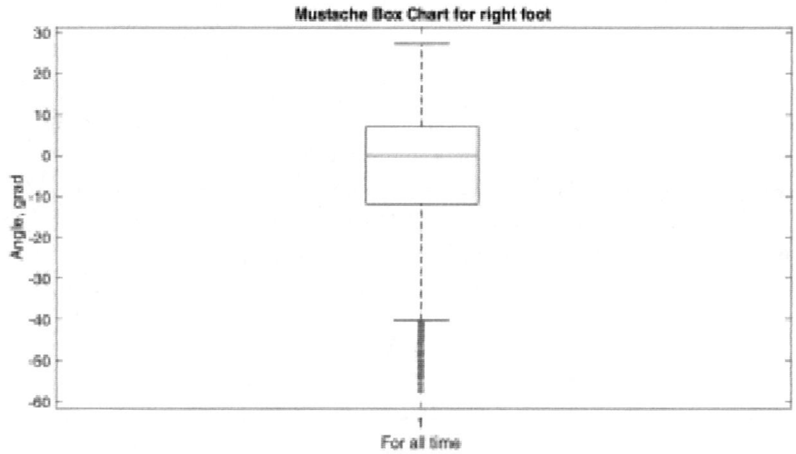

Figure 3. Daily deviation of the angle of the hip joint of the right leg when walking per day of the subject who underwent surgery on the knee joint

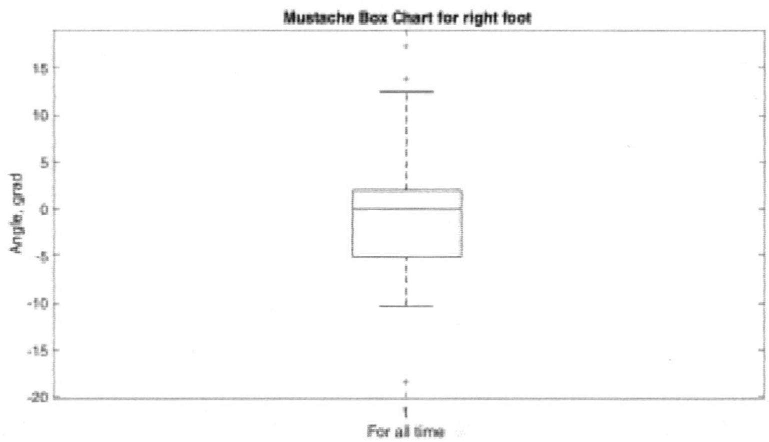

Figure 4. Daily deviation of the angle of the femoral joint when walking a healthy subject per day

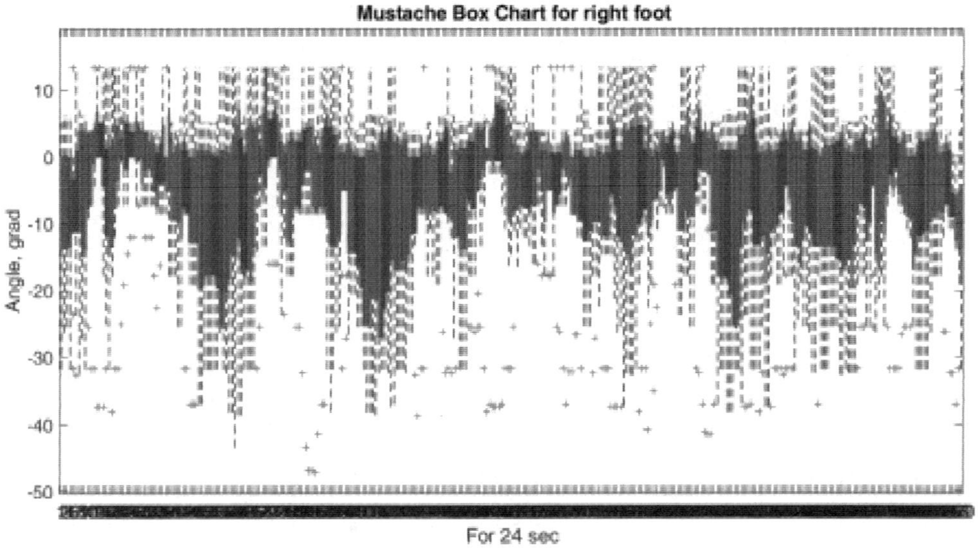

Figure 5. Diagram of the distribution of the angle of the hip joint of the right leg when walking the subject undergoing surgery on the knee joint

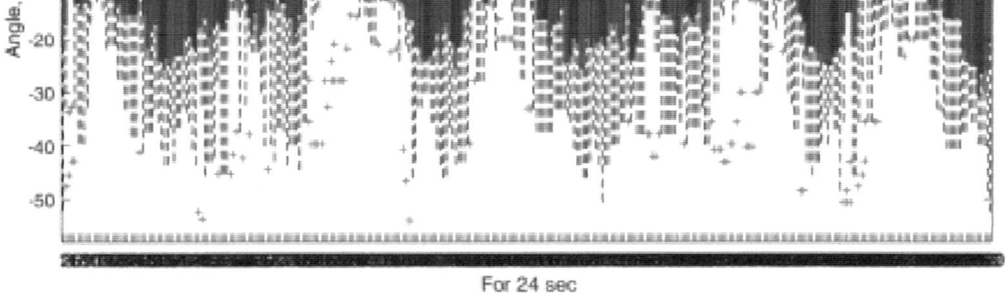

Figure 6. Diagram of the distribution of the angle of the hip left leg of the joint when walking the subject undergoing surgery on the knee joint

4. Conclusion

As a result of the research, the structure of the distribution of the angular parameters of the gait of a person was built, the events of the gait cycle were revealed, the significance of which was confirmed by similar studies in this area [21-25].

It should be noted that the above graphs were obtained in the course of studies of gait parameters using one accelerometer sensor. As can be seen from the presented results, the analysis of the data as a result of such an accelerometric research already allows you to get a fairly detailed picture of the gait (change in angle, walking speed, deviations in the functioning of the musculoskeletal system). However, the prospects for further research are aimed at finalizing data analysis algorithms using mining methods. The results obtained will allow us to form a preliminary structure of the neural network and evaluate its coefficients.

Using the approaches of accelerometry of the parameters of human movements will allow to register joint pathologies and evaluate the numerical characteristics of biomechanics in a dynamic mode. In the course of medical diagnostics, it is possible to assess the degree of joint damage and prescribe rehabilitation techniques using the provided graphic information.

References

[1] Niemier K, Ritz W and Seidel W 2007 Functional musculoskeletal diagnostics *Functional musculoskeletal diagnostics Manuelle Medizin* **45** **5** 123

[2] Andujo V, Fletcher I and McGrew C 2020 Musculoskeletal Preparticipation Physical Evaluation—Does it Lead to Decreased Musculoskeletal Morbidity? *Current Sports Medicine Reports* **19** 58-69

[3] Ludewig P, Kamonseki D, Staker J, Lawrence R, Camargo P, Braman J 2017 Changing our diagnostic paradigm: movement system diagnostic classification *Int. J. Sports Phys. Therapy* **12** 884-93

[4] Matveeva E and Spirkina E 2012 Some indicators of biochemical analysis of synovial fluid *Some indicators of biochemical analysis of synovial fluid Genius Orthopedics* **2012** 203

[5] Goto M, Hosako Y, Katayama M and Yamada T 1993 Biochemical analysis of rheumatoid synovial fluid after serial intra-articular injection of high molecular weight sodium hyaluronate *Int. J.l Clinical Pharmacology Research* **13** 161-6

[6] Kostrova O, Stomenskaya I and Timofeeva N 2017 Characteristics of synovial fluid under normal conditions and in some patthological processes *Acta Medica Eurasica* **4** 24-31

[7] Romakina N A, Fedonnikov A S, Kireev S I, Bakhteeva N Kh and Norkin I A 2015 Application of techniques of biomechanics in the status evaluation and pathology correction of locomotor system (review) *Saratov J. Medical Sci. Research* **1** 310–6

[8] National Clinical Guideline Centre. Osteoarthritis. Care and management in adults. London (UK): National Institute for Health and Care Excellence (NICE). 2014 Feb. pp 556 (Clinicalguideline; no. 177).

[9] Jansen B, Tan M and Bautmans I 2011 Accelerometer based gait analysis — multi variate assessment of fall risk with FD-neat Biosignals *Biosignals. Int. Conf. on Bio-inspired Systems and Signal Processing* (Rome: INSTICC Press) pp 138

[10] Bugané F, Benedetti M G, Casadio G 2012 Estimation of spatial-temporal gait parameters in level walking based on a single accelerometer: Validation on normal subjects by standard gait analysis *Computer Methods and Programs in Biomedicine* **108** 129–37

[11] Grecheneva A, Dorofeev N and Kuzichkin O 2016 Analysis of mistakes and errors application of the accelerometric phase method of biomechanical control *Int. Multidisciplinary Sci. GeoConf.* (Albena: Bulgary) vol 1 383-90

[12] Boutaayamou M, Schwartz C and Stamatakis J 2012 Validated extraction of gait events from 3d accelerometer recordings *IEEE Int. Conf. on 3D Imaging (IC3D)* (NJ: IEEE) p 4

[13] Che-Chang Y and Yeh-Liang H 2007 Algorithm Design for Real-time Physical Activity Identification with Accelerometry Measurement *33rd Annual Conf. of the IEEE Industrial Electronics Society* vol 3 pp 1-14

[14] Sant Anna A. and Wickstrom N 2010 A linguistic approach to the analysis of accelerometer data for gait analysis *Int. Conf.on Biomedical Engineering* (Anaheim, CA: ACTA Press) p 8

[15] Dorofeev N V, Grecheneva A V and Kuzichkin O R 2018 Principles of constructing of

intellectual systems for the diagnostics and the rehabilitation of the human spine *Procedia Comp. Sci.* **150** 270-9

[16] Yuming S, Sai G, Lin S and Weidong C 2017 Human Motion Classification Based on Range Information with Deep Convolutional Neural Network *4th Int. Conf. on Information Sci. and Control Engineering (ICISCE)* vol 1 p 8

[17] Chen P, Kuang Y and Li J 2016 Human Motion Capture Algorithm Based on Inertial Sensors *J. of Sensors* **2** 1-15

[18] Feng B and Xu G 2012 Semantic knowledge mining for human motion classification *J. of Information and Computational Sci.* **9** 2761-70

[19] Endres F, Hess J and Burgard W 2012 Graph-Based Action Models for Human Motion *7th German Conf. on Robotics* (German) vol 1 p 6

[20] Chan C and Chala G 2018 Impact of number of attributes on the accuracy of human motion classification *J. of Engineering Sci. and Technol.* **13** 1967-78

[21] Vorontcova O I and Lozovskaya M V 2017 Structure of gait cycle based on kinetic and kinematical parameters *J. of New Medical Technol.* **3** 9-15

[22] Borzikov V V, Rukina N N, Vorobyova O V, Kuznetsov A N and Belova A N 2015 Human Motion Video Analysis in Clinical Practice (Review) *Modern technologies in medicine* **7**(**4**) 201-10

[23] Likens A and Stergiou N 2020 A tutorial on fractal analysis of human movements In book: Biomechanics and Gait Analysis

[24] Geovanny P and Ernesto S-G 2018 Methodology for the registration of human movement using accelerometers and gyroscopes *15th Int. Conf. on Electrical Engineering, Computing Sci. and Automatic Control* (CCE) p 6

[25] Kielan P 2019 System for examination of human movement system dysfunctions *Przeglad Elektrotechniczny* **1**(**12**) 142-5

Developing models of IoT infrastructures to identify vulnerabilities and analyse threats

N A Proshkin[1], E S Basan[1], M A Lapina[2], A G Klepikova[3], V G Lapin[4]

[1] South Federal University, 2 Chekhov St., Taganrog, 347928, Russia

[2] North-Caucasus Federal University, 1 Pushkin St., Stavropol, 355017, Russia

[3] Belgorod State National Research University, 85 Pobeda St., Belgorod, 308015, Russia

[4] Stavropol regional clinical consulting and diagnostic center, 304 Lenina St., Stavropol, 355000, Russia

E-mail ele-barannik@yandex.ru

Abstract. This article describes the developing stands of typical IoT infrastructure Technologies, which are designed to identify vulnerabilities and analyse potential threats. Scenarios of attacks typical for such systems have been developed, and the consequences of implementing attacks have been determined. Attacker can use the developed scenarios, been motivated by different goals. The research is also aimed at increasing the awareness of IoT users and determining the necessity to compliance with the minimum set of requirements relevant articles in literature searches, great care should be taken in constructing both.

1. Introduction

In the modern world, there are more and more opportunities for learning, and it becomes easier to find the necessary knowledge. Every person will be able to find any information they are interested in on the Internet [1].

Along with this, in the age of information technology, such a direction as "Cybercrime" is particularly developed. Cybercriminals are people who have made every effort to find the necessary knowledge and tools to use their computer illegitimately and influence information systems.

Today, the range of devices and technologies that can be attributed to the "Internet of things" systems is actively expanding, primarily systems such as: "Smart home", "Smart city", "Smart greenhouses", "Smart farm", "Smart factory", etc., I.e., systems that can consist of sensors, cameras, actuators, in combination with information and telecommunications technologies and control systems [2]. The use of such systems in various spheres of human life gives a positive economic effect and makes additional benefits from various points of view. The direction related to the digitalization of the economy is one of the priorities for the Russian Federation.

All information about the devices listed above and the devices themselves are easily accessible. The use of such systems in various spheres of human life gives a positive economic effect and allows to get additional benefits from various points of view. For example, using the "Smart lighting" provides energy savings of 40%. The "Internet of things" is an information system that consists of various hardware, software and telecommunications components, each of which has threats and vulnerabilities, as a result of which it can be subject of attacks from an intruder, environmental influences, failures and other factors that can lead to malfunction of both the system itself and the object that this system manages. A study by Positive Technologies States: "On average, about 10 vulnerabilities are detected in home network devices (Wi — Fi and 3G routers) every month. According to PositiveTechnologies experts, most developers and suppliers of network devices do not attach proper importance to such concepts as "testing" and "security". Many serious vulnerabilities remain without updates, and if they do come out, users are in no hurry to apply them. For example, 87% of systems tested by Positive Technologies experts had vulnerabilities related to the lack of updates." Dmitry Yaroshevsky, senior solutions architect of the company "Kaspersky Lab", writes: "Deception of sensors, manipulation of data received in the program or shown to the operator, influence directly on the Executive mechanisms-successful attacks on any links in this chain can lead to disastrous consequences. Despite the reality of the threat, many users are negligent about basic cybersecurity rules. So, according to one of the studies, only 50% of the surveyed Russians changed their default passwords on the router, 61% check updates once a year or less often." In many ways, users of such systems are often the culprits of incidents related to the violation of the security of their "Internet of things" system themselves [3].

2. Description of Internet of things infrastructure stands

A. Automated dairy farm

The first example of a stand developed for information security analysis is "Automated dairy farm" (hereinafter referred to as "Smart farm"). The "Smart farm" model consists of temperature and carbon monoxide sensors, alarms, lighting, and an Arduino UNO microcontroller. Microcontrollers in such systems are used to control other devices and provide an interface for human interaction with the system. The structure of the remote farm management system, its topology, and physical and technological connections are shown in figure 1.

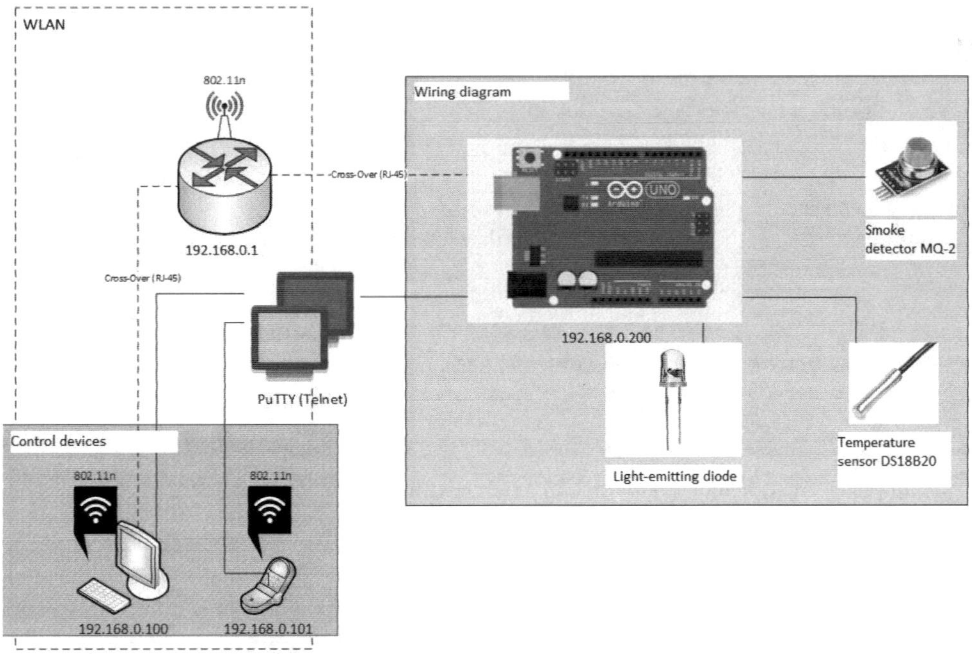

Figure 1. Block diagram of «Automated dairy farm»

The "Smart farm", connected to a wireless router via an Ethernet cable, is managed using control devices (PC or phone) located in the same WLAN (LAN) network with the Arduino UNO microcontroller, using the Telnet remote control protocol.

Analogous to such infrastructures are the systems of the companies "Dairy XL", "Iron Ox", the international agricultural Corporation "Lely" and others [4].

B. Automated oil pump

The second stand is "Automated oil pump" (hereinafter referred to as "Oil pump"). The "Oil pump" model consists of a stepper motor for adjusting the pump's swing speed, an Arduino UNO microcontroller, and a smoke generator that simulates the engine's excess power. Control of the "Oil pump" connected to the router via a cross-over cable is performed by control devices (PC or phone), located in the same WLAN (LAN) network with the Arduino UNO microcontroller, over the HTTP Protocol, by connecting to the management interface and supporting the system via the browser address bar. Analogues of such systems are: "AHM7500 series electric pumping oil mainline units" and others [5].

C. Smart house

The third stand is "Smart house". It includes several types of wireless communication: radio, Wi-Fi, telephone, temperature, humidity, pressure, carbon monoxide, motion, rain, sound, light controls, Russian microcontrollers Iskra and their Arduino counterparts, servos.

The Smart house connected to the router is managed via a mobile app that displays all the system's functionality.

The type of wireless communication is determined by the analysis and identification of the most secure type of communication.

Comparison of different countries-manufacturers of microcontrollers is related to the need to use domestic technologies.

Analogous to such infrastructures are many systems such as: "Xiaomi", "Yandex", "Rexant", etc.

3. Typical attacks on information systems and tools for attacks

A. DoS Attack

Denial of Service (DoS) attacks are a type of network attack [5]. A DoS attack causes interruptions in network services for users, devices, or applications.

There are two types of DoS attacks:

- Huge amounts of traffic;
- Packages with incorrect formatting.

The tool for carrying out the attack can be the utility "Netwox 76" under the Linux operating system.

B. Sniffing

This type of attack involves traffic tapping using the Wireshark sniffer program, which is used by network administrators to detect network problems, but a cybercriminal can use this program to commit illegal actions.

C. Spoofing

Spoofing is an impersonation attack in which attackers exploit a trust relationship between two systems.

Types of attacks using substitution.

- Spoofing the MAC address;
- IP address spoofing;
- ARP substitution.

The tool for carrying out the attack can be the utility "Netwox 40" under the Linux operating system.

D. Man in the middle (MITM)

An attacker becomes an intermediary by intercepting messages between devices on the network to steal information. A criminal can also manipulate messages and pass false information between hosts, since the hosts do not know that a message modification has occurred.

The tool for conducting an attack can be the "Netwox" utility under the Linux operating system.

4. Vulnerabilities of Internet of things infrastructure

Finding vulnerabilities in copies of the Internet of things systems allows to analyze a similar real system, identify potential vulnerabilities and, accordingly, potential threats.

It must be noted that in real systems, responsible information security specialists provide for all typical attacks on automated systems, thereby reducing the number of potential threats.

In particular, their security measures target network nodes such as routers, switches, and management devices of automated infrastructure.

But IoT Technologies, despite their simplicity, also have a number of vulnerabilities. It is difficult to influence the sensor modules from the outside, you must have access to control the microcontroller, but the wireless communication modules of "Internet of things", practically, do not have protection at all.

In this way, the basic infrastructure of IoT Technologies consists of two levels:

- Network devices;
- «Internet of things».

Therefore, it is advisable to separate vulnerabilities depending on the level of infrastructure components.

A. Vulnerabilities at the network device level

This layer, as mentioned earlier, includes: routers that provide remote connection and data transfer over the network; switches that allow to combine multiple devices into a single local network; and management devices that directly have access to the infrastructure.

This level is characterized by typical vulnerabilities and attacks, so it does not make sense to consider it in detail.

We will take a closer look at vulnerabilities at the Internet of things level

B. Vulnerabilities at the Internet of things level

This level includes microcontrollers, sensors, wireless and wired communication modules.

The sensors are only exposed to electromagnetic interference and do not act as a target. Microcontrollers are vulnerable only when interacting directly with them, that is why, they can be a subject to 0-day attacks. But the software or firmware of microcontrollers cannot be counted without special utilities and direct access to it. Therefore, the attacker does not appraise microcontrollers as a target.

The main target of such systems are wireless communication modules, as well as wired communication modules connected to the router.

Communication modules are as vulnerable as communication tools. They are vulnerable to eavesdropping, signal suppression, radio intelligence, etc. Most vulnerabilities are also related to the fact that IoT Technologies are connected to the Internet.

Let's look at the vulnerabilities of specific stands.

5. Vulnerabilities of IOT stands and attacks:

A. Vulnerabilities of "Auto farm"

"Auto farm" is connected to the local network via an Ethernet cable connection between the router and the Ethernet Shield.

The vulnerability of the Telnet remote access Protocol is that data is transmitted over this Protocol without encryption. In this way, you can conduct an attack on the confidentiality of information by using the Wireshark traffic sniffer to intercept packets and read data. The vulnerability of the router is that it does not have protection against sending ICMP packets, so an attacker can scan the network and identify devices inside this network. And also conduct a DoS or DDoS attack on the router, thereby disconnecting the control device from the system.

B. Vulnerabilities at the Internet of Things

This level includes microcontrollers, sensors, wireless and wired communication modules.
Sensors are exposed only to electromagnetic effects and do not act as targets.

Microcontrollers are only vulnerable to direct interaction with them and are susceptible to 0-day attacks. But the software or "firmware" of microcontrollers cannot be read without special utilities and direct access to it. Therefore, the microcontrollers do not see the attacker as a target.

The main goal of such systems are wireless communication modules, as well as wired communication modules connected to the router.

Communication modules are vulnerable as well as communications. They are vulnerable to listening, signal suppression, radio intelligence, etc. Also, most of the vulnerabilities are related to the fact that IoT-Technologies are connected to the Internet.

Consider the vulnerabilities of specific stands.

C. Vulnerabilities of the "Oil pump"

The "oil pump" is connected to the local network via a crossover cable connection between the router and the Ethernet Shield.

The vulnerability of the HTTP Protocol is that information transmitted in packets and headers of this Protocol can be used to get data about the web server or clients.

When connecting the router to an Internet provider, access to the web page for managing the "Oil pump" is not protected in any way, and a VPN is not used, so an attacker can find out the IP address of the system and connect to it from anywhere in the world.

There are also router vulnerabilities related to sending ICMP packets and attacks on system availability.

D. Vulnerabilities of the "Smart house»

The Smart house stand is a stand for analyzing the security of the majority of wireless communications technologies.

Microcontrollers are divided into:

- Central;
- Primary;
- Secondary.;

Primary and Central microcontrollers form a network of topology "Star", where the Central node is the Russian microcontroller Iskra Pro. They are connected to each other by a wired connection and use the UART Protocol for data exchange.

Sensors and communication modules are also wired to each secondary microcontroller and use UART, SPI, or I2C protocols, depending on the hardware of the device. Only the communication modules are connected to primary microcontrollers

Primary and secondary microcontrollers interact with each other via wireless communication. In the future, we will call this interaction "Pair".

Wireless communication modules for each "Pair":

- Radio communication module "nrf24l01" 2.4 GHz;
- "Troyka-module" Wi-fi module based on ESP-12 with 5 GHz ESP8226EX chip;
- GSM SIM800l cellular communication module without Bluetooth and GPS/GLONASS support.

To manage the Smart house infrastructure, a wireless connection to the Central microcontroller is used using the HC-05 Bluetooth module.

Vulnerabilities of the radio communication module are physical insecurity and the ability to listen to the communication channel. An attacker can intercept radio waves using a broadband software-defined radio receiver-transmitter "HackRF", and also, recognizing the frequency, send their data [6].

In this way, the confidentiality of information in the system will be violated without proper encryption of radio traffic.

Also, radio communication is subject of distortion or signal suppression by an attacker, which leads to a violation of the integrity and availability of information in the system.

The vulnerabilities of the Wi-Fi communication standard are similar to those of radio communications. In addition to them, the vulnerability of the "open hotspot" also applies. It appears when the Wi-Fi module does not have a hidden parameter in its configuration, which allows you to conduct a "Denial of Service" attack on the module itself. As well as attacks on vulnerabilities in the WEP and WPA security protocols.

Vulnerabilities of the cellular communication module are associated with the possibility of overflow of the incoming message buffer, as well as with suppression and jamming of cellular communication frequencies [7]. Since the cellular module uses telemetry to work in duplex mode, and there is no white list of numbers that the module can respond to, there is a vulnerability of the depletion of the money balance due to sending telemetry messages.

Vulnerabilities of the Bluetooth module are associated with vulnerabilities of the protocols LMP (Link management Protocol), L2CAP (logical link control and adaptation management Protocol), SDP (Service Discovery Protocol), TCS Binary (telephony control Specification - Binary), TCP and UDP.

This is not the final list of vulnerabilities, but an attacker can use them to attack the infrastructure using IoT Technologies, thereby exposing the entire system to threats.

Next, the scenarios of such attacks, analysis of the corresponding threats and their consequences, conducted at the stands of the Internet of things will be presented.

6. Attack Scenarios, threats and consequences

A. «Auto farm»

Scenario 1: A Cybercriminal learns that there is a milk production plant nearby that has remote access to control, determines the IP address of the router (target) and conducts a Denial of Service attack using the "Netwox 76" utility.

Possible consequences: during a DoS attack: all connected nodes lose connection. If you lose control of a plant, its owner may lose a large number of their assets, dairy products will deteriorate, and the sooner the owner reacts to the attack, the more money they will be able to save to restore the system.

The threat is related to the possibility of violating the availability of the system.

Scenario 2: A fired employee, been driven by revenge, connects to the internal network using old authentication data (or using a Brute force attack), and, already having access to the system, starts sending incorrect commands, overflow the incoming command buffer, and destabilize the enterprise. The operator who manages the plant by receiving and sending data, will receive distorted information or not receive it at all, and will not be able to send their commands.

Possible consequences: failure to take timely measures on the part of the owner will ensure the loss of his assets.

The threat is related to the possibility of violating the integrity and availability of the system.

B. «Oil pump»

Scenario 1: Infrastructure Operators use remote access to regulate engine power and limit its maximum value. An attacker using the «Netwox» utility arranges a DoS attack or, after infecting a large number of devices with viruses, makes a "Bot network" and arranges a DDoS attack at the moment when operators began to increase the engine power. When performing this attack, the connection is broken, so it is not possible to reduce the engine power remotely, it gradually approaches the limit and the engine breaks.

Possible consequences: Loss of assets.

The threat is related to the possibility of violating the availability of the system.

Scenario 2: An Attacker connected to the network by matching the password to the router using a Brute force attack. Using the «Wireshark» sniffer he captured packets with the HTTP Protocol, analyzed them, and found out the IP address of the system [8]. Since the infrastructure had a password system only at the router level, the attacker gained access to the system, increasing the power of the engine, thereby incapacitating it.

Possible consequences: Loss of assets.

The threat is related to the possibility of violating the system's privacy.

C. «Smart house»

Scenario 1: an attacker fraudulently entered the target's home by impersonating as a new neighbor. After establishing contact with the victim, he found out that the house is controlled with a phone, using Bluetooth. Knowing that the range of Bluetooth communication is not large, an attacker will take advantage of a vulnerability in the protocols of this wireless communication and perform a "Bluesnarfing" attack to steal data. Then, he can conduct a "Bluejacking" attack to send false alerts to the victim's phone and false commands to the system.

Possible consequences: an attacker will not be able to manage the Smart house remotely, but in a small radius from the house, if the owner is not in it, he will be able to set false infrastructure commands.

The threat is related to the possibility of violating the system's privacy, however, considering the fact, that the interaction distance is limited, this model is irrelevant.

Scenario 2: The Cybercriminal found out that the Smart house is controlled by sending common SMS. After learning the number of the Smart home, he will try to figure out its location using GPS [8]. If this attack fails, because the module cannot use this function, the attacker may start sending a large number of SMS messages to the Smart home number. And since now any mobile operator has a limited number of SMS messages, either in units per day, or in their cost, the funds or limit on the SIM card will quickly expire.

Possible consequences: you will constantly have to spend money in order to top up the balance of the smart house SIM card, and if the SMS limit is exhausted, data from the home will not be received.

The threat is related to the possibility of violating the availability of the system. To minimize the risks associated with the threat, it is enough to update the module's firmware by adding a list of «White numbers» to it.

Scenario 3: an attacker uses a powerful radio jammer located near the house will be able to disable all wireless communication modules.

Possible consequences: Loss of control of the system may cause jamming of doors, failure to turn on or off the light, etc.

The threat is related to the possibility of violating the availability of the system.

Scenario 4: a Cybercriminal uses a interceptor «HackRF» will be able to intercept a radio signal from a «Smart house» that extends over considerable distances, configure the" HackRF " to the frequencies that is used in the system, identify the carrier communication channel and compromise information, jamming the channel with interference or send false information.

Possible consequences: a Complete loss of control over the system will make it impossible to interact with it.

The threat is related to the entire face of the security cube: privacy, integrity, and availability, since the model of this threat is relevant.

7. Conclusion

To sum up, I would like to note the fact that the fight against cybercrime in the field of automated information systems using IoT Technologies should begin with the introduction of new information security measures for the «Internet of things» itself.

In IoT Technologies, all vulnerabilities are mainly related to the fact that IoT have access to the Internet. Here we should make a reservation – not "all", but "many", but there is also a share of vulnerabilities that are typical only for the «Internet of things», in particular for wireless communication modules.

Since IoT is a part of an automated system, their protection, from the point of view of information security, must be conducted properly.

The article presented primitive examples of real infrastructures. It is not possible to give an objective assessment of the security of the system based on them. However, real systems use devices that are as close to the devices in the stands as possible.

The development of such systems, the selection of components and their Assembly and configuration, allow us to give not a comprehensive, but individual

assessment of the security of each "Internet thing" that makes up the system, which is extremely necessary for the final assessment of the infrastructure audit.

8. Acknowledgment

This work was supported by the RFBR grant 18-07-00212 "Development of a method and decision protocol for detecting anomalous node behavior in group control systems of autonomous mobile robots".

References

[1] Basan A, Basan E, Makarevich O, Babenko L 2019 Investigation of the impact of active network attacks on a group of mobile robots. *Cybersecurity Issues* (1) ISSN 2311-3456 pp 35-44

[2] Makhdoom I, Lipman J, Ni W 2019 Anatomy of Threats to the Internet of Things *IEEE communications surveys & tutorials* **21(2)** pp 1636-1675

[3] Siboni S, Sachidananda V, Meidan Y, Bohadana M, Mathov Y, Bhairav S, Shabtai A , Elovici Y 2019 Security Testbed for Internet-of-Things Devices. *IEEE transactions on reliability* **68(1)** pp 23-44.

[4] Atamli A , Martin A 2014 Threat-based Security Analysis for the Internet of Things. *Proceedings of International Workshop on Secure Internet of Things.* pp 36-43.

[5] Zhou W, Yu B. 2018 A Cloud-Assisted Malware Detection and Suppression Framework for Wireless Multimedia System in IoT Based on Dynamic Differential Game. *China Communications* pp 209-223

[6] Carielli S, Eble M, Hirsch F, Rudina E, Zahavi R 2019 IoT Security Maturity Model: Practitioner's Guide Version 1.0 p 129

[7] National Institute of Standards and Technology. Framework for Improving Critical Infrastructure Cybersecurity 2018 Version 1.1.April 16 p 55

[8] Alam S, Mohammad M, Chowdhury 2011 Interoperability of Security Enabled Internet of Things. *Wireless PersCommun* pp 567-586 DOI 10.1007/s11277-011-0384-6

Development of Telecommunication System Units in Parallel Neural Network Basis

A A Smirnov[1], P A Sakhnyuk[2]

[1]North-Caucasus Federal University, Institute of Mathematics and Natural Sciences, Pushkina str. 1, Stavropol, Russia Federation, 355009

[2] National Research University Higher School of Economics, 20, Myasnitskaya str., Moscow, 101000

E-mail: shursun@mail.ru

Abstract: The article suggests the integration of a neural network as a parallel element base in a telecommunication system. In this case, the ability to learn or adapt to external conditions is applied as the main advantage. For telecommunication systems in conditions when it is possible, this ability will improve noise immunity, reliability, operability, etc. The article considers an example of the integration of a neural network into a discrete matched signal filter. It is noted that the use of parallel mathematical methods in signal processing leads to the maximum effect of increasing the quality parameters of such telecommunication elements.

Keywords: Neural Network, Matched Filter, Residue Number System, Learning Algorithms, Activation Function, Telecommunications.

1. Introduction

Nowadays, in the era of global informatization and multiple increases in the volume of information transmission, data transmission networks develop rapidly. The problems such as “Integration of various types of networks”, “Optimization of the information under existing channels”, “Adaptation of signals regarding the quality of channels”, etc. [1-5] arise on the agenda.

Along with the increase in the volume of information, a problem of the processing rate, speed and quality of transmission arises. One of the ways to increase these indicators is to enhance the clock rate or operating frequencies of telecommunications elements. However, at present, this direction has been exhausted by the technological limitations of the conductor dimensions in units of nanometers. Thus, the only way to increase the quality parameters is to parallelize the process of transmission and processing of information [6 -8, 13]. To construct systems for transmitting and processing parallel data, “parallel mathematics” is required, for example, a system of residual classes, a parallel element base, such as neural networks, is needed. Parallel communication channels are also needed based on various compression methods, algorithms, etc. This work is devoted to the introduction of mass parallelism in telecommunications. Let us consider the directions of possible integration of neural networks into telecommunication elements [2, 7, 8-11, 14, 15].

IOP Conf. Series: Materials Science and Engineering **873** (2020) 012019 doi:10.1088/1757-899X/873/1/012019

2. Possible directions of the neural network technologies integration

To determine the possibility of using neural networks in telecommunication elements, we consider the basic mathematics of neural networks and compare them with mathematical models of telecommunication elements [2, 4, 15].

2.1. Elemental base of neural networks

The basis for constructing neural networks is a neuron, as a physical prototype of a human brain's biological neuron. Functionally, it is represented in Figure 1, where x_i is neuron input vector (outputs of previous neurons); $w_{i,j}$ is weight vector; $F(u)$ is activation function (determined by the type of solvable problem).

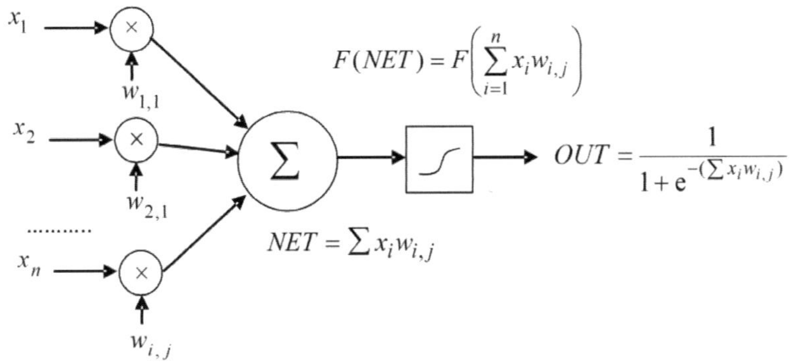

Figure 1. Neuron model

The activation function is considered as a nonlinear amplification characteristic of an artificial neuron that mathematically implements the following operation:

$$OUT = F\left(\sum_{i=1}^{n} x_i w_i\right). \quad (1)$$

The gain is calculated as the ratio of the increment of the output value to the small increment of the input value that caused it. It is expressed by the slope of the activation function curve. There are many activation functions, for example, a sigmoidal activation function (Figure 2), which is described by the expression:

$$F(u) = [1 + \exp(-u)]^{-1}. \quad (2)$$

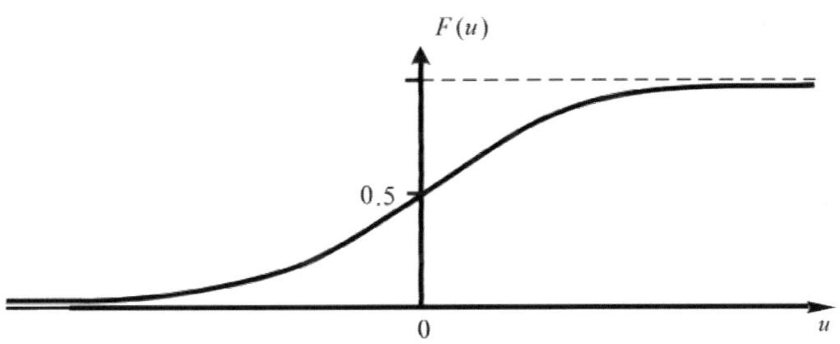

Figure 2. Sigmoidal logistic function

Another activation function adapted to the system of residual classes is a piecewise continuous function (Figure 3), which operates in the ring of integers:

IOP Conf. Series: Materials Science and Engineering **873** (2020) 012019 doi:10.1088/1757-899X/873/1/012019

$$F_2(u)=\left\lfloor \frac{10}{1+e^{-0,5(u)}} \right\rfloor . \tag{3}$$

This function allows smoothly to adjust the weight coefficients, eliminating the influence of random deviations in the received signal, i.e., it is more suitable for cases of constant interference.

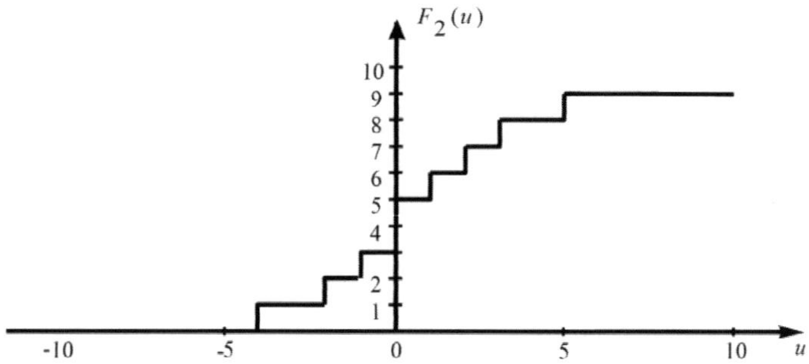

Figure 3. Sigmoidal piecewise continuous logistic function

In the adder, the input vectors of the neuron are multiplied by the vectors of weight coefficients, this sum is equal to: $NET = x_1w_1 + x_2w_2 + ... + x_nw_n = \sum_{i-1}^{n} x_iw_i$. Then the function takes the form:

$$OUT = F(NET) = F(\sum_{i=1}^{n} x_iw_{i,j}) = \left[1+\exp(-\sum_{i=1}^{n} x_iw_{i,j})\right]^{-1} . \tag{4}$$

One of the advantages of neural networks is their ability to self-learn. It means the ability to adapt weights w_{ij} to the actual conditions of propagation of radio waves. The difference between real conditions and the analytical model is most clearly expressed in those models where one of the input factors is the external environment, internal thermal noise, etc. The network is trained to give the desired set of outputs for some set of inputs. During training, the weights of the network gradually become such that each input vector produces an output vector. There are learning algorithms with and without a teacher. Learning with a teacher assumes that for each input vector there is a target vector representing the desired output. Comparison with the standard allows you to determine the error and adjust the weight coefficients of the neural network.

Learning without a teacher does not need a target vector for exits and does not require comparison with predetermined ideal answers. The training set consists only of input vectors. The training algorithm adjusts the network weights so that consistent output vectors are obtained, i.e. so that the presentation of sufficiently close input vectors gives the same outputs. Most modern learning algorithms grew out of the concepts of D. Habb, who in 1949 proposed the law of learning, which was the starting point for the learning algorithms of artificial neural networks.

A generalization of the neuron learning algorithm (delta rule) is formulated as

$$w_{i,j}(n+1) = w_{i,j}(n) + \Delta_{i,j} , \tag{5}$$

where $w_{i,j}(n+1)$ is weight value after correction, $w_{i,j}(n)$ is value weight i before correction, $\Delta_{i,j}$ is correction associated with *i-th* input x_i.

$$\Delta_{i,j} = \alpha\delta x_i , \tag{6}$$

where α is coefficient «learning rate», δ is the difference between desired or target output T and real output Y.

$$\delta = (T - Y) . \tag{7}$$

It follows that δ is multiplied by the value of each input x_i and this product is added to the corresponding weight. The learning coefficient is selected depending on the type of problem being solved and the type of signal. Obviously, the learning algorithm (5 - 7) can be implemented in the system of residual classes. For this, we will determine the weighting coefficients of the neural network according to (5), calculated on the basis of the corresponding mutually simple bases p_i and taking into account the proposed activation functions (Fig. 3). According to (6), when choosing an integer value of the learning speed and the initial values of the weight coefficients, the final learning result will also be integer.

$$w_{ij}(n+1) = \lfloor w_{ij}(n) + \alpha\Delta_{i,j} \rfloor, \qquad w_{ij}(n+1) \equiv (w_{ij}(n) + \alpha\Delta_{i,j}) \bmod p_i . \qquad (8)$$

In the matrix of weights, we get integers, not superior to the corresponding bases.

Note that it is impossible to see in neurocomputers a panacea for solving all problems. They are designed to solve difficult formalized and non formalized problems. As it turned out, the majority of such problems in the theory of telecommunications. Moreover, to formalize them, they are forced to resort to serious restrictions (the depth of feedback, the width of the signal spectrum, the use of standard radiation patterns, the limitation of the correcting capabilities of the code, etc.), to the detriment of quality.

Let us outline the main directions in the theory of telecommunications that require the use of neural networks and the development of the analytical foundations of their construction.

2.2. Neural networks in digital filtering

A digital filter is a device or signal processing algorithm implemented programmatically on an electronic computer. Mathematically, the operation of a digital filter is described by the equation in finite differences:

$$Sv(nT) = -\sum_{i=0}^{m-1} a_i \cdot Sv(nT - jT) + \sum_{j=0}^{k-1} b_j \cdot S(nT - jT), \qquad (9)$$

where a_m and b_k are coefficients, determined by signal processing tasks; $Sv(nT)$ is output values of discrete signal; $S(nT)$ is output signal; m are k are feedback depth coefficients. Under condition $m = k = 1$, a circuit that implements (9) can be performed by a single feedback neuron [14]. Moreover, multipliers of neural network coefficients will play as coupling amplifiers. As can be seen, expression (9) can easily be implemented in a neural network basis, since it contains the basic operations of this basis - addition and multiplication (4). An even greater effect is achieved when implementing digital filters in parallel mathematics.

2.3. Signal correlation filtering

The synthesis of correlation filters is based on the correlation properties of a signal having a large central response and small side lobes. The decision to receive a signal is made when the correlation response is exceeded:

$$R(k) = \sum_{i=1}^{n} h_i x_{i+k} , \qquad (10)$$

where: h_i is impulse characteristic; x_{i+k} are elements n*-th* bit signal with k*-th* shift, some threshold analogue of the activation function.

Correlation analysis in the residue number system is equivalent to correlation analysis in ordinary mathematics of a positional number system. The limitation in the development of correlation filter technologies is the number of admissible calculations (n) with simultaneous operation in real-time. In addition, there is the problem of creating continuous delay lines. The elimination of these problems is possible when implementing filters on a neural network basis inspired by the residue number system. This will allow for correlation analysis of any complexity signals in real-time. Analysis (10) shows that matched filters are a special case of neural networks. That is, such schemes can be implemented as a result of training a neural network, however, unlike a rigid circuit implementation, the latter are capable of adapting to real interference conditions.

2.4. The harmonic analyses, Fourier transform

The spectral analysis of discrete signals and filters is based on a discrete Fourier transform:

$$C(k\Omega) = \frac{1}{N}\sum_{n=0}^{N-1} S(nT)\exp(-j2\pi k)\,;\quad S(nT) = \sum_{k=0}^{N-1} C(k\Omega)\exp(\,jnk\Omega T)\,, \tag{11}$$

where: $S(nT)$ is original discrete signal; N is the number of counting in the signal cycle; T is discretization interval, $\Omega = 2\pi/(NT)$ is the main frequency of transform, $C(k\Omega)$ are coefficients of Fourier series.

The number of row coefficients is equal to the number of discrete signal samples (N). A limitation of existing circuits is the number of samples in signal N. The increase in the number of samples is limited by the capabilities of modern computing facilities. Therefore, fast Fourier transform algorithms are used. In this case, the analysis error either increases or the technical implementation of the analysis devices is complicated. An analysis of expression (11) shows that the discrete Fourier transform also contains the basic operations of neural networks (addition, multiplication). Therefore, the implementation of spectral analysis devices can easily be performed by neurocomputers and parallel mathematics.

2.5. Noiseless coding

Noiseless coding is based on the use of redundancy in the transmitted information. Information symbols are divided into intersecting groups and a verification symbol is assigned to each group so that each group has an even number of units. The mathematical form of this procedure is implemented by:

$$b_{l\,\text{ver}} = \sum_{i=1}^{k} \gamma_i a_i,\ \text{mod}2\,, \tag{12}$$

where a_i is information symbols; b_i is verification symbols; $\gamma_{i,l}$ is determined by l coding equation.

Analysis (12) shows that the coding scheme can also be implemented by a neural network in the residue number system based on modular operations. Such an implementation will allow receiving signals as a whole, combining the first and second decisive schemes.

2.6. Adapted communication channels

There is a direction of research on the creation of adaptive communication channels based on the "backward channel" method in real-time. The method is based on considering the effects of phase distortions in the front of the received wave in real-time and their compensation (for example, decameter channels by the inverse ionosphere method). This problem is formulated only for a multipath model of the communication channel.

Obviously, this problem relates either to non formalizable or difficult to formalize. Neurocomputers are designed specifically for solving such problems.

2.7. Phased array

The synthesis and operation of phased array is based on the common-mode addition of signals received or transmitted by individual emitters $S_i(t)\exp(-j\varphi_i)$. It can be written analytically as:

$$\dot{S}(t) = \sum_{i=1}^{n} S_i(t)\exp\{-j(\omega t + \varphi_i)\}\,, \tag{13}$$

where φ_i is phase offset at i-th radiator. The limitation in the implementation of phased arrays is the complexity of the analytical description of non-standard radiation patterns, adaptable to real conditions. For example, the formation of zeros in the direction of interference at the maximum signal-to-noise ratio. That is the solution of difficult formalized problems. Analysis (13) shows that phased arrays can be implemented in a neural network basis. Since the main operations in (13) are the sum and the product, their implementation in the system of residual classes will become even more efficient [16].

3. Modeling a discrete matched signal filter in a neural network basis

Existing discrete matched signal filters (Figure 4) has several disadvantages. This is low noise immunity due to the lack of consideration of the influence of constant signal distortions S_k caused by external additive and multiplicative noise, distortions caused by technical features of the implementation of delay lines, especially on the capacitive element base.

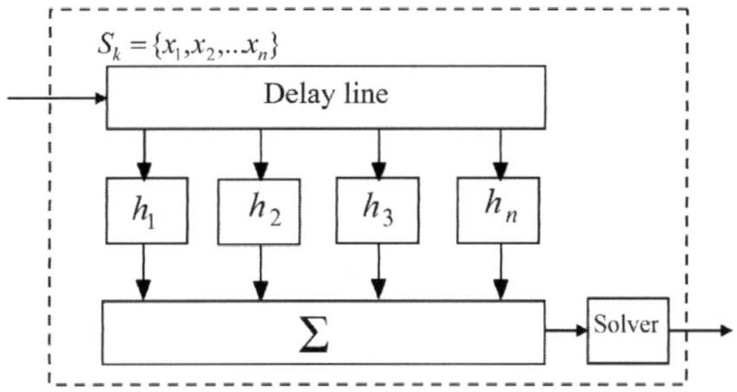

Figure 4. Filter matched to the signal

Another disadvantage is their rigid implementation and the inability to adapt to real signal distortions, which leads to a decrease in the correlation response and, therefore, to a decrease in noise immunity. In addition, discrete matched filters significantly lose their immunity when irreducible errors occur in the signal, since their multipliers h_i reflect the impulse characteristics of the expected signal and are fixed, and in case of distortion of the received signal, the response of such filters is significantly reduced, reducing their noise immunity.

The elimination of these shortcomings is possible only by adding neural network blocks to a discrete matched filter. Note that the structure of the matched filter has a parallel view, and the delay line is due to the need to receive a serial signal. The matched filter solves the problem of calculating the cross-correlation function $R(k)$ of the received signal $S = \{x_i\}$ and the impulse response $H = \{h_i\}$ of the expected signal depending on the shift k according to expression (10). The maximum $\max[R(k)]$ is determined at the time the signal is detected. Obviously, distortion of the received signal $S = \{x_i\}$ can significantly reduce the correlation response $R(k)$. In fig. 4 elements of the input discrete signal $S_k = \{x_1, x_2, \ldots x_n\}$ are delayed in the delay line, pass through the attenuators h_i or multipliers and, folding in the adder (Σ), give a response at the filter output. The value of the attenuators is determined by the impulse response of the signal. Therefore, the response of the autocorrelation function is formed at the filter output. Exceeding a certain threshold value by this response means signal detection. The Solver, which is set to the response threshold, decides to detect the signal. If you pay attention to the structure of the matched filter (Fig. 4), you can notice that it corresponds to the scheme of a trained neuron (Fig. 1) that implements function (10). In fact, you can replace the multipliers with a self-learning neural network. This will lead to increased noise immunity when working in communication channels with interference and significantly reduce the effect of internal interference. The neural network has pattern recognition properties, smoothing distortion, which allows you to compensate for various types of interference and distortion and increase noise immunity.

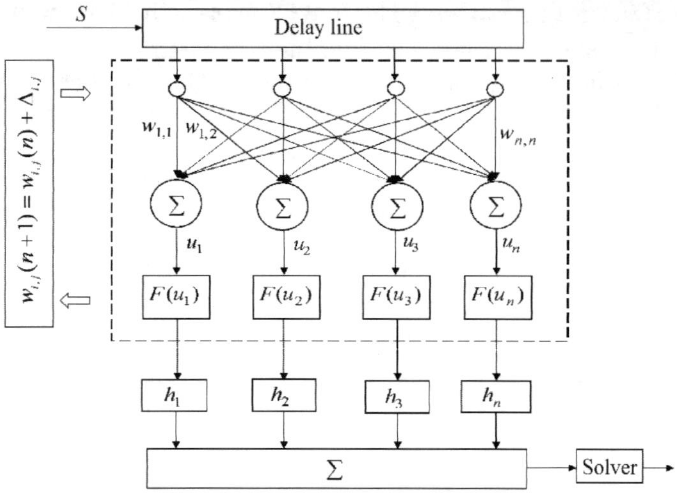

Figure 5. Adaptive filter matched to the signal

Figure 5 shows one of the options for such an implementation, where a neural network is introduced at the impulse response level. At the same time, attenuators acting as an impulse response h_i can be taken into account when choosing the activation functions $F(u_i)$ of a neural network. The role of the output adder and the deciding device, respectively, performs a common output neuron. A delay line is necessary to convert a serial signal in parallel. Obviously, if you use a parallel data format in the system of residual classes, then the delay line can be excluded.

For a sign-positive digital signal, we select the sigmoid function as the activation function (Figure 2). The solver, based on the Neumann-Pearson criterion, makes decision on the reception of the signal. This criterion takes into account the probability of a false alarm and can be taken into account in the output activation function. The value of the coupling weights is calculated according to the training algorithm of the neural network according to the Habb rule, for which the training set is the input signal. The learning algorithm, its learning speed and the activation function of the calculation unit can be selected depending on the specific types of signals, signal transmission speed and other tasks. The implementation of a neural network with training allows, due to its pattern recognition properties, to eliminate significant distortions of the input signal, eliminate fatal errors, eliminate the influence of internal interference of reception circuits and other interference, which gives the discrete matched filter adaptation properties. Finally, a mathematical model of an adaptive matched filter can be obtained by substituting expressions (2) and (4) into expression (10) for a conventional matched filter. Thus, we finally obtain:

$$R(k)=\sum_{i=1}^{n}\left(1/1+\exp\left[\sum_{j=1}^{n} x_{j+k} w_{i,j}\right]\right) h_i \ . \tag{14}$$

4. Conclusion

The resulting expression (14) can be directly used in modeling, as well as in the synthesis of adaptive matched filtering schemes. The introduction of mass parallelism in telecommunications becomes feasible with the integration of neural networks into telecommunication elements. Efficiency becomes maximum when applying simultaneously parallel mathematics of the system of residual classes.

References

[1] Schmidhuber J 2015 Deep Learning in neural networks: An overview *Neural Networks* **61** 85–117

[2] Takefuji Y 1992 A Neural Network Parallel Algorithm for Channel Assignment Problems in Cellular Radio Networks *IEEE Trans. Veh. Technol.* **41** 430–7

[3] Funabiki N, Takefuji Y and Lee K C 1992 A neural network model for traffic controls in multistage interconnection networks *IJCNN - International Joint Conference on Neural Networks* (Publ by IEEE)

[4] Smirnov A A, Bondar V V, Belokon L V, Saieg T kh., Mirzoyan M V and Smirnova O N 2018 The synthesis of the computing systems elements in neural network basis *Mod. High Technol.* 98–102

[5] Karras D A and Zorkadis V 2003 On neural network techniques in the secure management of communication systems through improving and quality assessing pseudorandom stream generators *Neural Networks* vol 16 (Elsevier Ltd) pp 899–905

[6] Smirnov A A, Bondar V V, Rozhenko O D, Mirzoyan M V. and Derzhaniya A D 2019 Mersenne numbers in the foundations of the system of residual classes when transmitting data in serial communication channels *The results of science and technology.* pp 87–94

[7] Aazhang B, Paris B P and Orsak G C 1992 Neural Networks for Multiuser Detection in Code-Division Multiple-Access Communications *IEEE Trans. Commun.* **40** 1212–22

[8] Hiramatsu A 1989 ATM communications network control by neural network *International 1989 Joint Conference on Neural Networks* (Publ by IEEE) pp 259–66

[9] Tavanaei A, Ghodrati M, Kheradpisheh S R, Masquelier T and Maida A 2019 Deep learning in spiking neural networks *Neural Networks* **111** 47–63

[10] Herzog S, Tetzlaff C and Wörgötter F 2020 Evolving artificial neural networks with feedback *Neural Networks* **123** 153–62

[11] Smirnov A A and Sakhnyuk P A 2018 Neural network implementation of an optimal receiver in binary communication channels *Des. Technol. Electron. tools* 3–9

[12] He G, Tang P and Pang X 1996 Neural network approaches to implementation of optimum multiuser detectors in code-division multiple-access channels *Int. J. Electron.* **80** 425–31

[13] Smirnov A A, Sakhnyuk P A, Sayeg T K, Vasilchenko A S and Yaroshuk A A 2019 Evaluation of noise immunity of communication channels with fading during neural network integration *Des. Technol. Electron. tools* 3–9

[14] Smirnov A A, Sakhnyuk P A and Senik V V 2004 Synthesis of digital signal processing circuits based on training of non-network circuits *Neurocomputers, Dev. Appl.* 91–4

[15] Brown T X 1989 Neural network for switcing *IEEE Commun. Mag.* 72–81

Development of a model for detecting security incidents in event flows from various components in a network of telecommunication service providers

D I Parfenov[1], I P Bolodurina[1], M A Lapina[2]

[1]Orenburg State University, 13, Pobedy Ave., Orenburg, 460018, Russia

[2]North-Caucasus Federal University, 1, Pushkin St., Stavropol, 355017, Russia

E-mail parfenovdi@mail.ru

Abstract. In the framework of this study, a technical solution was developed that makes it possible to detect network security incidents with a high probability using data arrays about device statuses, network events, and information stored in system logs. A model for identifying attacks on a network has been developed, using behavioral analysis and allowing the identification of suspicious network activity An algorithmic solution has also been built that allows aggregating data in a single store based on Cassandra and correlating events from specified sources using gradient boosting of decision trees in the CatBoost implementation. During the computational experiment, the study of the proposed hybrid solution for the accuracy of identification of individual types of attacks was conducted. It is proved that the proposed approach can effectively detect and repel attacks by reducing the response time to security incidents.

Keywords: Security incidents; intrusion detection; network of telecommunication service providers; network monitoring

1. Introduction

The volume of services provided by providers using publicly available data networks is increasing annually. The modern borderless is the main network architecture for the organization of the technology of the Internet of things (IoT). In turn, the borderless network architecture allows you to associate each object or device with network capabilities. Today, devices included in such a network can have completely different purposes, from household items and video cameras to automation systems, sensors, medical equipment, and other vital real-time systems [1]. Ensuring network security in such difficult conditions is one of the main tasks of telecommunication service providers. At the same time, I take into account the flexible and fairly dynamic topology, as well as the wide range and convergent nature of the services provided, the issue of trust between devices when sharing communication channels is no less important. It is important for users to be sure that the data network to which their network devices are connected has a fairly high degree of security. A feature of modern communication networks is the presence of many heterogeneous devices that use a variety of data transfer protocols. Each protocol has its own mechanisms and security measures.

Traditionally, to protect the network perimeter, telecom operators use approaches based on the use of intrusion detection systems (IDS) and intrusion prevention systems (IPS), as well as firewalls [2]. Such methods are effective if the attack on the network comes from the external perimeter. However, as noted earlier, in the modern conditions in which telecom operators work, the boundaries of networks are blurred. In such circumstances, each client network device is a potential source of cyber attacks. In such circumstances, traditional control systems are not effective enough.

Generally, there are three basic approaches to detecting attacks in the networks.

The most common are signature-based approaches. Signature intrusion detection systems detect attacks when the behavior of a system or network matches the attack profile stored in their own IDS databases. If any activity of the system or network matches the saved templates, a warning will be issued. Signature-based intrusion detection systems are accurate in detecting known threats, and their technique is easy to understand. However, this approach is ineffective for detecting new attacks and other variants of known attacks, since a matching signature for these attacks has not yet been established.

Anomaly-based approaches are also quite common. They allow you to almost instantly compare the current traffic profile with the profile of normal behavior. This approach is effective in detecting new attacks, in particular those attacks that involve resource abuse. A significant drawback of this behavior is that any changes in traffic profiles, including legitimate ones, are considered an intrusion. Therefore, using exclusively this approach to detect attacks has a high level of false positives. To build a normal behavior profile, statistical methods or machine learning algorithms are usually used.

Specification-based approaches typically include groups of rules and thresholds that determine the expected behavior for network elements, such as nodes, protocols, and routing tables. Specification-based approaches detect intrusions when network behavior deviates from specification definitions. Therefore, specification-based detection has the same purpose as the anomaly-based approach: detecting deviations from normal behavior. The key difference between these two approaches is the following: in the approaches based on specifications, a human specialist must manually determine the rules for each specification. Manual specifications typically provide lower false positives than anomaly-based detection. In addition, specification-based detection systems do not require a training phase, as they can start working immediately after setting up the specification. However, manually defined specifications may not adapt to different environments and may be time consuming and error prone.

Thus, it has been established that existing trips have a number of significant drawbacks, and today there is no comprehensive solution providing a mechanism for organizing the protection of more devices from a single point. Therefore, in the framework of this study, it is proposed to use a hybrid solution, which includes: the analysis of network traffic and the search for abnormal activity on network devices, as well as the analysis of data logs generated on network equipment. Such an approach will allow using ideas of detection based on signatures, specifications and anomalies, which will maximize the advantages of these approaches and minimize their disadvantages.

2. Related work

Scientists around the world offer various options for approaches to detecting and identifying attacks in networks.

As noted earlier, existing signature mechanisms are not enough to identify zero-day attacks. Researchers [3] offer an approach to combat this type of attack, based on the use of a modified support vector system (SVM) method. To increase the efficiency of Enhanced SVM, it preliminarily forms a profile of legitimate traffic, and for training SVM, a self-organizing characteristic map (SOFM) is used. The authors also apply a packet filtering scheme based on a passive TCP / IP fingerprint to cut off packets of the same type. In addition, the authors use a feature selection method using a genetic algorithm to obtain optimized information from raw traffic.

Study authors [4] propose the construction of two-tier models for classifying traffic anomalies based on Naïve Bayes in versions of KNN classifiers with the determination of the confidence coefficient, as well as linear discriminant analysis to reduce the dimension. Experimental results with the NSL-KDD dataset have shown significant gains in the speed of anomaly detection by reducing computational time.

Modern networks change dynamically over time and to adapt security systems to work in such conditions, a number of researchers use deep learning methods. In article [5] researchers proposed models for detecting anomalies based on various structures of deep neural networks. Training samples from the public datasets NSLKDD, NSLKDDTest + and NSLKDDTest21 were used as the main data set. A feature of the computational experiment is the use of a GPU, which significantly accelerates the learning process of the proposed model. A comprehensive review of deep learning methodologies for identifying anomalies in networks was conducted by the authors of the study [6]. They conducted extensive comparative computing experiments to compare the efficiency and accuracy of network traffic analysis.

As part of the study [7] the authors propose a new approach to identifying anomalies based on the center of the cluster and the nearest neighbor (CANN). The experimental results of the authors show that the CANN classifier works better in terms of classification accuracy, detection frequency and false positives.

As part of the study [8] the authors propose a model of the support vector machine (SVM), combining the analysis of the kernel principal component analysis (KPCA) with the genetic algorithm (GA). The authors propose using the multilayer SVM classifier as the basic classifier for attack / non-attack. At the same time, KPCA is used as a preprocessor to reduce the dimension of feature vectors and reduce training time. The genetic algorithm is used to optimize penalty functions. Experimental results show that the proposed model provides a higher accuracy of forecasting, a higher rate of convergence.

Modern network intrusion detection systems must be able to handle large and rapidly changing data. Therefore, the authors of the study [9] propose the use of ensemble solutions and a distributed approach and implementation of intrusion detection systems. For such systems, data exchange between distributed nodes is critical. The authors review the studies related to the performance of such solutions.

To search for attacks, the authors of the study [10] propose a distributed environment for extracting functions and preparing data for extracting functions from raw network traffic. The results are used to calculate during clustering, and anomalies are detected based on the outlier search.

In view of the high dynamics of changes in modern networks, the prediction of the development of attacks in priority. The authors of the study [11] propose a mechanism for detecting anomalies based on the statistical procedure, analysis of the main components, and metaheuristics of optimization of the ant colony. Based on the data obtained, the main traffic characteristics are obtained, from which they subsequently form a profile. The resulting typical profile is used in the future to compare forecast traffic values, due to this, anomalous activities are detected.

A review of existing approaches to solving the problem of searching for attacks in networks showed that most of the approaches considered to one degree or another use anomaly search, a signature approach or specifications, to identify attacks. The main significant drawback of the approaches considered is the high computational complexity, which is unacceptable for real-time systems. An equally important drawback of research is an experimental study that affects no more than 5-6 varieties of attacks. Nevertheless, it is worth noting that most of the solutions presented show a fairly high accuracy in determining attacks, on average from 50 to 90%.

3. Materials and Methods

Intrusion detection can be considered as one of the following machine learning tasks.

The task of forecasting. This task is aimed at predicting future opportunities and trends. Having at its disposal a prior information about the nature of upcoming events and upcoming changes, it is possible to prepare the necessary solution in advance. One of the main forecasting methods is regression analysis, with which you can predict the possibility of an attack. Regression analysis allows predicting similar behavior from collected attack logs.

The task of classification. Similar tasks are aimed at dividing into classes of objects (network behavior) according to certain and predefined signs. Classification allows you to highlight everything that does not fit into standard classes, i.e., to identify an intrusion. Thus, the classification helps the security administrator determine the direction of protection and analysis. To solve the classification problem, teaching methods with a teacher are used.

The task of finding associative rules. This task is aimed at identifying hidden relationships between data. A relation from a time stream is called a sequence rule. If patterns are found in such sequences, it is possible with some probability to predict the occurrence of events in the future, which makes it possible to make more correct decisions. This analysis method can determine abnormal network behavior by analyzing user behavior or the operation of network devices.

4. Informal statement of the problem

An urgent task is to develop models and mechanisms for ensuring network security. To do this, it is necessary to develop a model for identifying attacks on the network using behavioral analysis based on monitoring the characteristics (attributes) of network behavior. Under the behavior of the network we understand the totality of the characteristics (attributes) of the network at a certain point in time.

Identification involves matching current network behavior with known behavior (normal behavior, behavior under attack). Thus, the task of identifying attacks reduces to the task of classifying network behavior.

In the framework of this study, the classification problem can be formulated as follows.

Let there be given many behavior classes of the network C (normal behavior, behavior during an attack); many signs of the behavior of the network K; many investigated network behaviors X_c.

Of the many signs of the behavior of the network K we single out a subset $K_I \subseteq K$ – the set of the most informative signs. That is, each behavior is represented by a feature vector K_I of dimension k.

Assuming that, X_{cK_I} – the set of studied objects represented by a feature vector K_I, $|K_I| = k$.

Let, with the help of some rule, R be constructed a training set - a set of labeled vector-signs of network behavior, i.e. $\Psi_{X_c^L} = \left\{\left(x_i, \bar{c}\right)_i\right\}_{i=0}^{M} = \cup_{C_i \in C} \cup_{x \in C_i \cap X_c^L} \left(x, \lambda(C)\right)$, where $X_C^L \subseteq X_c$ is plurality of training vectors, and $x_i \in C_{i_{\bar{C}_i}}$ is attitude of belonging.

It follows that to solve the problem of identifying attacks on a network, it is necessary to solve the following optimization problem based on a training sample $\Psi_{X_c^L}$.

For this, we additionally introduce the set $G = \left\{K_{I_1}, ..., K_{I_k}\right\}$, where $K_{I_i} \subseteq K_I$, $|K_{I_i}| = i$, as well as an auxiliary function $\psi : G \to \left\{X_{C_{KI_1}}, ..., X_{C_{KI_k}}\right\}$, where $X_{C_{KI_i}}$ is investigated network devices represented by a vector of length i.

It is required at a given level of identification efficiency σ to minimize the number of used signs $k \in K_I$, i.e.

$$\Lambda(a, X_c, \sigma) = \left\{|K_{I_i}| \,|\, K_{I_i} \in G,\ \Omega(a, \psi(K_i)) \geq \sigma\right\} \to \min_{i \in \overline{1,k}},$$

where $\Lambda(a, X_c, \sigma)$ is function to evaluate the minimum set of behavioral signs; σ is some given positive number.

To implement the identification and classification of attacks on the network in the framework of this study, an appropriate model is constructed.

5. Formalization and analysis of the attack identification model

The developed model of node attack identification is based on the definition of a typical attack profile. An attack profile usually includes the formation of a whole chain of nodes connected by the same attack. Such a process requires significant computing resources. However, resource monitoring systems already have the necessary profiles of legitimate traffic behavior, as well as historical records of recorded attacks, on the basis of which it is possible to build profiles of illegitimate traffic behavior in the network.

We will identify attacks based on tracking changes in the main characteristics of network nodes. Suppose that at time *t* in the network of telecommunication service providers on some node 2, an attack was carried out previously carried out on node 1 (Figure 1).

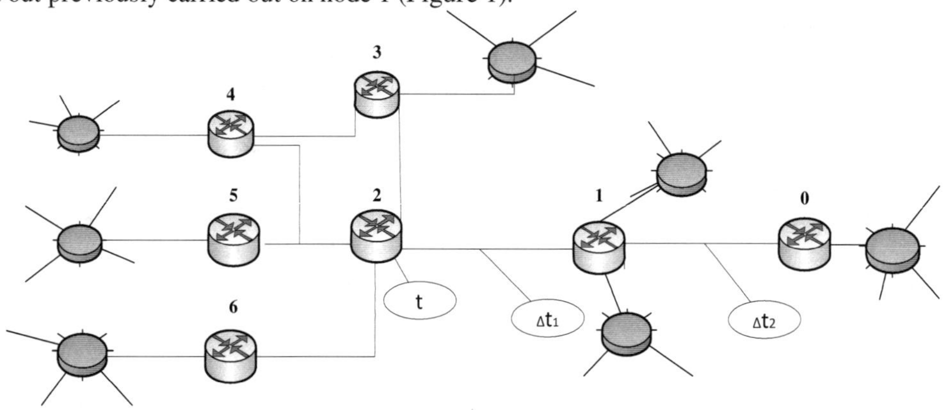

Figure 1. Telecommunications Service Provider Experimental Network Segment Diagram

In addition, it is known that on each of the attacked nodes {0, 1, 2} a chain of changes in the states of the nodes was found, which characterizes the profile of the recorded attack: $S_i \xrightarrow{t^1} S_j \xrightarrow{t^2} S_k \xrightarrow{t^3} S_m \xrightarrow{t^4} S_r \xrightarrow{t^5} \ldots \xrightarrow{t^T} S_p$,

It is clear that each transition from one state to another is accompanied by a change in characteristics:

1. The amount of traffic K_1 passing through the node at time t_i;
2. Indicator CPU K_2,%;
3. K_3 node response time, ms;
4. The number of lost packets K_4,%;
5. The state of RAM K_5,%;
6. Throughput K_6,%;
7. Statistics of violations of inputs K_7,%.

To describe a typical attack profile based on a change in the characteristics of a node, we pass from discrete values to a description format of the form:

+1 if K_i increased during the transition from one state to another;

-1 if K_i decreased during the transition from one state to another;

0 if K_i not changed.

Thus, the profile of the attack corresponding to the parameters will look like:

$$\begin{array}{c} \\ K_0 \\ K_1 \\ K_2 \\ K_3 \\ K_4 \\ K_5 \\ K_6 \end{array} \begin{array}{c} \begin{matrix} S_i \to S_j & S_j \to S_k & \ldots & \ldots \to S_m \end{matrix} \\ \begin{bmatrix} k_{01} & k_{02} & \ldots & k_{0p} \\ k_{11} & k_{12} & \ldots & k_{1p} \\ k_{21} & k_{22} & \ldots & k_{2p} \\ k_{31} & k_{32} & \ldots & k_{3p} \\ k_{41} & k_{42} & \ldots & k_{4p} \\ k_{51} & k_{52} & \ldots & k_{5p} \\ k_{61} & k_{62} & \ldots & k_{6p} \end{bmatrix} \end{array},$$

where $k_{ij} \in \{-1;0;+1\}$.

This approach allows you to quickly identify low-intensity and high-intensity attacks of the same type to determine regardless of the values of the parameters themselves, focusing on the nature of the processes.

Then the task of detecting attacks on related nodes {3, 4, 5, 6, etc.} can be determined much earlier, before it damages network performance. The solution scheme can be represented as the following sequence of steps.

To identify the nodes associated with the identified, in which the values of the characteristics as a whole differ from the norm. This stage will reduce the dimension of the problem.

Among the nodes that have a deviation from the norm in the characteristics, it is necessary to analyze the coincidence of the chain of states (partial). Due to the fact that we consider the attack on these nodes to be incomplete. For example, nodes 4 and 5 have at state time t chains of states $S_i \to S_j$ and $S_i \to S_j \to S_k$ respectively.

To directly confirm the conduct of the same type of attack, it is sufficient to classify the attack according to the profile of the change in characteristics. However, to improve the accuracy of determining attacks, in the framework of this study, an algorithm for collecting and analyzing logs generated on network equipment was developed.

6. Algorithm for collecting and analyzing logs for system logs of network devices

The algorithm for collecting and analyzing logs of network devices is an additional mechanism for detecting suspicious activities in the network. At the same time, to obtain complete and reliable information, it is necessary to carry out not only collection, but also filtering of aggregated data. This is necessary in order to generate incidents only for extremely dangerous events. At the same time, such a solution should work in real time with minimal delays. The solution to this problem is a combination of these factors and a large number of devices becomes difficult to implement. To reduce the computational complexity and increase the performance of the algorithmic solution, it is proposed to use the Cassandra distributed database, which allows collecting logs of network devices in a single repository. The algorithm is also based on a previously developed mathematical decision-making model based on the identification of related sets of events by generating sets of metadata at the preliminary stages of information receipt (in the signature analysis system, quick classifier, anomaly identifier).

Thus, the algorithm for collecting and analyzing logs of network devices can be represented as the following sequence of steps.

Step 1. Information is collected from security elements and network nodes in the syslog (log of system, services and applications) and SNMP (node status, performance counters) formats through a server that saves raw data for long-term storage in a relational distributed database Cassandra. The database stores information about system logs of applications and services operating in the network of a telecommunications service provider which do not write to syslog. In this case, a separate, private collection client is installed on the node, which publishes the data in the Apache Kafka sections. The collection server then reads the data from Kafka and sends it to Cassandra. NetFlow / sFlow network traffic data is also placed in Cassandra.

Step 2. Raw data is extracted from the Cassandra distributed database and events are generated from them using a set of structuring rules. Events have a hierarchical structure, each new node in the hierarchy adds additional fields. A tree of nodes is created for security elements of a particular type, which take into account all possible security events. The resulting structured set of events is saved back to Cassandra. To work in streaming mode, this step uses the Apache Spark framework.

Step 3. The structured data must be processed and the events in the network correlated with the state of the devices from which the data was received. Based on the data obtained, it is necessary to generate incidents with an intrusion detection system. For these purposes, it is proposed to use Spark Streaming in combination with the use of gradient boosting of decision trees in the CatBoost implementation. The advantage of this approach is the ability to identify new types of attacks that are a modification of existing ones, which existing signature methods cannot cope with.

It is worth noting that the originality of the developed algorithm is the possibility of its operation in two modes - in the mode of immediate processing as new events occur and in the execution mode on demand (for a retrospective analysis of events and investigation of incidents). For data analysis according to the presented algorithm using Python-based tools, a prototype module for data mining of system logs of network devices logs is implemented.

7. Experimental results

To assess the effectiveness of the sets when detecting suspicious network activity, real data is needed to describe the operation of the network of telecommunication service providers in which various attacks took place. Therefore, in the study, the CICIDS2017 data set developed at the University of New Brunswick was used to evaluate intrusion detection in the framework of this study. To study the effectiveness of these systems in this work, the following types of attacks were studied: DOS Slowloris, DOS Hulk, DOS Slowhttptest, DOS Goldeneye, Heartbleed, Web Brute Force, Web XSS, Web Sql inject, Infiltration - Dropbox download, Meta exploit Win Vista. On the proposed data set, a study of the accuracy of the proposed solution was conducted. The results of analysis are presented in Table 1.

Table 1. Network Attack Identification Experimental Data

	fp	tp	fpf	tpf
DOS Slowloris	988	4012	0,01643	0,100635
DOS Hulk	5124	5430	0,101641	0,236837
DOS Slowhttptest	4929	5625	0,18361	0,377932
DOS Goldeneye	6100	4454	0,285051	0,489653
Heartbleed	4599	5955	0,361532	0,639025
Web Brute Force	7601	2953	0,487935	0,713096
Web XSS	7748	2806	0,616783	0,78348
Web Sql inject	7776	2778	0,746096	0,853162
Infiltration – Dropbox download	7327	3227	0,867943	0,934106
Meta exploit Win Vista	7941	2627	1	1

The result of applying the proposed solution to the developed solution for identifying attacks in the network of a telecommunications service provider, taking into account the monitoring of system logs to recognize suspicious activities on network devices, is the correct identification of critical situations in 90% of the cases considered, which confirms its sufficient effectiveness.

8. Conclusion

In the framework of this study, the problem of ensuring network security has been solved; it is reduced to the problem of identifying attacks carried out on a telecommunications service provider network using an analysis based on monitoring the characteristics of network behavior and the information from system logs of network devices.

During the study, existing approaches to the search for anomalies were examined and a hybrid solution was proposed that allows identifying individual attacks in the general flow of events, as well as comparing attacks with events occurring in the system logs of network devices.

Acknowledgment
The study was carried out with the financial support of the RFBR in the framework of scientific projects No. 18-07-01446, No. 20-07-01065, as well as a grant from the President of the Russian Federation for state support of leading scientific schools of the Russian Federation (NSh-2502.2020.9) and a grant from the President of the Russian Federation Federations for state support of young Russian scientists - candidates of sciences (MK-860.2019.9).

References

[1] Zabrodina L, Parfenov D, Bolodurina I, Torchin V and Zhigalov A 2019 Development of a model of cyberattacks identification based on the analysis of device states in the network of a telecommunications service provider *Proceedings of the 2019 International Multi-Conference on Engineering, Computer and Information Sciences (SIBIRCON)* (IEEE) pp 0675-0680.
[2] Bolodurina I, Parfenov D and Torchin V 2018 Development and investigation of adaptive firewall algorithm to protect the software-defined infrastructure of multi-cloud platforms *Proceedings of the 2018 IEEE 12th International Conference on Application of Information and Communication Technologies (AICT)* (IEEE) pp 1–5.
[3] Shon T and Moon J 2007 A hybrid machine learning approach to network anomaly detection *Information Sciences* **177(18)** (Elsevier) pp 3799-3821.
[4] Pajouh H H, Dastghaibyfard G and Hashemi S 2017 Two-tier network anomaly detection model: a machine learning approach *J Intell Inf Syst* **48** (Cham: Springer) pp 61-74.
[5] Naseer S, Saleem Y, Khalid S, Bashir M K, Han J, Iqbal M M and Han K 2018 Enhanced network anomaly detection based on deep neural networks *IEEE Access* **6** (IEEE) pp 48231-48246.
[6] Kwon D, Kim H, Kim J, Suh S C, Kim I And Kim K J 2019 A survey of deep learning-based network anomaly detection *Cluster Comput* **22** (Cham: Springer) pp 949-961.
[7] Lin W C, Ke S W and Tsai C F 2015 CANN: An intrusion detection system based on combining cluster centers and nearest neighbours *Knowledge-based systems* **78** (Elsevier) pp 13-21.
[8] Kuang F, Xu W and Zhang S 2014 A novel hybrid KPCA and SVM with GA model for intrusion detection *Applied Soft Computing* **18** (Elsevier) pp 178-184.
[9] Ashfaq R A R, Wang X Z, Huang J Z, Abbas H and He Y L 2017 Fuzziness based semi-supervised learning approach for intrusion detection system *Information Sciences* **378** (Elsevier) pp 484-497.
[10] Bhuyan M H, Bhattacharyya D K and Kalita J K 2016 A multi-step outlier-based anomaly detection approach to network-wide traffic *Information Sciences* **348** (Elsevier) pp 243-271.
[11] Fernandes Jr G, Carvalho L F, Rodrigues J J and Proença Jr M L 2016 Network anomaly detection using IP flows with principal component analysis and ant colony optimization *Journal of Network and Computer Applications* **64** (Elsevier) pp 1-11.

IOP Conf. Series: Materials Science and Engineering **873** (2020) 012021 doi:10.1088/1757-899X/873/1/012021

Application of cloud technologies for effective management

S G Shmatko, D V Shlaev, A A Sorokin

Stavropol State Agrarian University, Zootekhnichesky Lane, 12, Stavropol, 355017, Russia

E-mail sshmatko@yandex.ru

Abstract. In modern conditions, when the flow of collected and accumulated information in the organization is growing rapidly, cloud services have become a popular solution for storing huge amounts of data. But not only data storage, but also its processing has become one of the advantages of cloud technologies implementing. Getting up-to-date analytical information for management and analyzing it for further use in maximizing profit for the organization is the most important factor in migrating to the cloud. All the work experience accumulated by the organization over many years becomes digital, which guarantees the success of management decisions. Therefore, cloud technologies also solve the problem of reducing the complexity of IT systems, their maintenance costs and increasing business dynamism. The paper studies the possibility to improve the efficiency of the organization's customer service department by automating its analytical activities based on the capabilities of cloud services and big data processing technologies. In order to achieve the aim, the authors solved the problem of analyzing the activities of the organization's customer service department using cloud technologies.

Keywords: cloud technologies, data analysis and processing, Python, implementation of digital technologies

1. Introduction

Nowadays, the use of cloud technologies for data analysis and processing is gaining popularity. Many organizations see a huge potential of these solutions and implement them to improve the efficiency of using the information they have collected over many years. The global market for cloud services is growing every year and is strengthening its position [1, 2, 5]. With the growth of this market, organizations spending on technologies related to cloud services is growing.

However, not all organizations get the maximum benefit from using of cloud technologies. This is due to a lack of understanding of the cloud deployment model, consideration of alternatives for implementation, and identification of data migration risks to the cloud.

Therefore, the management of any organization when making a decision to switch to cloud technologies needs to create requirements for using the cloud service and indicators that will be used to monitor the effectiveness of their use. Sometimes it is necessary to get the experience and understanding of all their capabilities before implementing technologies. For these purposes, you can use cloud service rates that do not require payment, but that allow you to evaluate all the positive and negative aspects of their use.

The purpose of the study is to show how it is possible to improve the efficiency of the organization's customer service department by automating its analytical activities based on the capabilities of cloud services and big data processing technologies.

In accordance with this goal, the authors solved the problem of analyzing the activities of the organization's customer service department using cloud technologies.

The practical significance of the research lies in the possibility of applying the analysis of the structural unit's activity in order to justify strategic decisions when approving the organization's development plans and selecting cloud services for their further use.

Today, there are many providers of cloud environments, platforms, storage and other services. In this regard, active attempts are made to compare them with each other. Due to the variety of services provided, it is quite difficult to figure out what is better.

Currently, the main cloud infrastructure providers are IBM, Microsoft, Amazon, Google and other companies. Each of the companies has a whole range of services. There are also small firms that provide cloud technology, but the price of these services, data protection and the provision of free work is preferable for large companies. Therefore, the analysis of the organization's customer service department was carried out using IBM cloud technologies plan Lite. This is enough to test the platform's capabilities and get the necessary experience to choose the placement model and manage the development process.

2. Using of cloud services capabilities

The use of IBM Watson Studio starts with a connection and download data. Data is loaded after creating a new Jupyter Notebooks [3, 4]. Creation is performed using the "New Notepad" command.

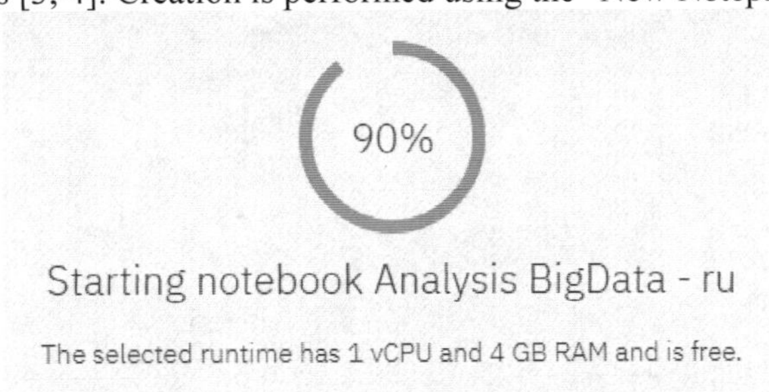

Figure 1. Loading data to Jupyter Notebooks Lite plan

After creating a new Notepad, you must add a database for analysis in *.csv format. Then we will generate the necessary code for connecting the database by the system:

```
# @hidden_cell
# The following code accesses a file in your IBM Cloud Object Storage. It includes your credentials
.
# You might want to remove those credentials before you share your notebook.
client_f05faf9965ecf8ffbae6fb71 = ibm_boto3.client(service_name='s3',
  ibm_api_key_id='JHsgqCHQQ7Bd2Y3NIZmH9UTbSJlNfzO1eW0X',
   ibm_auth_endpoint="https://iam.bluemix.net/oidc/token",
   config=Config(signature_version='oauth'),
   endpoint_url='https://s3.eu-geo.objectstorage.service.networklayer.com')
body = client_f05faf9965e245cf8ffbae6c1406fb71.get_object(Bucket='electro-donotdelete-pr-gugo
r7kp1pzg9s',Key='electro.csv')['Body']
# add missing _iter_ method, so pandas accepts body as file-like object
if not hasattr(body, "_iter_"): body._iter_ = types.MethodType( _iter_,body)
```

To solve the problem with data cleaning and processing, Watson Studio has developed a Data refining tool that allows you to organize independent data preparation. Watson Studio includes over 35 interfaces for connecting to data sources from IBM Cloud, other vendors ' clouds, applications, and on-premises systems [6].

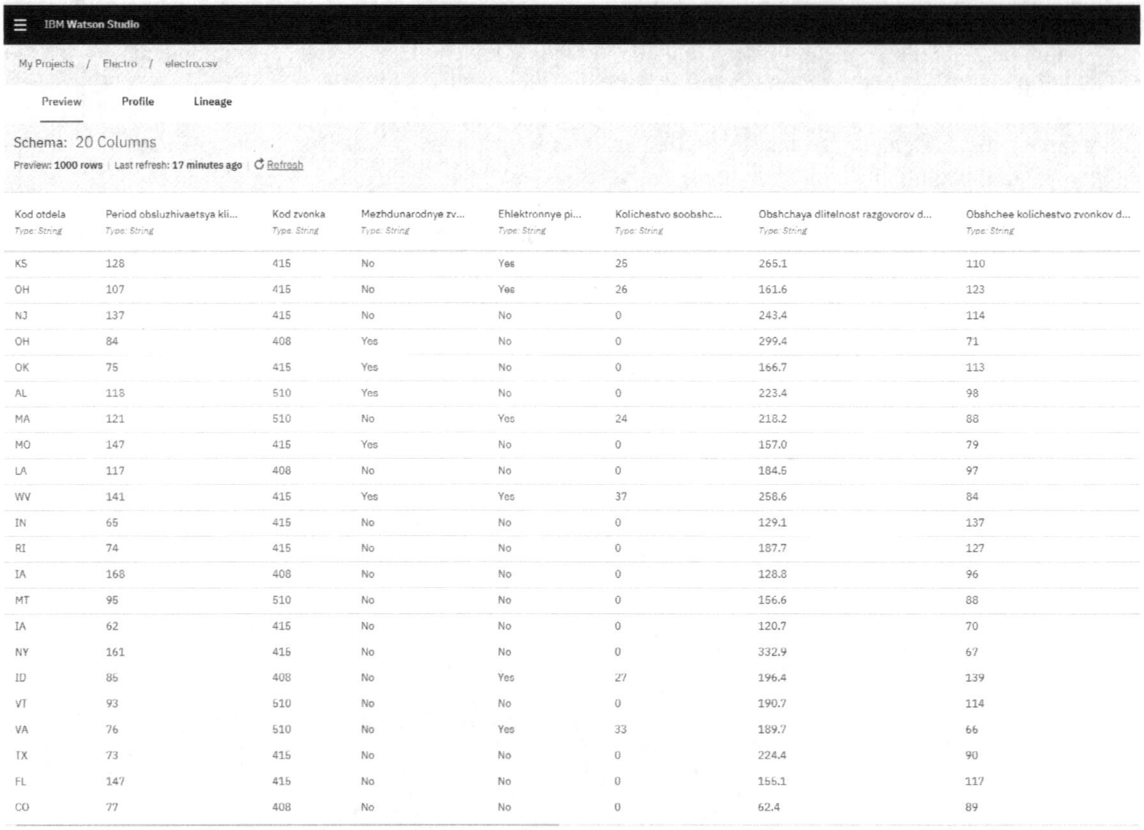

Kod otdela Type: String	Period obsluzhivaetsya kli... Type: String	Kod zvonka Type: String	Mezhdunarodnye zv... Type: String	Ehlektronnye pi... Type: String	Kolichestvo soobshc... Type: String	Obshchaya dlitelnost razgovorov d... Type: String	Obshchee kolichestvo zvonkov d... Type: String
KS	128	415	No	Yes	25	265.1	110
OH	107	415	No	Yes	26	161.6	123
NJ	137	415	No	No	0	243.4	114
OH	84	408	Yes	No	0	299.4	71
OK	75	415	Yes	No	0	166.7	113
AL	118	510	Yes	No	0	223.4	98
MA	121	510	No	Yes	24	218.2	88
MO	147	415	Yes	No	0	157.0	79
LA	117	408	No	No	0	184.5	97
WV	141	415	Yes	Yes	37	258.6	84
IN	65	415	No	No	0	129.1	137
RI	74	415	No	No	0	187.7	127
IA	168	408	No	No	0	128.8	96
MT	95	510	No	No	0	156.6	88
IA	62	415	No	No	0	120.7	70
NY	161	415	No	No	0	332.9	67
ID	85	408	No	Yes	27	196.4	139
VT	93	510	No	No	0	190.7	114
VA	76	510	No	Yes	33	189.7	66
TX	73	415	No	No	0	224.4	90
FL	147	415	No	No	0	155.1	117
CO	77	408	No	No	0	62.4	89

Figure 2. Clearing and processing data of customer requests to the customer service department

You can configure and use data analysis using the following languages: Python, R and Scala. Watson Studio also includes a development environment RStudio that simplifies data analysis using the R language.

Watson Studio supports package management. The conda environment is used for configuration. Computing environments are considered as project resources and can be used by all users for research [6].

IOP Conf. Series: Materials Science and Engineering **873** (2020) 012021 doi:10.1088/1757-899X/873/1/012021

Anaconda Environment

Summary

Environment	Anaconda Environment
Creator	
Language	Python 3.5
Hardware configuration	16 vCPUs and 64 GB
Software configuration	Default Python 3.5

Software configuration details

```
name: DSX-Python35
channels:
- defaults
dependencies:
- alabaster=0.7.10=py35h6fb19ab_0
- anaconda-client=1.6.5=py35hd900ed1_0
- anaconda-project=0.8.0=py35hb597f32_0
- asn1crypto=0.22.0=py35h0d675fe_1
- astroid=1.5.3=py35h1d0c565_0
- astropy=2.0.2=py35h2d2a8a6_4
- babel=2.5.0=py35ha5fc564_0
- backports=1.0=py35hd471ac7_1
- backports.shutil_get_terminal_size=1.0.0=py35h40844db_2
- backports.weakref=1.0rc1=py35_0
- beautifulsoup4=4.6.0=py35h442a8c9_1
- biopython=1.69=np113py35_0
- bitarray=0.8.1=py35h14c3975 1
```

Figure 3. Configuring package management

Watson Studio dashboards simplify data visualization in applications. The analyst can independently create visual models to search for new ideas. Interactive dashboards provide real-time data visualization. Data analysis and visualization tools help to find patterns in the presented data. You can also use filters and data navigation tools. You can integrate dashboards into the application context for maximum convenience for analytics [6].

To perform queries and analyze data, we will use the built-in Python programming language version 3.5, since this version is available for the Lite plan in IBM Watson Studio. Let us start by connecting the necessary libraries:

```
import requests, json, re
from io import StringIO
import io
import types
import pandas as pd
from botocore.client import Config
import ibm_boto3
```

Data analysis uses the Pandas library. This is a Python library that you can use to analyze data. Data arrays are usually stored in a table format, which is very convenient to use in analysis. Using SQL queries, such data is pre-processed for further use in the cloud environment.

To view the correctness of loading data for analysis, we will use the read_csv method and for their output we use the head method.

```
dt = pd.read_csv(body)
dt.head()
```

In Jupyter notebooks, Pandas dataframes are output as tables. The Pandas library outputs a small number of columns and rows, so if you need to output all the data, you can use the set_option function:

pd.set_option('display.max_columns', 48)
pd.set_option('display.max_rows', 752)
Target variable: loyalty Index, binary attribute (1-complaints).
The dataset has 3983 rows and 7 columns. Display the column names:
print(dt.columns)
Index(['Period obsluzhivaetsya klienta', 'Kod zvonka', 'Ehlektronnye pisma', 'Kolichestvo soobshchenij', 'Obshchaya dlitelnost zvonkov', 'Chislo obrashchenij', 'Indeks loyalnosti'], dtype='object')
Column names in Latin are not readable in the environment, so all column names were transliterated in accordance with GOST 7.79-2000.

The database uses the following feature types: bool, int64, float64 and object. The info method is used to analyze the data table for omissions in its filling. No data omissions were detected after using this method.

Now for further data analysis, we need to change the column type "Indexes loyalnosti" using the astype method and convert it to int64:
dt['Indeks loyalnosti] = dt['Indeks loyalnosti'].astype('int64')
The next step in the analysis involves calculating the main statistical characteristics for each numeric attribute. This can be done using the describe method (Table 1).
dt.describe()

Table 1. The main statistical characteristics of the data

Characteristics of data	The company customer service	Call code	Message ammount	Duration of calls	E-mails	Number of requests	Loyalty index
count	3983	3983	3983	3983	3983	3983	3983
mean	101.06	437.18	8.10	10.24	4.47	1.56	0.14
std	39.82	42.37	13.69	2.79	2.46	1.31	0.35
min	1	8652	-	-	-	-	-
25%	74	8652	-	8.50	3	1	-
50%	101	86554	-	10.30	4	1	-
75%	127	86544	20	12.10	6	2	-
max	243	86544	51	20	20	9	1

Now we need to consider the distribution of data by the target variable of our study. To do this, we will use the value_counts method.

dt['Indeks loyalnosti'].value_counts()

0 3500

1 483

Name: Indeks loyalnosti, dtype: int64

So 3500 of 3983 the organization's customers are loyal, since the value of the Index loyalnosti variable is 0.

For further study of the data you need to sort the DataFrame by the value of the Obshchaya dlitelnost zvonkov attribute:

dt.sort_values (by = 'Obshchaya dlitelnost zvonkov', ascending = False).head()

To answer one of the main questions of our research and find out what percentage of disloyal clients in the presented dataframe, we write the following expression:

dt['Indexes loyalnosti']. mean (). # result: 0.12126537

12.12% is a bad indicator for an organization. This means that urgent measures must be taken to identify areas of work that cause such consequences for the organization as a whole.

It is very convenient and appropriate in this case to perform indexing by a single column. Therefore, using indexing, we can determine the average values of numeric attributes that relate to disloyal clients and draw conclusions that interest us.

dt[dt['Indeks loyalnosti'] == 1].mean()

#input

Period obsluzhivaetsya klienta	102.6645
Kod zvonka	437.8178
Kolichestvo soobshchenij	5.115942
Obshchaya dlitelnost zvonkov	206.9140
Ehlektronnye pisma	4.163561
Chislo obrashchenij	2.229814
Indeks loyalnosti	1.000000

Now we can find out valuable information for the organization's management about the average duration of a call on the phone of a disloyal client using the following construction:

dt[dt['Indeks loyalnosti'] == 1]['Obshchaya dlitelnost zvonkov'].mean() # input: 206.91407867494823

3. Results

According to the data grouping by the columns we are interested in, we can conclude that, despite the fact that the percentage of disloyal clients is only 12.12%, this category spends 6 points more on various requests to customer service departments. In other words, one loyal customer has an average of 20 seconds to contact the departments, while disloyal customers have an average of 4.2 minutes.

For further analysis, you can group the data by the value of the total number of calls from all customers, which will later help you study the data for the workload of customer service departments and identify uneven distribution of service time.

4. Conclusions

Based on the organized and conducted analysis, the authors believe that the use of cloud data analysis technologies allows you to quickly get meaningful information for the organization's management and contributes to its effective development in the digital world [7, 8]. This analysis can be configured to run automatically based on uploading the required dataframe to the cloud. Conducting regular data analysis will provide strategic advantages to the organization and in current world, the use of cloud technologies is an important factor in achieving its goals.

References

[1] Chin S, Vos J, Weaver J, Chin S, Vos J and Weaver J 2019 JavaFX, the Web, and Cloud Infrastructure *The Definitive Guide to Modern Java Clients with JavaFX* (Apress) pp 367–409

[2] Balis B, Orzechowski M, Pawlik K, Pawlik M and Malawski M 2020 Cloud Infrastructure Automation for Scientific Workflows pp 287–97

[3] Perkel J M 2018 Why Jupyter is data scientists' computational notebook of choice *Nature* **563** 145–6

IOP Conf. Series: Materials Science and Engineering **873** (2020) 012021 doi:10.1088/1757-899X/873/1/012021

[4] Hamrick J B 2016 Creating and Grading IPython/Jupyter Notebook Assignments with NbGrader (Association for Computing Machinery (ACM)) pp 242–242
[5] Yang J and Chen Z 2010 Cloud computing research and security issues *2010 International Conference on Computational Intelligence and Software Engineering, CiSE 2010*
[6] Cecil R R and Soares J 2019 IBM Watson Studio: A Platform to Transform Data to Intelligence pp 183–92
[7] Kaur T 2019 Cloud Computing: A Study of the Cloud Computing Services *Int. J. Res. Appl. Sci. Eng. Technol.* **7** 1933–8
[8] Hashem I A T, Yaqoob I, Anuar N B, Mokhtar S, Gani A and Ullah Khan S 2015 The rise of "big data" on cloud computing: Review and open research issues *Inf. Syst.*

Technology development for the production of thermoplastic composites with agricultural fillers by compounding method on co-directional twin screw extruder

A M Kuzmin, E A Radaikina
Mordovia Ogarev State University, Mechanics and Power Engineering Institute, 7 Rossiyskaya, Saransk, 430904, Russia

E-mail kuzmin.a.m@yandex.ru

Abstract. The paper is devoted to the development of a continuous process for the production of thermoplastic composite materials based on low-pressure polyethylene and agricultural plant waste products by the components compounding on a Rheomex PTW 16 PolyLab laboratory twin-screw extruder. The processing modes of the developed compositions are given. The effect of melt degassing on the quality of the obtained granulate is noted.

Keywords: continuous process technology, thermoplastic composite, polyethylene, agricultural origin fillers, dispenser, compounding, twin-screw extruder, degassing, vacuum treatment, granulate.

1. Introduction

The main problem for thermoplastic composite materials with agricultural origin filler creating is the problem of uniform distribution of components by mass when they are mixed and compounded. This process in practice is carried out firstly by periodical method in rotary and roll mixers and secondly by continuous one in single-screw pin extruders and co-directional twin-screw extruders. Recently, to obtain the specified class of composite materials, twin-screw extruders of co-directional rotation are more often used [1, 2]. They provide high-quality high-speed compounding of components and obtain granulate composites with high uniformity of components distribution.

The mixing effect in twin screw extruders is provided by screws located parallel to the cylinder and rotating towards each other or in the same direction. In the opposite rotation of the screws, they work like rollers, while the material is fed along the screws. The shift and grinding of the material occur both between the turns of the screws and in the gap between the screws and the cylinder body. The value of material reverse flow increases with increasing distance between the screws and enhances the mixing effect. When the screws rotate in one direction, if they fully engaged, large shear deformations are achieved. It allows to mix highly viscous materials with high content of fillers [3].

A number of researchers [4–7] speak of advantages of twin-screw extruders in the process of compounding thermoplastic polymers with various modifiers. Due to their high productivity and good mixing properties, twin-screw machines can be used to implement a direct (single-stage) extrusion scheme of composites, in which the mixing of the components and the extruding of the resulting compound through the spinneret is carried out as a single stage. In this case, the extruders should be equipped with an effective degassing device (vapors and gases removal) and have wear-resistant surfaces.

The two-stage process for the production of wood-polymer composites consists of the pre-production of compound granulate with its subsequent processing by heat pressing, and injection molding or extrusion.

The possibility of processing composite materials with a thermoplastic polymer matrix and lignocellulose fibers on a twin-screw extruder was shown in this study [8]. It was noted that the particle size and fiber length significantly affect the compounding process.

Compositions of polypropylene and sunflower husk with organoclay modification were studied in this study [9]. Compounding of the components was carried out on a Leistritz Micro-18/GL-40D twin-screw extruder. It has been shown that the introduction of a small amount of modifier such as organoclay (5%) contributes to the improvement of the mechanical characteristics of composites. This improvement is due to the uniform distribution of particles in the polymer matrix, which has been proved by a scanning electron microscope using.

The purpose of this experimental study was the laboratory testing of a continuous process for producing thermoplastic composites filled with barley straw and wood flour on a twin-screw extruder with co-directional rotation, as well as studying the physical and mechanical characteristics of the obtained composites.

2. Materials and Methods

To obtain the composite material, such filler was used as:

- barley straw with a dispersion of 200 ... 350 microns, a moisture content of 8.7% and a bulk density of 0.14 g/cm3;
- wood flour with a dispersion of 180 ... 250 microns, humidity of 7.9% and bulk density of 0.16 g/cm3.

Grinding of a agricultural origin fillers was carried out on a rotary knife mill. Humidity was determined by thermogravimetric method by a moisture analyzer "Evlas-2M".

Low pressure polyethylene (HDPE 273-83) manufactured by Kazanorgsintez according to GOST 16338-85 was used as the polymer matrix.

As a result, samples of compounds of two compositions were obtained:

composition 1: 55% HDPE + 45% barley straw;

composition 2: 55% HDPE + 45% wood flour (powder).

Compounding was carried out on a Rheomex PTW 16 PolyLab twin-screw extruder. The appearance of the extruder with peripheral devices is shown in Figure 1.

The line includes the HAAKE RheoDrive 7 OS, the Rheomex PTW 16 PolyLab measuring extruder-compounder (modular 16 mm twin-screw co-directional extruder with an L/D ratio of 40:1), a cooling bath, a granulator and a metering system including volumetric single-screw DRS 28 for polymer granules, MT 1 volumetric twin-screw metering unit for fibrous fillers, Vertical Stuffer Feeder screw feeder for powder fillers, vacuum pump and Accel 500LC circulation cooler.

The equipment starts as follows.

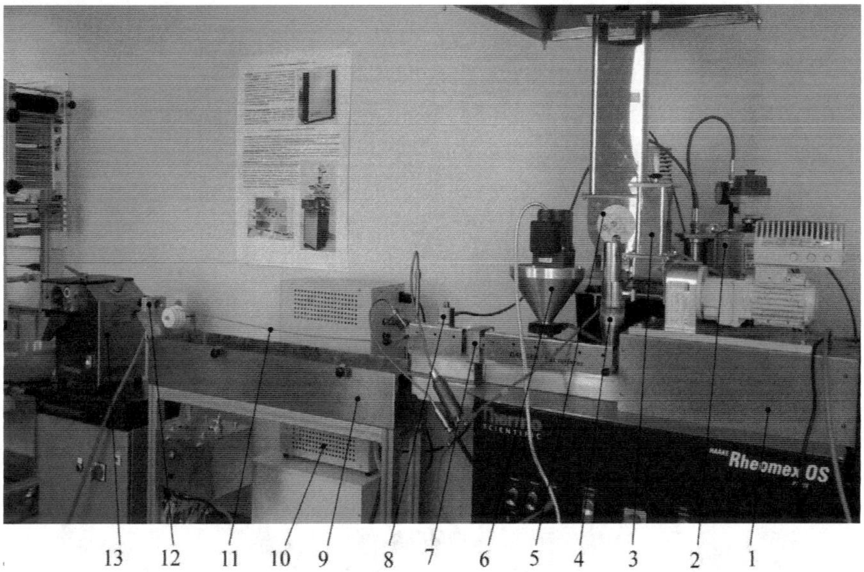

Figure 1. Appearance of a twin-screw extruder-compounder Rheomex PTW 16 with peripheral devices: 1 - extruder body; 2 - vacuum pump; 3 - agricultural origin fillers dispenser; 4 - boot device; 5 - polymer granulate dispenser; 6 - supply of powder filler; 7 - heated case with screws; 8 - adapter of the degassing system; 9 - a cooling bath with a temperature control system; 10 - thermostat; 11 - extrudate (strand) with air cooling; 12 - device for air drying and strand cooling; 13 - granulator

The cylinder was opened and the screws were cleaned of a protective lubricant with detergent. Then, the extruder heating zones were turned on and the temperature controllers were installed at the required temperatures (PolySoft OS program). The extruder reached operating temperatures within 10 minutes. By turning it on at a low speed (5 rpm), it was found that the screws rotate freely. Next, the PolySoft OS program and the drive unit were turned on, then, with free rotation of the screws at a frequency of 50 rpm, dispensers were switch on, which began to feed components (polyethylene granules and agricultural origin er fillers) into the loading device of the extruder. Further, the screw speed was increased to the recommended value at which the torque on the extruder drive does not exceed 60% of the maximum permissible value (130 Nm).

At the first stage, the masterbatch “HDPE 273-83 + agricultural origin fillers” has been obtained. Based on the metering calibration schedules, the relative rotational speeds of the metering screws were determined. For the MT 1 dispenser (due to its low productivity), the maximum (100%) rotation frequency was set, the rotational speed of the DRS 28 dispenser was calculated based on the ratio of filler to polymer.

The processing modes of thermoplastic composites are shown in Figures 2-3.

IOP Conf. Series: Materials Science and Engineering **873** (2020) 012022 doi:10.1088/1757-899X/873/1/012022

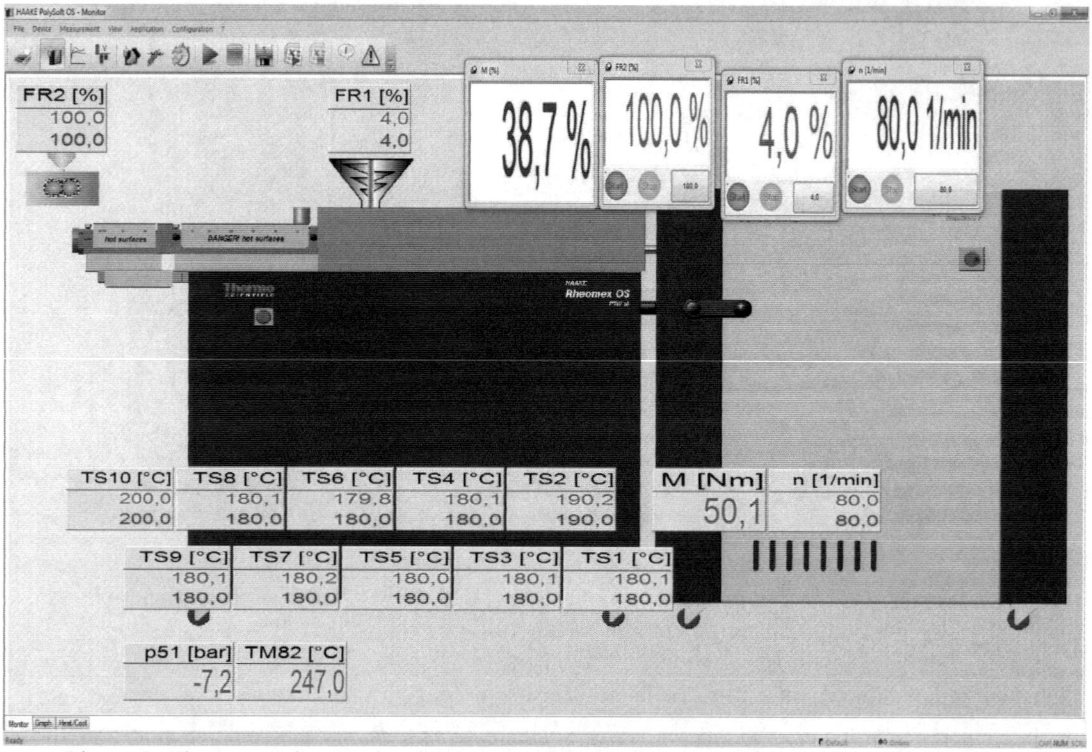

Figure 2. The processing mode of the composition "55% HDPE 273-83 + 45% wood flour"

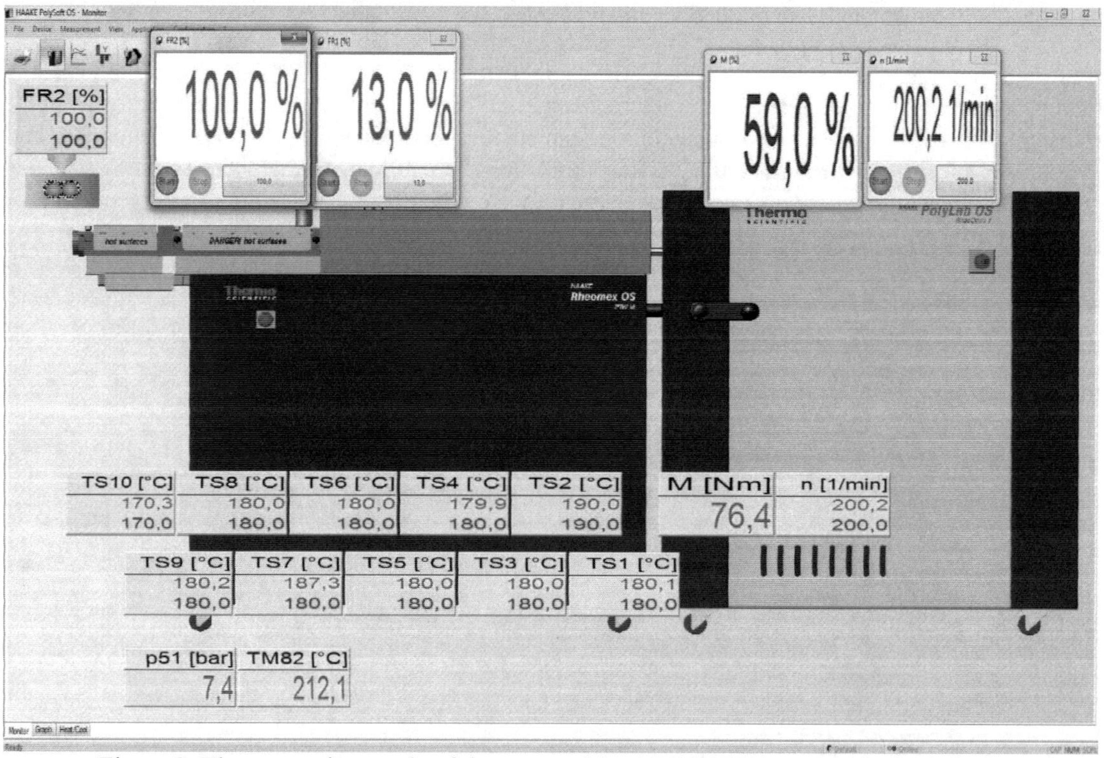

Figure 3. The processing mode of the composition "55% HDPE + 45% barley straw"

Figures 4-6 present some features of the granulation process.

Figure 4. The output of the strand (extrudate) from the extrusion head into the cooling bath

Figure 5. The output of the strand (extrudate) from the cooling bath to the drying device

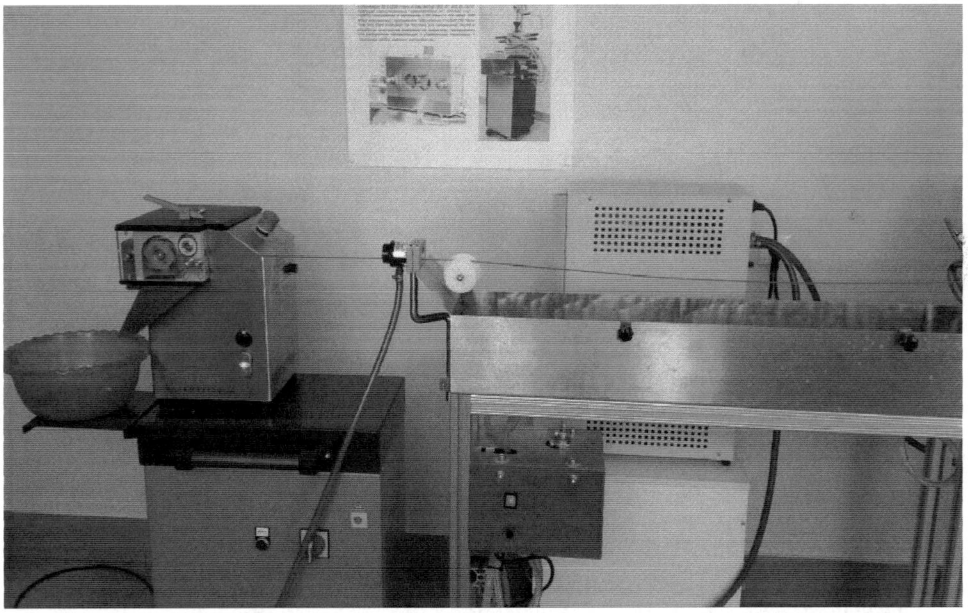

Figure 6. The output of the strand into the granulator (air cooling of the extrudate)

To carry out physical and mechanical tests by hot pressing on a Gibitre hydraulic press, plates with a size of 200×200×1 mm were formed, from which the plates were cut.

The density of the composites was determined on an H-200L densitometer with an ultrahigh resolution of 0.001 g/cm3 according to GOST 15139 - 69.

The values of water absorption of the samples were determined according to GOST 4650 - 80 (method A, water exposure for 24 hours).

Physical and mechanical tests of the prepared samples were carried out on a UAI-7000 M tensile testing machine at a temperature of 23 ± 2°C and a clamp speed of 1 mm/min. The tensile strength and tensile modulus were calculated according to GOST 11262 - 80 and GOST 9550 - 81. The measurement error for all tests was not more than 10%.

3. Results and Discussion

Table 1 shows the physical and mechanical characteristics of the obtained thermoplastic composites.

Table 1. The values of the physical and mechanical characteristics of the investigated compounds

Composites	Tensile Strength, MPa	Tensile modulus, MPa	Relative extension, %	Moisture absorption,%	Density, g/cm^3
Composite 1	13,5	1830	7,4	5,1	1,15
Composite 2	12,8	1750	6,9	5,3	1,12

Figure 7 shows typical tensile curves for the resulting thermoplastic compositions.

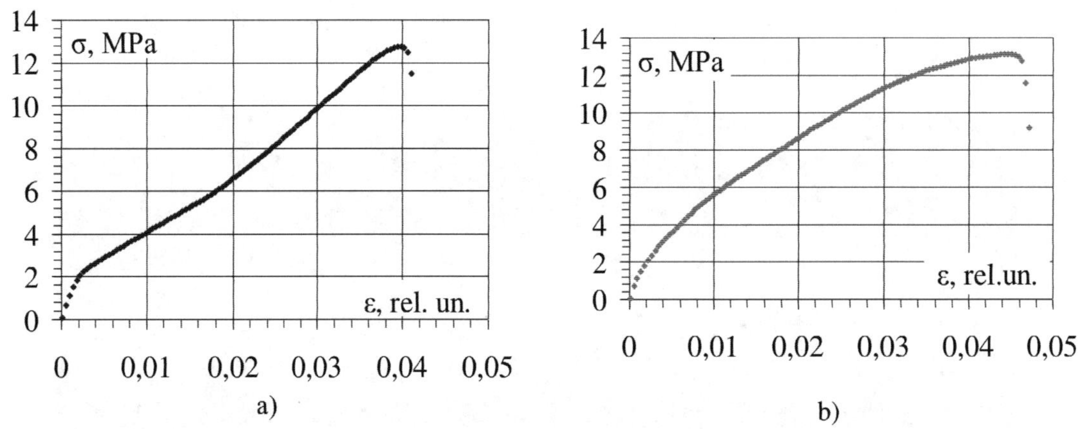

Figure 7. Typical tensile curves of thermoplastic composites filled with 45% wood flour (a) and 45% barley straw (b)

The values of physical and mechanical characteristics shown in Tab. 1 and Figure 1are consistent with the data of other researchers, which gives reason to believe that it is possible to use fillers from wood flour and barley straw to obtain composite materials for engineering purposes.

Thermoplastic composites with such fillers should, if possible, be processed without preliminary drying, but the resulting products should meet high operational and technological requirements.

To a large extent, this set of requirements depends on the porosity of the initial granulate, the reason for which is air capture, moisture in the feedstock, as well as partial decomposition of thermoplastics and agricultural origin fillers under high temperatures in the extruder with the release of volatile substances. In practice, this means that gas inclusions must be removed from the extruder before the forming head.

To combine extrusion with the operation of removing volatile substances released during processing, the extruders should be equipped with devices for melt degassing. The gaseous components are sucked off in all cases in the areas of expansion of the molten mass through holes in the cylinder of the extruder [1, 3]. An extruder Rheomex PTW 16 is equipped with a similar device such as vacuum pump 2 and adapter 8 (Figure 1).

Figure 8 shows micrographs of a slice of composite granules obtained with melt degassing and without degassing in extruder.

As follows from Figure 8, the granulate obtained without degassing does contain pores and cracks, leading to a decrease in physical and mechanical properties and moisture resistance of the products.

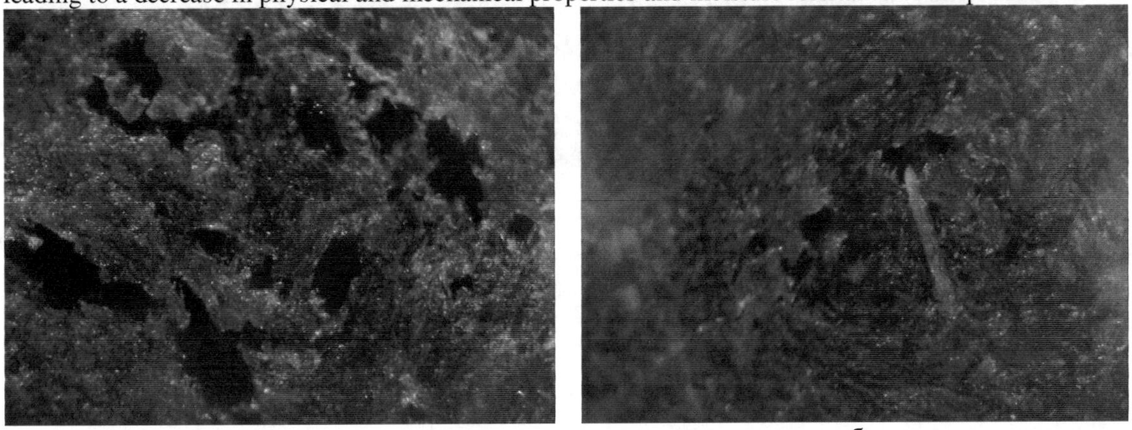

а б

Figure. 8 Photos of the structure of transverse sections (200-fold increase) of the masterbatch compound granulate obtained without degassing (a) and with degassing (b) of the last extruder mixing zone

4. Conclusions

Thus, in the course of the research, a two-stage technology for producing granulate of thermoplastic composite materials filled with wood flour and finely dispersed barley straw by continuous compounding of components on a twin-screw extruder of co-directional rotation was developed.

As a result of the degassing processes study, it was found that degassing of the melt of the compound with a vacuum value of 0.8 ... 0.9 bar can significantly reduce the porosity of the granulate.

References

[1] Klesov A A Wood-polymer composites [trans. from English.] / A. A. Klesov. *St. Petersburg:NOT Publishing House* 2010. 736 p.

[2] Kuster Moro M and Sposina Sobral Teixeira R and Sant'Ana da Silva A and Duarte Fujimoto M. and Albuquerque Melo P and Resende Secchi A and Pinto da Silva Bon E 2017 Continuous pretreatment of sugarcane biomass using a twin-screw extruder (*Industrial Crops and Products*) 97 pp. 509–517.

[3] Kohlgrüber K 2008 Co-rotating twin-screw extruders: fundamentals, technology, and applications *Hanser Publishers*. 365 p.

[4] Kim H and Oh K and Seo Y 2019 Rheological and Mechanical Properties of a Novel Polyamide 6 Synthesized by Anionic Polymerization of ε-caprolactam in a Twin-Screw Extruder (*Polymer*) 177 pp. 196-201.

[5] Lee J K and Han C D 2000 Evolution of polymer blend morphology during compounding in a twin-screw extruder (*Polymer*) 41(5), pp. 1799–1815.

[6] Singh R and Singh S and Hashmi M S J 2017 Polymer Twin Screw Extrusion With Filler Powder Reinforcement (*Reference Module in Materials Science and Materials*) pp. 1-15.

[7] Liu S P and Hwang S S and Yeh J M and Hung C C 2011 Mechanical properties of polyamide-6/montmorillonite nanocomposites — Prepared by the twin-screw extruder mixed technique (*International Communications in Heat and Mass Transfer*) 38(1), pp. 37–43.

[8] Berzin F and Beaugrand J and Dobosz S and Budtova T and Vergnes B 2017 Lignocellulosic fiber breakage in a molten polymer. Part 3. Modeling of the dimensional change of the fibers during compounding by twin screw extrusion (*Composites Part A: Applied Science and Manufacturing*) 101, pp. 422–431.

IOP Conf. Series: Materials Science and Engineering **873** (2020) 012022 doi:10.1088/1757-899X/873/1/012022

[9] Sui G and Fuqua M A and Ulven C A and Zhong W H 2009 A plant fiber reinforced polymer composite prepared by a twin-screw extruder (*Bioresource Technology*) 100(3) pp. 1246–1251.

Simulator of the navigation equipped with LIDAR of the mobile robot based on the neural network

R V Faizullin

Kalashnikov Izhevsk State Technical University, Faculty of Digital Economics, Studencheskaya St. 42, Izhevsk, 426069, Russia

E-mail rf85@mail.ru

Abstract. The paper considers the issue of using the robot simulators equipped with LIDAR (Light Detection and Ranging). The simulator application allows decreasing the experiment costs, their time and risks or negative aftereffects of accidents, but, at the same time, the simulated environment should identical to the real one. In practice, the simulated environment can have differences and there comes the issue of estimating the robot ability to adapt to new conditions. Neural networks are used to teach the robots, due to which the robots reach the target in the shortest or the fastest way, at the same time, there should be no accident/collision. The initial series represents the multiple time series. To teach the neural networks to control the robots, the paper proposes to introduce the additional index in the form of the minimal distance to the nearest object in the actual time moment. The approach novelty in using the new penalty function for teaching the neural network is the speed index of approaching the nearest object/obstacle. The neural networks were taught on the maps without dynamic objects, but tested in the conditions when several mobile robots were used on the map, thus, they became the dynamic obstacles for each other. The accident risk for robots without the proposed penalty function during testing was 4.5% against 0.3% - with the penalty function.

Keywords: mobile robot, LIDAR, laser sensors, collision, neural network, simulator.

1. Introduction

The robotics development is one the technological development trends in general. The desire to computerize part of the functions is widespread and perspective. Computerization allows decreasing the influence of subjectivity and human factor, and opens the way for improving the productivity of system activity. At the modern development stage robotics allows solving many problems in different knowledge areas.

One of the characteristics of a robotic system is its ability to interact with the environment. Interaction with the environment assumes the ability to identify the surroundings and capability of self-localization in the random environment. Several ways of solving similar problems are known [1, 2, 3], but their success depends directly on the ideal conditions and cost of such systems. The high cost of the sensor part of the known autonomous vehicles (e.g., self-driving cars) is the obstacle to their widespread occurrence. For this reason, great many papers dedicated to the development of autonomous motion algorithms using a relatively inexpensive element base have been recently published [4, 5].

Taking into account the application scale and complexity of robots, the development of applied robotics generates interest in the modern world. In reality, the data collection or system testing can be expensive, insecure or impracticable. The simulator application allows improving the security and decreasing the cost of robot testing.

The development of simulator to visualize the action spot comes down to three components. The first is the sensor simulation, which describes how the sensor interacts with the action spot (space). The second is the generation (visualization) of the action spot necessary to construct simulated worlds corresponding to the reality. The third is the simulator quality assessment based on the comparison of real and simulated data. The availability of simulated spaces allows imitating the process of robot navigation in this space. The application of computer-aided learning and artificial intelligence for robot control allows teaching it to reach the targets, hurdle obstacles, do optimal routing.

The use of 2D LIDAR by mobile robots and solution of the problem for creating navigation systems based on these data is an urgent task. The attempts are made to accomplish this task, the description of which can be found in the following papers [2, 3, 6, 7].

At present, there is great number of modifications of Simultaneous Localization and Mapping method (SLAM method) adapted for various sensor systems. But the basis of all SLAM modifications is Iterative Closest Point algorithm (ICP algorithm). ICP algorithm and its widespread variations are given in the paper [8].

Let us consider the mobile robot equipped with single beam laser scanner LIDAR. LIDAR is Light Identification Detection and Ranging ("detection, identification and ranging with the help of light"), it is the technology of obtaining and processing information about distant objects with the help of active optical systems using the light absorption and dissipation phenomenon in optically transparent media.

LIDAR data serve as the source for forming the information about the environment. If the sensor transfers the information about the distance from the sensor to the objects around it and there is no other information, the mobile robot action spot is formed only on the available information. The data set is not a perfect method to describe the action spot. The data on the distances "sensor – observation object" can be distorted. The distortion can be caused by the measurement, rounding off and other errors. This fact does not mean that we cannot form the information about the action spot but we should take into account the possible error in the incoming data. The main task of the data collection about the robot action spot is the action strategy formation, i.e. formation of the options of the system reaction to certain characteristics of the environment.

2. Research objective

In order to solve a certain problem of the system, the formation of its mathematical formulation becomes necessary. Let us formulate the problem as follows. A mobile robot (MR) should not approach another object closer than some distance. Two options should be singled out from the point of actual situations. The robot can collide with a stationary (immobile) or mobile object. Since we may not know about the behavior of mobile objects in the surroundings, we should consider the environment as the one with a random component.

The task of preventing the mobile robot collision with objects of the environment can be split into two sub-tasks: identification of the obstacles in front of the robot and movement routing to bypass the obstacles. To accomplish the first sub-task it is necessary to make the following steps:

1. To obtain the data about the distances to the surroundings.
2. To extract the relevant information (possibly being the obstacles).
3. To process the data and obtain information about the object, its danger, trajectory.

The task of object recognition can be significantly simplified, not taking into account the height and shape of the objects in vertical plane. For this, it is sufficient to check the obstacles not by the whole picture in the vertical plane but only along one line located at a certain height from the floor. In our experiments this height was determined by the MR height.

Modern algorithms of MR control based on LIDAR data obtained by computer-aided learning methods allow making the control trouble-free.

Despite the attempts to develop trouble-free control systems, they are still rather urgent. Even with the information from different sensors (radar, LIDAR and cameras), the optimal solution has not been found so far.

The work relevance is connected with broad capabilities of using the results. Let us consider the case of the car trouble-free driving. A human factor is the main reason of car road accidents. The system, which could predict collisions, could significantly decrease a number of road deaths. We can use different sensors but not all of them work ideally. The problems in the sensors operation can be caused by the weather conditions. Even if the sensors obtain the correct information, it should be correctly processed, and the car should have time and possibility to take measures for avoiding an accident.

Apart from the data from the sensors, cars use the wireless automobile communications. There are several types of such communications: vehicle-to-everything (V2X) communications, vehicle-to-vehicle (V2V), vehicle-to-infrastructure (V2I), and vehicle-to-pedestrian (V2P). The investigations demonstrated that the combination of V2V and V2I technologies can help to avoid about 80% of all car accidents [9].

In this paper [10], the development of a vehicle collision warning system based on multisensors and V2X communications is presented. On-board sensors including radar, LIDAR, and camera systems that have already been adopted in production vehicles are chosen for this work such that by adding V2X communication devices to the vehicle, it is possible to evaluate the benefits of introducing V2X communications to today's vehicles in terms of road safety.

However, simple models are also necessary. If the model is simple, it requires less resources and works faster and clearer for the user. In the paper [11] the method to estimate the risk of collision with an obstacle is proposed. The simulated data on the moving object coordinates are added to the LIDAR data. The trajectory prediction allows defining the collision possibility. The collision fact is predicted, the time before the collision is estimated with the help of the analytical method. The use of additional information about the object physical properties improved the accuracy of the results modeling and collision risk estimation, and decreased the computational and memory load on the results processing system.

Neural networks are successfully used to develop the algorithm of robot control [1, 2, 3, 4, 6, 7], however, there are no proofs that the robot route will be ideal. Apart from the optimality, the majority of papers do not consider the possibility of force majeure situations or situations, which were not envisaged at the stage of neural network teaching. Such situations can be connected with the sensors errors; in case of LIDAR, there can be problems with reflecting and transparent objects, the robot can change its behavior by itself, because, for example, its battery is getting down. The robot can collide with the object, which it has not seen before (in the teaching process), or collide with the dynamic object with unpredictable behavior. Those robots, which consider the possibility of dynamic object availability and predict its trajectory [9, 10, 11], do not take into account the random trajectory changes.

Thus, we come across the necessity of developing such neural network and selecting such development criteria for the robot to cope with situations not envisaged at the simulation/teaching stage.

3. Methodology

Let us consider the change in the action spot state as the change in the distances from the mobile robot location to the objects detected by LIDAR sensor.

The collision detection system [11] is targeted at determining if and when the collision between the mobile system and environment can occur to notify the motion planning system (or a human operator) that the current motion trajectory needs to be adapted to avoid the collision. Various approaches exist for detecting the presence and range of an obstacle; at present, they are widely used in mobile systems with beam ranging and swept area predictions being the most common.

Beam ranging uses a concept of directing some sort of energy towards the obstacle and evaluating the reflection. Most common are: acoustic waves (sonars), radio waves (radars) and light (LIDAR), with ranging devices providing the information on the distance between the device and an obstacle. Single or multiple sensors can be attached to a mobile system, covering the area of the predicted motion.

As the mobile system is moving along the predicted trajectory, its bounding box sweeps over an area in the environment (also called configuration space). In order to determine the possibility of collision between the mobile system and environment, each measurement must be compared against the predicted path of the system. It is possible to predict this swept area in advance and check whether any of the detected obstacles are within this area. In this case the time to collision can be estimated based on the predicted moment the mobile system's bounding box touches the obstacle.

A lot of dynamic objects make the task more complicated. A mobile robot has limited resources. This is caused by the limited computational and memory power and limited energy storage. So, the necessity to seek a simplified method of collision prediction arises.

Under collision between objects we understand the situations when the distance between them equals zero. The robot with LIDAR can detect the distance to the obstacle with the accuracy over 99%. The experimental results [12] demonstrate that LIDAR can measure the distances from 0.12 up to 10.5 m with the error around 0.9%. The object color and environment light intensity do not influence the measurements obtained with the help of LIDAR. An autonomous mobile robot can avoid colored objects of different sizes. The robot faced the problems of distance detection only with transparent objects.

So, it follows that we should solve the problem of predicting the multiple time series. Actually, the data flow from LIDAR is the multiple time series.

Let us set the task of multiple time series prediction.

Let us designate $X = \left[x^{(1)}, \dots, x^{(s)}\right]^T$ is the given s-dimensional time series. Now we formulate the design matrix from the series segments:

$$\begin{pmatrix} x_0^{(1)} \cdots x_{n-1}^{(1)} \\ \vdots \\ x_0^{(s)} \cdots x_{n-1}^{(s)} \end{pmatrix} = X_{0\div(n-1)} \quad (1)$$

Let $X = \left[x_n^{(1)}, \dots, x_n^{(s)}\right]^T$ is the value of X series in the time moment n. Now we build up the prediction $\widehat{X}$ of X series in the point X_n. Let us do it k times for different teaching samples $X_{train}^{i} = X_{i\div(n+i-1)}, i = 0, \dots, (k-1)$. We get k predictions $\widehat{X} = \left[\widehat{X}_n, \widehat{X}_{n+1}, \dots, \widehat{X}_{n+k-1}\right]$.

The prognostic model looks as follows:

$$\widehat{X}_{t+1} = f(\widehat{w}, X_t, X_{t-1}, \dots, X_{t-L+2}) \quad (2)$$

$$\widehat{w} = \operatorname{argmin} S(w, X, \widehat{X}_n, \widehat{X}_{n+1}, \dots, \widehat{X}_{n+k-1}) = S(w, X, \widehat{X}) \quad (3)$$

where the loss function:

$$S(w, X, \widehat{X}) = \sum_{i=0}^{k-1} \mathcal{L}(X_{n+i}, \widehat{X}_{n+i}) = \sum_{i=0}^{k-1} \mathcal{L}(X_{n+i}{}^{(1)}, \widehat{X}_{n+i}{}^{(1)}) \quad (4)$$

In case of LIDAR, which transfers us the data on the distances of the objects in 360 directions in each time unit, s=360. Depending on the prediction model, we should also choose the number of time periods, which influences the current value. We select n periods for matrix (1), then our computations have to be carried out with the matrix with the dimensions s* n. The complexity of problem solution is caused not only by the larger size of the data analyzed but also by the availability of complex bonds in them. It means that the data will change depending on where to and along which trajectory the mobile robot moves, as well as on the behavior of dynamic objects in the robot vision area.

For dynamic objects not to detain mobile robots, the robots should be able to predict the behavior of the dynamic objects. The given task is being actively accomplished, for example, for self-driving cars. In such tasks other automobiles and pedestrians are considered as dynamic objects [13]. The prediction model should take into account future interactions with an unknown number of dynamic objects. It is necessary to bear in view that dynamic objects can also change trajectories as a result of our behavior or changes in the behavior.

The model teaching based on the simulated data is a widespread practice, though it has a number of shortcomings. It is easier, faster and cheaper to obtain the simulated data.

In our paper, we consider the task of selecting the teaching process of the neural network based on simulated data. The program operation starts from the level loading location of all obstacles, MR starting point, as well as the area, in which the stop should be made. After that the first generation with random parameters of neurons weight is generated. The neurons are represented in the memory as the samples of class "Neuron". The class contains the array of numbers: neuron weight, mutation function randomly changing up to 1/3 values from this array, and activation function "Activation()". One specimen is represented as a sample of AI class. AI class consists of three layers (arrays) of neurons, mutation method "mutate()" calling the mutation function of all neurons in all layers with the chance of 40%, crossing-over function "crossBreed()" creating one new specimen out of two, and decision-making function "makeDecision()". The decision-making function of the specimen activates all neurons one by one in all layers producing 4 Boolean values at the output. These values indicated what action should be taken at the given time moment based on the environment data, whether it is necessary to speed up or slow down, whether it is necessary to turn left or right.

After the first generation is initialized, the evolution cycle starts. We create some number of specimens. Each specimen is a different set of weights of neurons and, consequently, a separate way of MR reaction to the environment data.

The game cycle starts:

For all specimens of the first generation the environment analysis is initiated by data from the LIDAR, i.e. the data on the distances to the objects by all 360 degrees. After that the function makeDecision()" is called upon for each specimen of the first generation. Based on the decision, the shift of the robot (robot version) fixed to it is calculated; taking into account the current speed, acceleration and its turning angle the specimen will get the scores for the distance covered within one frame. The frame is renewed. The accident risk is calculated for each robot; if it collides with the obstacle, it stops, it is marked with the attributes "crashed" and "finished", and gets the scores by the adaptability function: for the distance to the target, as well as the penalty (punishment) is the number of scores is decreased by 20%. Further, the car marked as "finished" does not participate in decision-making, miscount of collisions or shift (the game is over for it). If the time of one game cycle exceeds the set time, the car is marked as "finished", and we calculate the scores with the help of the adaptability function. Then it is estimated if there are still other cars without the attribute "finished". If there is at least one available, the game cycle is iterated again. The specimen is given scores if it reaches the target. If even none specimen has reached the target and if they have covered different distances, those who covered longer distances will have more scores.

When all the cars are finally marked as "finished", the time for creating the new generation comes. The new generation is formed by cross-breeding the best specimens of the previous generation. That is, the couple of parents from the best specimens of the previous generation is selected, and the function "crossBreed()" is called upon. The specimen obtained becomes a new member of the next generation, and the method "mutate()" is called upon for it.

Let us teach several neural networks selecting different criteria when awarding the specimens. Let us further consider the capabilities of the neural networks, which were taught on the static data, to get adapted to the situations with the emergence of dynamic objects. This situation will be an example for estimating the capabilities of neural networks to non-standard situations.

4. Results

During the investigation it was revealed that the main aim of teaching the robots is to reach the target, the distance and time criteria are used for optimization. Although the time spent by the robot depends on the distance, the criteria are not identical. If we optimize only the distance covered, the robot stop will not influence the result, therefore, the time is more important. On the other hand, we understand that the distance covered mostly influences the energy spent. The time spent by the robot for covering the distance depends on the robot average speed. Finally, these three indexes are the most widely spread in the process of robot teaching [14, 15, 16].

The collision prediction based on solving the prediction task of zeros emergence in the problem as (1), is a neutral task, however, we propose the model simplification. The task complexity is connected with the fact that, on the one hand, the multiple time series reflects the distance to the obstacles around, and, on the other hand, this distance changes with time not only based on the location of obstacles but also on the changes in the robot location. From the information capacity point, the information which of 360 points that are described by the LIDAR data will become zero is not that important as the prediction of time (moment) when it happens. What is the difference from what side the hit comes, if we do not know when? Let us consider the possibility of using the information on the proximity of the nearest object instead of the information on the remoteness of the objects from all sides. Thus, it is proposed to transform the multiple series into the one-dimensional one. In our case, we are interested in the nearest point, therefore, the minimal value ($\min x_n$). Since random factors influence the collision fact and its time, the prediction is only some expected value with some probabilistic characteristics.

Actually, the task comes down to extrapolation. If the schedule describing the change in the minimum distances from MR to the obstacle can be described by some curve (straight line or parabola), it is enough to understand if this curve takes the zero value in the future time moments.

The latest observations of the considered value are more important than the old ones, therefore, we cannot use, for instance, the model of simple moving average. Thus, there comes the task that the newer data are the most important. It was also considered that the expected collision time or expected distance to the nearest object/obstacle can have some error. So, it becomes necessary to modify the neural network teaching process and introduce the penalty not only for collision but also for the "proximity/approximation" to the object. Here, we should consider two moments. First of all, the danger of approaching changes non-linearly, and, second, the fact of approximation cannot be dangerous in some cases, for instance, when the robot moves along the wall but close to it. We understand that if we give penalty for the approximation to the object/obstacle in each time moment, the robot will not move close along the wall, though, it can be an optimal decision (the shortest way). The penalty was proposed not only for the proximity to the object/obstacle, and for the speed proximate to the object/obstacle, i.e. the speed of changing the minimum distance was considered. So, if the robot moves along the wall, it does not get additional penalties for the wall proximity apart from those it got in the moment of approximating them. Let us repeat that the penalties are increased non-linearly.

Thus, we achieved the sufficient decrease in the potentially dangerous situations.

The neural network was tested as follows. We have trained the neural network using the methods described in the previous section. We found out that after 10,000 iterations, the robot finds the target well and does not cause accidents. If the robot did not make an accident in the last 300 tests, we believe that the robot has been trained to drive safely. The neural networks taught in different ways were used for the robot control simulator in the new environment. If the teaching was carried out only for different maps, but without dynamic objects, the possibility of robot adaptation to new conditions was initiated at the map of several robot simulators. The robots "competitors" played the role of dynamic objects. Dynamic nature of the environment is the reason of the external faults or exogenous events for robot that are not the fault of the system [17]. These faults are usually unpredictable [18].

The neural networks taught to 1) minimize the distance covered to the target, 2) minimize the tine to reach the target, 3) with the combined target were compared. Also the comparison was made in parallel with penalty function for the speed to decrease the minimum distance to the obstacles and without the penalties.

During the experiments it was revealed that neural networks, which have penalties for speed slowing down to the minimal distance to the object during teaching, demonstrated the results better than those, which did not have penalties while teaching. Mobile robots, which did not know that the danger is in the fact that the speedily shortened distance to the nearest object is speedily decreasing not to get into accidents, get the target better. Robots controlled by neural networks without a penalty function, collided, especially when a large number of robots were used on the map and their concentration was high. Роботы проверялись на разных картах с разным числом роботов-«конкурентов». The accident rate was calculated based on the results of 300 tests in each condition (by different maps and by different number of robots - "competitors" on maps). The maximum accident rate is mentioned below. The accident occurrence for a robot without the proposed penalty function is 4.5%, against 0.3% - with the penalty one.

5. Conclusion

Thus, the conclusions that neural networks allow controlling robots and are able to learn making the routes to the target are obtained. The optimization of the route target reach can be made by time and distance. The use of the average speed is the combination of previous approaches as the average speed is the ratio of the distance covered to time. During the investigation it was revealed that the robot should consider the minimal distance to the nearest object. To teach the neural networks to control robots, the paper proposes to introduce an additional index as a minimum distance to the nearest object in the given time moment. The approach novelty is in the use of a new penalty function to teach the neural network is the speed index to approach the nearest object/obstacle. The neural networks were taught on the maps without dynamic objects but they were tested in the conditions when several mobile robots acted on the map, thus, they became dynamic obstacles for each other. The accident of the robot accident without the penalty proposition when testing was 4.5%, against 0.3% with the penalty one. The proposed testing approach to root control can be used for other models. The use of the proposed penalty function is also appropriate for teaching with more diverse situations. Since the robot could adapt well to the emergence of dynamic objects, it can be ascertained that the robot can somehow avoid the prediction of forecasting and avoiding emergencies. The proposed penalty function can be used to predict the accidents moments. These suggestions are explained by the fact that the initial data were the ones with LIDAR data, and during teaching there were no dynamic objects on the map, consequently, the program could know how to learn to escape the collisions predicting the dynamic object trajectory. The investigation of other possibilities of using the other penalty function can be directed by future investigates.

Acknowledgment
The work was carried out with the financial support by Kalashnikov Izhevsk State Technical University within the framework of the grant no. FRV/20-55-25.

References

[1] Lozano-Perez T 2012 *Autonomous robot vehicles* (Springer Science & Business Media)

[2] Csaba G, Somlyai L and Vamossy Z 2018 Mobil robot navigation using 2D LIDAR *2018 IEEE 16th World Symposium on Applied Machine Intelligence and Informatics (SAMI)* doi:10.1109/sami.2018.8324002.

[3] Cheng Y and Wang G 2018 Mobile robot navigation based on lidar *2018 Chinese Control and Decision Conference (CCDC)* doi:10.1109/ccdc.2018.8407319

[4] Yeon A S A, Kamarudin K, Visanathan R, Mamduh S M, Kamarudin L M, Zakaria A and Shakaff A Y M 2015 Feasibility analysis of 2D-SLAM using combination of Kinect and laser scanner *Jurnal Teknologi*, 76(12).

[5] Gong Z, Li J and Li W 2016 A low cost indoor mapping robot based on TinySLAM algorithm *2016 IEEE International Geoscience and Remote Sensing Symposium (IGARSS)* pp 4549-4552

[6] Carlos E Aguero, et al. 2015 Inside the virtual robotics challenge: Simulating real-time robotic disaster response. *IEEE Transactions on Automation Science and Engineering*, 12(2) pp 494–506

[7] Shah Sh, et al, 2018 Airsim: High fidelity visual and physical simulation for autonomous vehicles *Field and Service Robotics* (Springer) pp 621–635

[8] Chetverikov D, Svirko D, Stepanov D and Krsek P 2002 The trimmed iterative closest point algorithm *Object recognition supported by user interaction for service robots* (IEEE) vol 3 pp 545-548.

[9] Najm Wassim G, et al. 2010 *Frequency of target crashes for intellidrive safety systems* (United States. National Highway Traffic Safety Administration)

[10] Baek, Minjin, et al. 2020 Vehicle Trajectory Prediction and Collision Warning via Fusion of Multisensors and Wireless Vehicular Communications *Sensors* 20.1 288

[11] Bosnak M and Skrjanc I 2017 Efficient Time-To-Collision Estimation for a Braking Supervision System with LIDAR *3rd IEEE International Conference on Cybernetics (CYBCONF)* doi:10.1109/cybconf.2017.7985775

[12] Hutabarat D, Rivai M, Purwanto D and Hutomo H 2019 Lidar-based Obstacle Avoidance for the Autonomous Mobile Robot *12th International Conference on Information & Communication Technology and System (ICTS)* doi:10.1109/icts.2019.8850952

[13] Rhinehart N, McAllister R, Kitani K and Levine S 2019 Precog: Prediction conditioned on goals in visual multi-agent settings *Proceedings of the IEEE International Conference on Computer Vision* pp 2821-2830

[14] Gao M and Tian J 2007 Path planning for mobile robot based on improved simulated annealing artificial neural network *Third International Conference on Natural Computation (ICNC)* vol 3 pp 8-12

[15] Al-Sagban M and Dhaouadi R 2012 Neural-based navigation of a differential-drive mobile robot *12th International Conference on Control Automation Robotics & Vision (ICARCV)* pp. 353-358

[16] Algabri M, Mathkour H, Ramdane H and Alsulaiman M 2015 Comparative study of soft computing techniques for mobile robot navigation in an unknown environment *Computers in human behavior* (Elsevier Ltd)vol 50 pp 42-56

[17] James P Gunderson and Louise F Gunderson 2008 *Robots, Reasoning, and Reification.* (Springer Science & Business Media).

[18] Bhat P, Thoduka S and Plöger P 2019 A Dependency Detection Method for Sensor-based Fault Detection.

The impact of integrated modelling on the technical and economic indicators of the oil production process

G V Kutergina, E N Yapparov

Perm State National Research University, Department of Economics, 15, str. Bukireva, Perm, 614990, Russia

E-mail gkutergina@psu.ru

Abstract. The article outlines the findings of the reviewed list of tasks to be solved at different stages of oil fields development and operation with the use of integrated modelling technology. Operational and technological models are the most popular nowadays. However, there is no unified field model that would include all the elements of integrated planning. Systematisation of the solved problems and directions for reducing operating and investment costs at different stages of oil field development and operation remains relevant with the use of new modelling technology. Considering the possibilities of a series of integrated modelling programs (Mbal, Prosper, Gap) for optimising oil production process allows us to group the main problems using integrated modelling technology at the following stages: design and strategic decision-making, evaluation and optimising individual activities in the process of operational management, monitoring the development and oil field operation. Based on the analysis of open sources, we have concluded that the introduction of new 'smart' technologies significantly impacts technical and economic performance of oil and gas companies, including the level of oil recovery, reduction of drilling time, reduction of operating costs and profit growth, increasing the stability of design solutions and project efficiency in general.

Keywords: integrated modelling, intelligent field, oil production modelling, technical and economic parameters of oil production, list of integrated modelling tasks.

1. Introduction

The growing cost of developing new oil and gas fields under conditions of fluctuating demand for hydrocarbons forces subsoil users to develop and implement new technologies that reduce production costs. Field design plays an important role in this process, enabling optimisation of technological modes and ensuring an appropriate level of technical and economic performance.

The major driver of ensuring the search for optimal solutions in the area of field development and exploitation is the acceleration of data processing and application of intellectual technologies, which are extensively used by global foreign companies (Shell, BP, Exxon Mobil, Schlumberger, etc.) in new previously unprofitable projects, as well as to increase oil recovery at existing fields. Among Russian companies, Oil Company Rosneft and Gazprom are the most successful in implementing digital technologies. Rosneft is developing 10 digital fields, while Gazprom is developing 7, including one unmanned marine field, and LUKOIL is developing 5 of them [1].

The analysis of global and domestic experience of intelligent field technology implementation allows researchers to conclude that it is of great interest to the oil and gas industry of Russia as a tool for managing high-tech projects for oil and gas field development, ensuring competitiveness and high industry profit [2]. The significance of applying new information technologies grows as oil producers switch to developing fields in remote areas with a high level of reserve recovery complexity.

Modern integrated solution technologies in the oil and gas industry are united by the following categories: 'intellectual field', 'smart field', 'digital field'. These concepts are relatively new yet often used by experts when discussing the problems of improving the efficiency of hydrocarbon production and field development. The stages and directions of their development are intensively discussed by the scientific community [1; 3-8]. In the meantime, researchers focus on engineering and technical issues:

1) justification of the name and configuration of the components included in the 'intellectual field' system, based on the introduction of hardware and software package into production processes of oil and gas producing enterprises, a set of 'business processes for operational management of the field, technical development project based on the most advanced technologies and fully automated infrastructure [2];

2) analysis of relevant trends in the development of digital and intellectual technologies for the oil and gas industry, promising solutions for geology, drilling, development and operation of intellectual fields in Russia and abroad [1, 9];

3) development of systems for control of large volumes of oilfield information, its processing and display in comprehensible form based on the implementation of information technologies and high-tech equipment providing cybernetic control of individual field components (intellectual well, intellectual oilfield, integrated modelling of work planning)[9];

4) analysis of the innovative programs for the development of oilfield services in major oil and gas companies: Rosneft, Gazprom, Gazpromneft and Zarubezhneft [10], establishment of digital twins for production enterprises using modern simulation tools [11] etc.

Few works consider the following:

– opportunities to leverage integrated field modelling for effective business planning of field development [16],

– characteristics of technical and economic indicators of oil and gas field exploitation at various stages of development associated with the assessment of innovative oil production technologies progresses [13];

– methods of modelling indicators' calculation for the field development investment project and its efficiency parameters as a part of integrated modelling technology, economic and management criteria for the selection of projects for the implementation of smart field technology, financial and economic model of oil and gas asset [2;14].

2. Problem Statement

Implementation of integrated field modelling technology by Russian companies is constrained by the high capital costs associated with the cost of equipment currently purchased from foreign service companies, which requires a detailed analysis of the limitations and opportunities of new technologies in order to improve the technical and economic parameters of oil and gas production.

The review of publications has shown that only separate instruments of field economic modelling for calculation of basic engineering package parameters for field development are currently applied. The main disadvantage is the absence of a single field model, which would include all elements of integrated planning. Unfortunately, the scientific publications have paid insufficient attention to these issues, which makes it urgent to systematise the list of tasks to be solved and directions to reduce operating and investment costs through the optimisation of technical and economic parameters and indicators of oil production at different stages of development, as well as operation of oil fields using integrated modelling technology.

3. The Integrated Model of Oil Production: Opportunities and Limitations

Oil production is a complex production process, which includes geological exploration, drilling and construction of wells, well repairs, as well as purification of produced oil from water, sulphur, paraffin and other impurities. As a rule, the life cycle of oil fields is several decades long.

Optimisation of technological modes of production well exploitation includes the creation of favourable conditions for oil inflow from the formation, the selection of downhole equipment, and the establishment of its mode of operation in line with the productive characteristics of wells and other factors, which creates the necessary conditions for improving technical and economic performance of oil production. For example, reducing the suspension depth of pumps in wells to optimal values leads to a reduction in the frequency of underground repairs and a decrease in the repair price [15].

Integrated modelling allows to automate data collection, filtering, storage, and processing, describe physical processes, predict hydrocarbon production, and visualize key parameters for management decisions. Availability of an integrated model provides an opportunity to make a reasonable choice of development field of project options, reduce operating and investment costs, energy consumption by infrastructure facilities located in the fields, and increase the speed of management decision-making.

System analysis and optimisation of well and downhole equipment operation modes requires a large amount of calculations, which is impossible without development and use of appropriate information models adapted to the conditions of specific fields. Therefore, over the past years, there has been a trend towards more active use of methods of mathematical modelling of processes related to oil and gas production in the oil industry. This is facilitated by the active introduction of modern well research technologies, information systems enabling the registration of diverse data on the state of field development facilities, formations and wells [9].

Commercial and technological models are the most common. They are based on displacement characteristics, fluid filtration equations, and low-parametric regression models of oil, liquid, and water production. At the same time, the actual conditions of oil and gas production are characterised by insufficient adequacy of models, errors in registration of geological and technological parameters, deficiency, heterogeneity, incompleteness of initial data that characterise the energy state of oil reservoirs[1]. In such circumstances, the use of traditional methods to identify technological parameters of development often does not ensure the required solution accuracy. To overcome the complexity of the oil and gas production process, limited information, and ensure the stability and accuracy of solutions, it is recommended to use integrated identification systems that record information related to previously accumulated experience, provide the ability to use expert assessments, models of technological parameters of development within a set of models that reflect the properties inherent in real systems [11].

The integrated model presents related models of reservoirs, wells, and engineering infrastructure of oil field, which allows optimising the process of oil production as a whole. The increase in prediction accuracy is ensured by taking into account the limitations associated with the reservoir and the collection network when calculating the well operation and production mode. The inherent advantages of the integrated model are as follows: taking into account the impact on the indicators characterising well exploitation, collection networks, identification of bottlenecks within collection networks, etc.

At the same time, the integrated model is characterised by the slowdown of calculations in comparison with local models; its implementation requires more significant computing power and labour costs, the requirements to the initial data increase. Table 1 shows the limitations of the hydrodynamic and integrated models, which demonstrate the advantages of the integrated model in terms of managing factors for optimising technical and economic indicators of field development.

[1] Materials used: Nizhelsky D V Presentation of the project ‘Conducting Comprehensive Research to Configure the Well Model and Optimize the Operation of Wells Without Technological Losses’, Perm, 2018

Table 1. Comparison of Limitations of Integrated and Hydrodynamic Models[2]

Limitation Nature	Models	
	Integrated	Hydrodynamic
Optimization of well equipment operation	+	-
Taking into account the performance limitations inherent in a compressor station that provides complex gas treatment	+	+
Optimisation of the ground collection system	+	-
Taking into account the specifics of well operation methods	+	-
Taking into account maximum pressure limits in the collection system	+	-
Taking into account limitations inherent in well equipment	+	-

Integrated modelling provides a way to see the possibilities, simplify and speed up work, reduce errors and shorten the time to correct them. One of the most popular integrated modelling programs in the world comprise Mbal, Prosper, Gap [16] which, when operated as a complex, allow for the analysis of the field during the entire period of exploitation. If we consider this software as modelling tools, we can describe the main features of each program for solving problems that arise in the oil production process.

Mbal is an analytical tool used for dynamic modelling of fluid distribution in a reservoir (material balance). This program provides the ability to determine the main parameters required for modelling the volume of oil and gas reserves, and the mechanism of production. It also allows you to adapt the models taking into account the geological and field data at the stage of field development.

Mbal is actively used to form a preliminary model of dynamic processes occurring in the reservoir prior to hydrodynamic modelling, which allows for the following [17]:

- adaptation of the reservoir model based on available field data to determine hydrocarbon reserves and the displacement mechanism;
- forecast of production dynamics;
- analysis of field development scenarios;
- determination of the daily contract amount of gas;
- modelling the dynamics of gas condensate field operation in the depletion and recycling modes;
- analysis of the production drop curve;
- construction of a multi-collector field model, Monte Carlo modelling, linear flooding modelling, modelling of gas production in dense rocks and methane production from coal seams.

Prosper is a program for modelling most types of oil and gas well configurations, including design and optimisation of operation modes and production forecasting. The program provides the ability to [15]:

- distribute production between wells;
- design and optimise well completion for horizontal, multi-layer, and multi-hole wells;
- to design piping and tubing, to carry out the optimisation;
- monitor the operation of wells, identify the need for repairs;
- design wells with centrifugal, hydraulic, jet, cavitation pumps, gas lift wells, and optimise them;
- determine the temperature of the fluid in relation to pipelines and wells;
- calculate pressure drop in relation to fittings, conduits, and wells;
- investigate the continuity of the fluid flow in the well.

The GAP program is an optimiser software that simulates surface product collection system. GAP integrates the reservoir model (MBal) and well model (PROSPER) in dynamic mode, focused on optimising gas and oil collection networks. This program makes it possible to create a production model for oil, gas and gas condensate fields, as well as gas and water injection systems, field modelling taking

[2] Based on the work [12]

into account elements in the form of productive reservoirs, wells, ground-based pipelines, and liquid injection systems. The GAP program features fast and powerful optimisation algorithm that provides the following features [16]:

- designing surface pipelines network,
- integration of models of gusher and with mechanised extraction wells, including gas lift, centrifugal, cavitation, hydraulic, jet and rod pumps;
- optimisation of fields with mixed well types;
- optimisation of closed oil collection systems,
- calculation of parameters of wellhead fittings to provide a plan for monitoring the development of the field,
- production forecast.

GAP program allows to integrate MBAL and PROSPER, which makes it possible to model and optimise the production process as a whole [2]. Monitoring the oil production process using integrated modelling capabilities allows solving a wide range of practical tasks at different stages of oil field development and exploitation, and management levels.

3.1. Field design, strategic decision-making level.

At this stage, the presence of an integrated model allows to reasonably choose certain options for the development of the project, the concept of field development, and solve problems related to the justification of the effectiveness of implementing certain components of smart field technology. Typical issues addressed by integrated modelling at this stage [19]:

- choice of the appropriate concept of field development;
- justification of the effectiveness of implementing certain components of smart field technology;
- optimisation of measures to intensify development, optimise well operation (selection of pumps, optimisation of gaslift, selection of string diameter, etc.);
- optimisation of surface network operation (changes in pipeline diameter, pipeline re-routing).

3.2. Evaluation and optimization of individual activities, operational management level.

At this stage, the integrated model can solve a wide range of tasks related to optimizing measures to intensify development (hydraulic fracturing, transferring wells to another category, etc.), optimizing the operation of wells (selection of pumps, optimization of gas lift, selection of column diameter, etc.), as well as optimizing the operation of the surface network (changes in the diameter of the pipeline, re-routing of pipelines, modernization of the booster compressor station, etc.).

3.3. Monitoring of field development.

At this stage, the integrated model can solve a wide range of tasks related to optimizing measures to intensify development (hydraulic fracturing, transferring wells to another category, etc.), optimizing the operation of wells (selection of pumps, optimization of gas lift, selection of column diameter, etc.), as well as optimizing the operation of the surface network (changes in the diameter of the pipeline, re-routing of pipelines, modernization

The integrated model displays the field operation based on the available views of the formation, wells and equipment. Therefore, periodic updating of the model and its comparison with the available actual data allows to quickly identify changes in the state of the model components, which are reflected in the form of discrepancies between actual and calculated data. Thus, for example, violation of tubing tightness will lead to discrepancies between the calculated and actual pressure losses in the well. Thus, the integrated model can serve as an important indicator of problem occurrence.

When external constraints (e. g. oil or gas production target) are encountered, the integrated model automatically determines the optimum operating points for the equipment, taking these constraints into account so as to minimise network pressure losses and energy consumption for the operation of pumps and compressors. When using the active components of a smart field system (e. g., controllable connectors or flow control systems), the integrated model can be used to develop the optimal algorithms for controlling them depending on sensor readings.

Due to technologies related to integrated modelling, it provides increased efficiency and savings in the design and implementation of investment projects, current operating costs, including:

– reduced costs and shorter R&D and industrial testing periods due to more accurate field modelling and reserve forecasting;

– reduction of investment costs, profit growth and decrease of the payback period for capital investments when commissioning new industrial facilities due to development project optimisation;

– reduction of energy consumption costs for infrastructure facilities located in the fields based on optimisation of equipment usage modes;

– minimisation of costs for repair, maintenance and operation of fixed assets by optimising their operation and maintenance modes, increasing the productivity of production wells, reducing the number of oil production wells downtime and all types of emergency incidents (including leaks and emissions); reducing oil, gas and water losses by optimising separation, dewatering, desalination and early detection of gusts of oil and gas collection networks;

– reduced staff costs due to automation and increased efficiency of management decision-making, increased productivity and labour safety;

– increase in production volumes due to an increase in the oil recovery coefficient as technologies develop [20; 21].

The publications of various authors provide estimates of expert organisations that confirm the significant impact of new ‘intelligent’ technologies on the technical and economic results of oil and gas companies.

Linnik Yu N, Balashova A D and Bolshakova O I with references to foreign sources [22] argue that the widespread introduction of integrated modelling technologies and intelligent fields will increase global oil recovery from 30 to 50% [1 . 39; 20]. Based on the analysis of data from six foreign projects (oil and gas companies: Brunei Shell Petroleum, Saudi Aramco, Statoil, ONGC, PEMEX, Saudi Aramco), the introduction of smart field technologies will result in increase in average annual profit from 15 to 20% [13; 23–26].

The significant impact of integrated modelling technology on technical and economic indicators of oil field development is also demonstrated by the statistics of Russian companies. Thus, as a result of the introduction of digital technologies, Rosneft has developed Digital Centre for analysis and visualization of geologic formations, Digital shelf, Integrated Field Model, Production Robotisation, and Digital Worker. The company plans to reduce the drilling time by 5%, increase the efficiency of well capital repairs by 20%, achieve a reduction in operating costs by 2–3% per year [21], and increase profits by more than 20% [9] by 2022.

NOVATEK benefited in 40% reduction of metal consumption in terms collection system due to the new technology compared to the traditional approach to design; according to preliminary estimates Gazpromneft-Khantos LLC (Gazpromneft subsidiary) may reduce operating costs for the development of the field by 15% due to the launch of the digital double system at the Yuzhno-Priobskoye field [1].

Application of integrated modelling allows to transfer the revealed uncertainty of parameters at early stages in the category of project risks, to carry out their estimation and to receive possibility to make decisions on prevention of risks or essential reduction of their negative influence. It improves stability of design decisions and efficiency of the project as a whole.

4. Conclusion

This article provides an overview of publications and practices of integrated modelling in oil and gas industry in terms of opportunities to improve the technical and economic performance of the oil production process. It is urgent to develop a classification of the main technical and economic parameters and factors to reduce operating and investment costs of the oil producer when using technology intelligent field to assess the effectiveness of its implementation.

References

[1] Linnik Yu N and Kiryukhin MA 2019 *Digital technologies in the oil and gas complex* (University Bulletin) 7 pp 37-40. DOI: 10.26425 / 1816-4277-2019-7-37-40

[2] Berezina AA 2015 *Economic and managerial criteria for selecting projects for the implementation of intellectual field technology* (Internet journal "SCIENCE") 7 (1). DOI: 10.15862 / 20EVN115

[3] Rosendahl T and Hepsø V 2013 *Integrated Operations in the Oil and Gas Industry: Sustainability and Capability Development* (Business Science Reference) (an imprint of IGI Global).

[4] Al-Balushi M et al., Real-Time 2014 *Surveillance:* How *System Integration Allows One Company to Minimize Deferment, Optimize Production, Maximize Test Unit Capacity, and Track the Operating Envelopes of its Wells,* SPE-167857-MS.

[5] Talabi O A et al. 2016 *Integrated Asset Modelling: Modernizing the Perspective for Short-Term Forecasting and Production Enhancements,* SPE-182496-MS.

[6] Vliet van JP M and Male P T 2017 *Well, Reservoir and Facility Management - Process, Practice and Impact*, SPE-185880-MS.

[7] Huiqiang Shen et all, 2017 *Design of an Integrated Production Planning System Framework Based on Simulation Using a Production Information Model, Proceedings of the Twenty-seventh* (International Ocean and Polar Engineering Conference), San Francisco, CA, USA, June 25-30.

[8] Volkov S et al., 2016 *Optimization of Oil and Gas Production Based on Integrated Planning*, SPE-181955-MS.

[9] Vlasov A I and Mozhchil A F 2018 *Technology overview: from digital to intelligent field* (In Russ.) (PRONEFT. Professionalno o nefti) no. 3(9), pp. 68-74. DOI: 10.24887/2587-7399-2018-3-68-74

[10] Beloshitsky A I 2019 *Technological trends in the development of the oilfield services industry* (Oil and gas business) 17 (2) pp 50-55. DOI: 10.17122/ngdelo-2019-2-50-55

[11] Makarov VL et al. 2019 *Development of digital doubles for manufacturing enterprises* (Business Informatics) 13 (4) pp 7-16.DOI: 10.17323/1998-0663.2019.4.7.16

[12] Apasov RT 2018 *Integrated modelling - a tool for improving the quality of design solutions for the development of oil rims of multi-layer oil condensate fields* (Oil Industry) 12 pp 46-49. DOI: 10.24887/0028-2448-2018-12-46-49

[13] Cherepovitsyn A E and Kraslavsky A 2016 *Study of the innovative potential of an oil and gas company at different stages of field exploitation* (Notes of the Mining Institute) 222 pp 892-902. DOI: 10.18454/PMI.2016.6.892

[14] Bogatkina Yu G and Eremin NA 2019 *Intelligent modeling technologies for calculating economic indicators for the evaluation of oil and gas fields* (Bulletin of Tula State University. Earth Sciences) 3 pp 344-355. eLIBRARY ID: 40870199

[15] Ushmaev OS et all 2016 *Integrated modeling as a tool for assessing the impact of well operating modes and a surface collection network on the development of an oil rim* , SPE182007

[16] Petroleum Engineering And Structural Geology Software [Electronic resource] – URL https://www.petex.com

[17] Ismagilov R P et al. 2014 *Integrated model for integrated management of field development and development* (Oil industry) 12 pp 71-73. https://elibrary.ru/item.asp?id=22704011

[18] Grishagin AV 2009 *On the problems of integration of the reservoir - well - development - economics system using the example of the development project of the West Kommunarskoye oil field* (Scientific and Technical Bulletin of NK ROSNEFT) 1 pp 30–35. eLIBRARY ID: 15200634

[19] Acosta L M et al. 2005 *Integrated modeling of the Furrial Field Asset applying risk and uncertainty analysis for the decision taking*, SPE 94093

[20] Balashova A D and Bolshakova O И 2019 *Influence of business digitalization on oil recovery coefficient and increase in the efficiency of oil and gas companies* (University Bulletin) 5 pp 73-79. DOI: 10.26425/1816-4277-2019-5-73-79

[21] Katysheva E G 2019 *Economic and institutional problems of digital transformation of the oil and gas complex of Russia* (Digital Economy and Industry 4.0: Trends 2025 Proceedings of the scientific and practical conference with international participation. Edited by A.V. Babkina) pp. 175-184.DOI: 10.18720/IEP/2019.1/25

[22] *New Realities in Oil and Gas: Data Management and Analytics* 2017 (White paper Cisco public) https://www.cisco.com/c/dam/en_us/solutions/industries/energy/docs/OilGasDigitalTransformationW hitePa per.pdf (19.03.2019).

[23] Al-Dhubaib T A 2011 *Intelligent fields:industry's frontier and opportunities* (Proceedings of conference:SPE Middle East Oil and Gas Show Conference, Manama, Bahrain, September 25-28) pp 9-11.

[24] Al-Mulhim W A 2010 *Mega I-Field Application in the World* (Proceedings of conference: SPE intelligent energy conference and exhibition, Utrecht, The Netherlands. March 23-25) pp 6-9.

[25] Sankaran S 2009 *The Promise and Challenges of Digital Oilfield Solutions: Lessons Learned from Global Implementations and Future Directions* (Proceedings of conference: SPE digital energy conference and exhibition held. Houston, Texas, 7-8 April)13 pp 10-12.

[26] Soma R 2006 *Service oriented data composition architecture for integrated asset management* (Proceedings of conference: SPE Intelligent Energy Conference and Exhibition, Amsterdam, The Netherlands. April 11–13) pp. 1–2.

Development of haptic gloves with vibration feedback as a tool for manipulation in virtual reality based on bend sensors and absolute orientation sensors

V I Petrenko, F B Tebueva, V O Antonov, A A Apurin, U V Zavolokina

North Caucasus Federal University, Pushkin street, 1, Stavropol, 355009, Russia

E-mail ant.vl.02@gmail.com

Abstract. In the article, the existing developments of haptic gloves were considered, an analysis of scientific publications on this topic, which showed a variety of technical solutions. The main problems of the study area are indicated and two methods for the implementation of haptic gloves, which have a lower cost in comparison with analogues, are proposed. The first method uses nine axial sensors of absolute orientation, which allow to accurately determine the position of the joint in space, and the second method uses bend sensors, which are easier to use but less accurate. To implement the haptic gloves, software was developed, a model was created in the Unity cross-platform development environment.

Keywords: haptic glove, virtual reality

1. Introduction

It becomes necessary to perform tasks in places where the presence of a person is difficult, risky or even impossible. In such dangerous situations, the use of anthropomorphic manipulators is effective. At this stage of the development of science and technology, only robots that realize the remote presence of an operator by using the copying type of control can completely replace a person when performing complex tasks in a dynamic environment. With copy control, the robot repeats the actions of the operator [1]. The basis of this control method is the simultaneous formation of laws of motion for all degrees of mobility of the anthropomorphic manipulator through a copy type master [2].

For remote control of the manipulator, a haptic glove worn by an operator can be used. Haptic gloves significantly expand the possibilities of plausible immersion, allowing you to feel virtual objects by affecting the tactile sensations of a person for greater dexterity and control during operations. This can be an advantage not only during rescue operations, but also in the energy and construction industries, medicine, and industry.

The aim of this work is to develop haptic gloves. To achieve this goal, literary sources with presented existing haptic gloves were analyzed. Methods for implementing haptic gloves on absolute orientation sensors and bend sensors, as well as software architecture, are proposed.

2. Analytical review of current developments of haptic gloves

As part of the work, an analysis has been made of systems and mechanisms developed in recent years that are used in the construction of haptic gloves. There is a general trend for all developments, which consists in the desire to reduce the weight of the structure for the possibility of more comfortable and long-term use, as well as moving towards a cheaper design.

The work [3] describes the development of a haptic glove with a pneumatic drive to bring closer the sensations of interaction with a virtual person to a natural person. To improve the perception of human skin, a pneumatic drive system with silicone cavities is proposed, they are located at the fingertips and are integrated into the glove in the center of the palm. Air pressure is used to simulate the softness of the skin with different textures. At the moment, the device consists of only three mechanisms: on the thumb and forefinger, as well as in the center of the palm. Wi-Fi support is provided.

In [4], authors proposed a system that recognizes the angle formed by the movement of a human fist to control the Scorbot-Er 4U robotic arm. Technically, this is implemented using an accelerometer in a passive haptic glove that measures the inclination of the hand, then the information is processed by Arduino and transmitted to MATLAB (application package for solving technical computing problems) for further sending the command to the robot.

The article [5] describes the design and optimization of a five-finger mechanism for a tactile glove that uses a worm-geared motor. The glove is a wireless and autonomous mechatronic system that is mounted above the outside of the palm and provides tactile force feedback for each finger. The total weight of all elements of the mechanism is 310 g.

In [6], the ExoTen-Glove haptic glove with integrated force sensors and small-sized DC motors is described. The weight of the glove is 360 g, it is worn on the whole hand, using five fingers.

The RML Glove with a reverse response is put on two fingers (forefinger and middle finger) and weighs 180 g, it also uses a worm motor [7].

The source [8] shows the haptic glove CyberGrasp, driven by a direct current electric motor, worn over the entire wrist and using five fingers. The weight of the device is 450g.

The DEXMO glove, presented in [9], is driven by servos, has a weight of 320 g and is also worn on five fingers.

The article [10-11] describes the development of the haptic gloves and the interface to them The Rutgers Master II. The design works with pneumatic cylinders - these are mechanical devices that use the force of compressed gas to create force in a reciprocating linear motion - weighs 185 g and is worn 4 fingers: thumb, forefinger, middle finger and ring finger.

The Wolverine device described in article [9] consists of a base that is mounted on the thumb and three rods attached to it, each of which has a sliding mount for the tips of the forefinger, middle finger and ring fingers. Each sliding mount has a brake that can lock the corresponding rod. In this way, the braking system can generate motion capture accuracy.

In [12-15], the authors analyze the problems of copying control of robot manipulators using haptic gloves. Which is also an important aspect of glove structure development.

The feedback force for each of the above devices is 80, 35, 12, 29.4, 16, and 106 N, respectively.

Table 1. Device comparison

	Drive unit	Wireless	Number of fingers	Max response force, N	Weight, g
ExoTen-Glove	twisted string system	no	5	80	360
RML Glove	worm-geared motor	yes	4	35	180
CyberGrasp	DC electric motor	no	5	12	450
DEXMO	servos	yes	5	29.4	320
The Rutgers Master II	pneumatic cylinders	no		16	185
Wolverine [11]	brake gear	yes	3	106	55

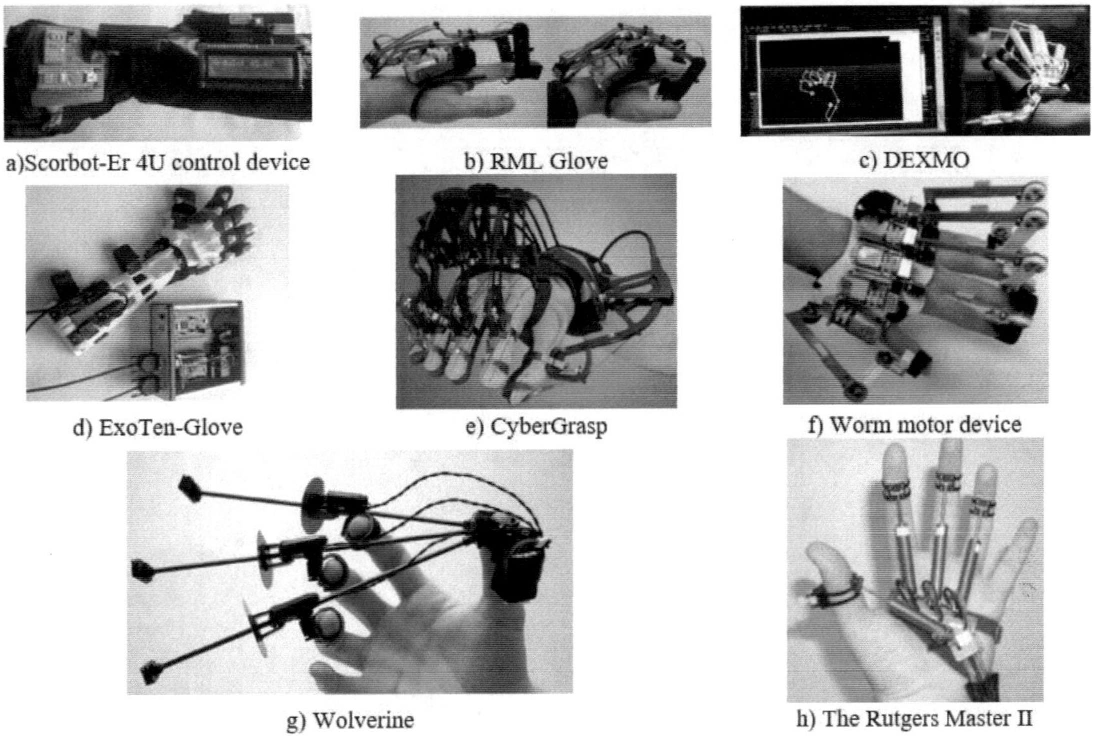

Figure1. Existing Haptic Gloves

As a result of the analysis of haptic gloves, it was revealed that existing solutions have several disadvantages. So, the worm-geared motors used in the RML Glove, cause significant losses in transmitted power, their disadvantages also include the tendency of thread turns and teeth to seize. RML Glove and CyberGrasp are expensive because they use expensive drives. The remaining devices considered are not available for purchase, since they are presented only in laboratory copies.

This article proposes two methods for implementing haptic gloves that are cheap but no less accurate.

3. Architecture

Figure 2 shows a block diagram of the software.

Figure 3 shows a diagram of a haptic glove on inertial measuring modules for one finger. The absolute orientation sensor allows to more accurately measure the position of the object in space thanks to three sensors with three axes for simultaneous measurement of tangential acceleration, rotational acceleration and the strength of the local magnetic field, also the built-in microprocessor eliminates a significant amount of mathematical calculations, this will reduce the delay time and increase the accuracy of execution target operations. Each module is located on the phalanx of the finger and fixes its movement in space. To more accurately determine the movement of the hand, one module is installed on the hand. Information about the movement is transmitted to the Arduino, where it is processed and using the Wi-Fi module is transmitted to the computer. The glove is powered by a lithium polymer battery. Using vibration motors, feedback is implemented. The figure shows:

1. Arduino Mega;
2. Lithium polymer battery;
3. ModuleWi-Fi;
4. Logic level converter;
5. Vibration motor;
6. Inertial measuring module.

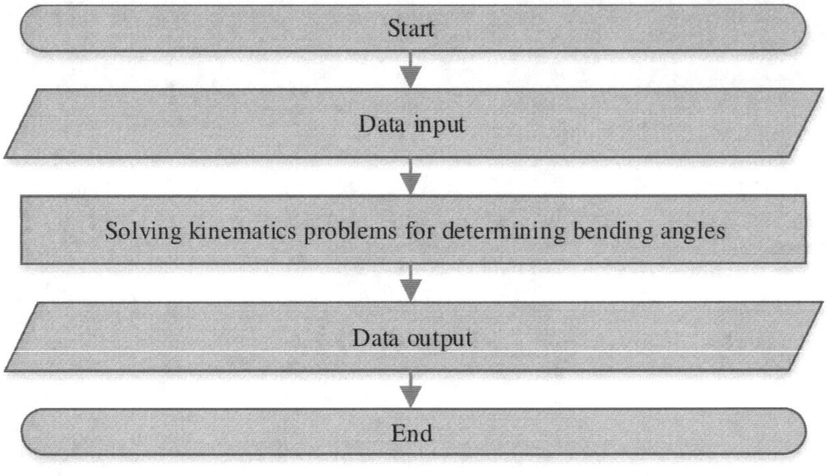

Figure 2. Software block diagram

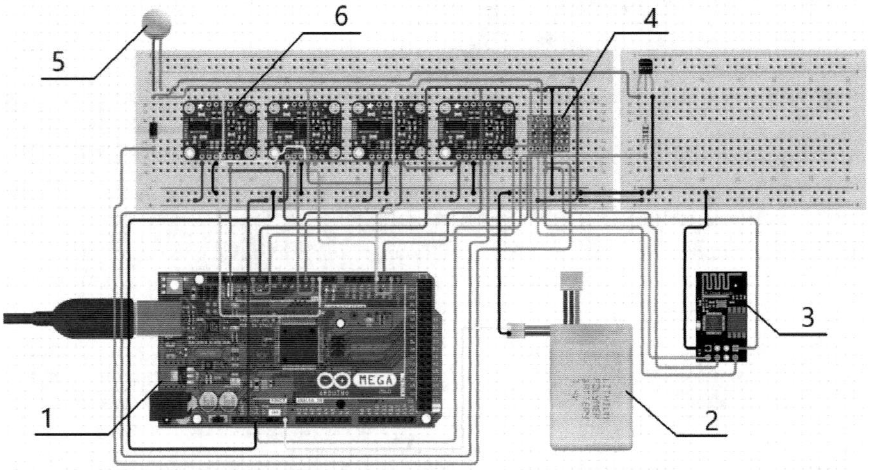

Figure 3. The electrical circuit of the haptic gloves with absolute orientation sensors

Figure 4 shows a diagram of a haptic glove on bend sensors for five fingers. The difference from the first option is the use of bend sensors to determine the movement and bend angles of the operator's fingers. Depending on the degree of bending, the sensor returns a number from 0 to 1023. An inertial measuring module helps determine the movement of the operator's hand. The figure shows:

1. Arduino Mega;
2. Lithium polymer battery;
3. Module Wi-Fi;
4. Logic level converter;
5. Vibration motor;
6. Inertial measuring module;
7. Bend sensor.

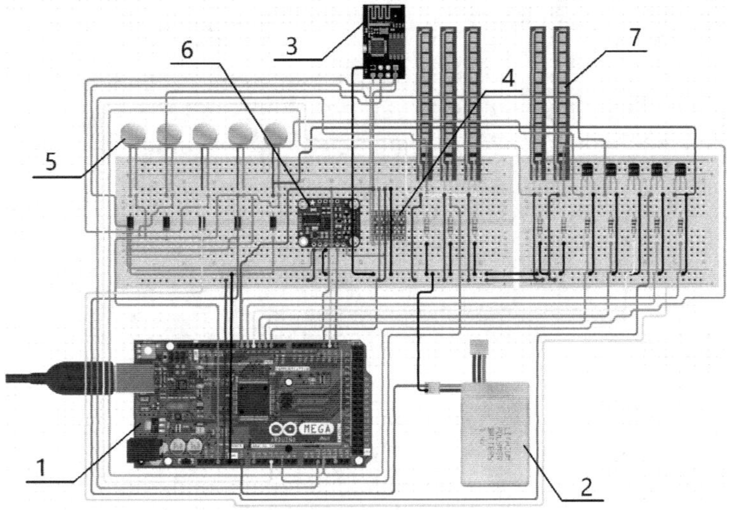

Figure 4. Haptic glove wiring diagram with bend sensors

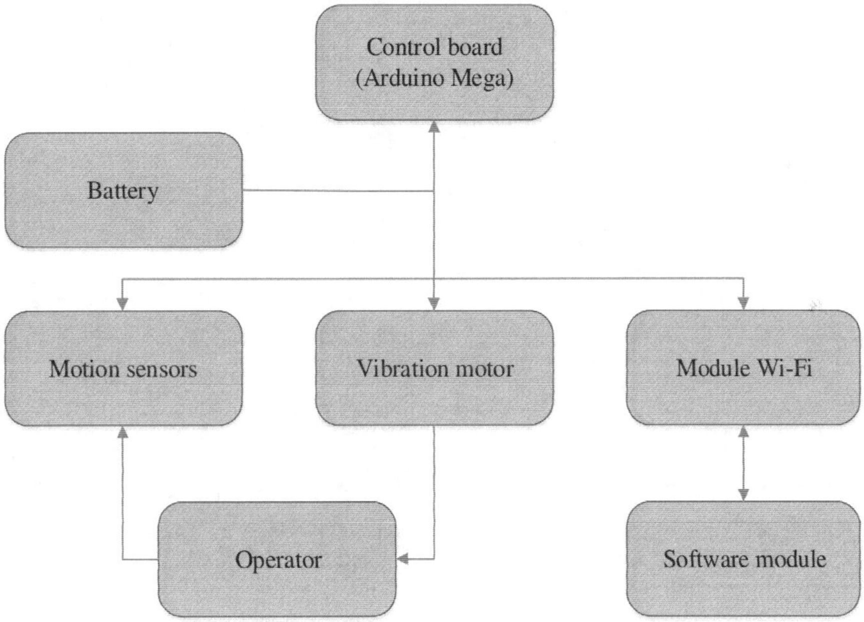

Figure 5. Functional diagram of the haptic glove

Advantages relative to analogues are light weight, ergonomics and ease of implementation. The approximate price is 11 thousand rubles.

The disadvantage is the impossibility of tracking the movement of the elbow.

4. Software

To implement the presented haptic gloves, software was developed and a model was created in the Unity cross-platform development environment (figure 6). Scripts are written in C #. Hand model rigging was done in Blender v2.81. The software allows you to display in real time each registered movement of the fingers and hands of the operator.

IOP Conf. Series: Materials Science and Engineering **873** (2020) 012025 doi:10.1088/1757-899X/873/1/012025

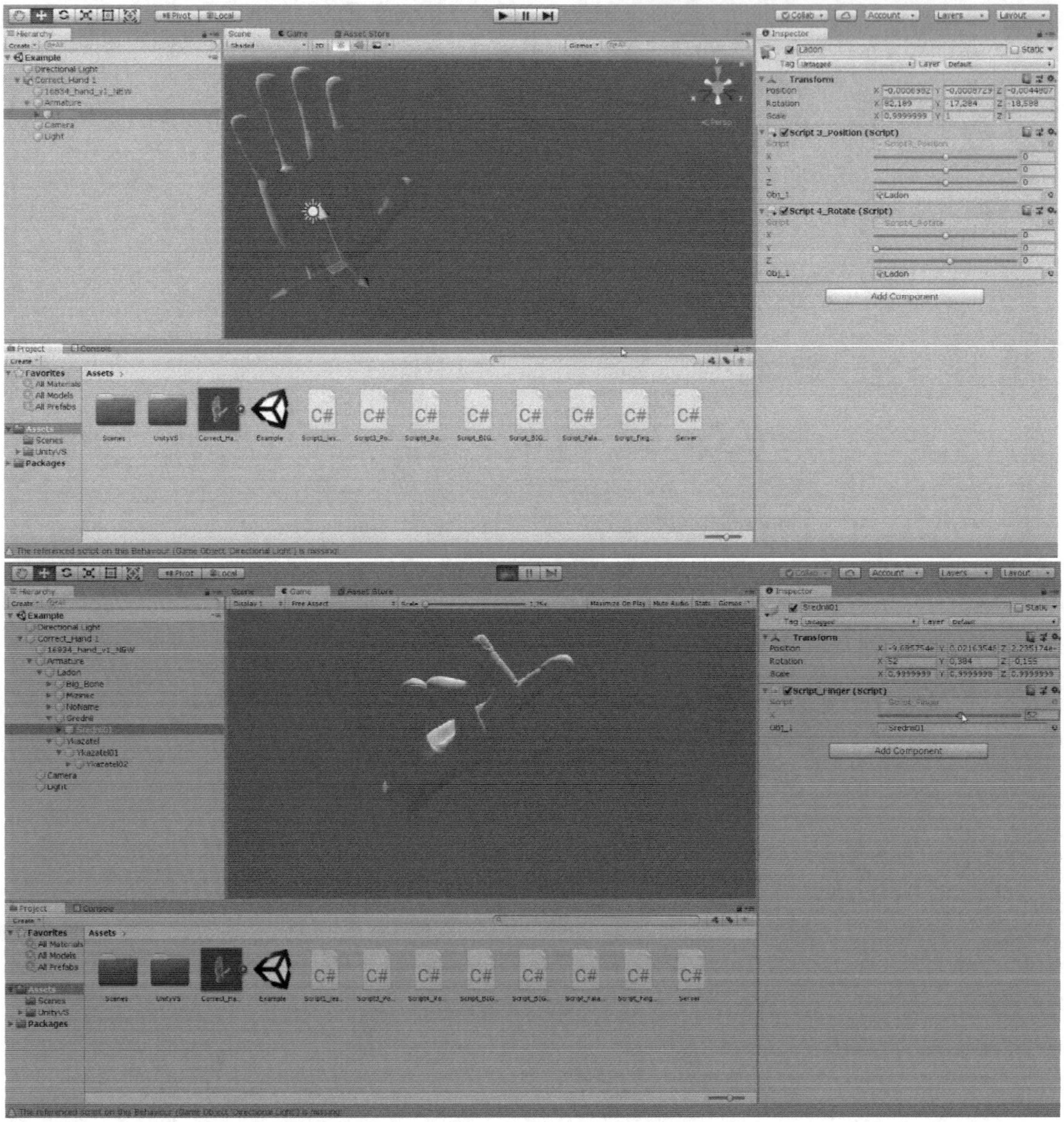

Figure 6. Virtual model

Using the Wi-Fi module, the haptic glove connects to the computer on which the software is installed. Absolute orientation sensors transmit measurements of tangential acceleration, rotational acceleration, and local magnetic field strength, and bend sensors transmit values from 0 to 1023, depending on the degree of bending.

5. Conclusion

Based on the results of the analysis of current developments of haptic gloves, problems such as high cost and significant losses in transmitted power were identified. It is also noted that most of the devices examined are presented exclusively as laboratory specimens and are not available for purchase.

As possible solutions to the above problems, two methods for the implementation of haptic gloves based on bend sensors and absolute orientation sensors and software developed for them were proposed.

IOP Conf. Series: Materials Science and Engineering **873** (2020) 012025 doi:10.1088/1757-899X/873/1/012025

These developments additionally solve the issue of reducing the weight of the structure and increase the ergonomics of the device.

References

[1] Petrenko V I, Tebueva F B, Gurchinsky M M, Antonov V O and Pavlov A S 2019 *Predictive assessment of operator's hand trajectory with the copying type of control for solution of the inverse dynamic problem* (SPIIRAS Proceedings) pp. 123-147.

[2] Petrenko V I, Tebueva F B, Antonov V O and Ryabtcev S S 2018 *Development of an algorithm for constructing the spatial position of the joints of the operator's hand on the basis of solving the inverse task of kinematics* (Proceedings of the 2017 Instrumental Engineering in the XXI Century. Integration of Science, Education and Production) pp. 733–740.

[3] Talhan A, Kim H, Kumar S, Raza A and Jeon S 2018 *Pneumatic actuated haptic glove to interact with the virtual human* (AsiaHaptics).

[4] Dos Anjos G A, Santos G A, Diaz Amado J A and Soares Marques J E 2016 *Haptic glove development for passive control of a robotic arm (*IEEE XXIII International Congress on Electronics, Electrical Engineering and Computing (INTERCON)) pp. 1-6.

[5] Ma Zh and Ben-Tzvi P 2015 *Design and optimization of a five-finger haptic glove mechanism* (Journal of Mechanisms and Robotics).

[6] Hosseini M, Sengül A, Pane Y, Schutter J and Bruyninckx H 2018 *ExoTen-Glove: A force-feedback haptic glove based on twisted string actuation system* (IEEE International Conference on Robot and Human Interactive Communication) pp. 320 – 327.

[7] Ma Zh and Ben-Tzvi P 2014 *RML Glove – An Exoskeleton Glove mechanism with haptics feedback* (IEEE/ASME Transactions on Mechatronics) 20 pp. 641-652.

[8] CyberGrasp *CyberGlove System* URL: www.cyberglovesystems.com (date of the application: 29.01.2020).

[9] Gu X, Zhang Y, Sun W, Bian Y, Zhou D and Kristensson P O 2016 *Dexmo: An inexpensive and lightweight mechanical exoskeleton for motion capture and force feedback in vr* (Proceedings of the 2016 CHI Conference on Human Factors in Computing Systems) pp. 1991–1995.

[10] Bouzit, M, Burdea G, Popescu G and Boian R 2002 *The Rutgers Master II-New Design force-feedback glove* (IEEE/ASME Transactions on Mechatronics) 7 pp. 256 - 263.

[11] Choi I, Hawkes E W, Christensen D L, Ploch C J and Follmer S 2016 *Wolverine: A wearable haptic interface for grasping in virtual reality* (IEEE Intelligent Robots and Systems) pp. 986–993.

[12] Petrenko V I, Tebueva F B, Gurchinsky M M, Antonov V O, Shutova Yu A 2018 *Solution of the dynamics inverse problem with the copying control of an anthropomorphic manipulator based on the predictive estimate of the operator's hand movement using the updated Brown method* (IOP Conf Series: Materials Science and Engineering) p. 450.

[13] Bogdanov A A, Permyakov A F and Kutlubaev I M, *Basic Principles of an Exoskeleton Construction for an Anthropomorphic Manipulator Control* 2019 (International Multi-Conference on Industrial Engineering and Modern Technologies (FarEastCon), Vladivostok, Russia) pp. 1-4.

[14] Magrini E, Ferraguti F, Ronga A J, Pini F, De Luca A and Leali 2020 *Human-robot coexistence and interaction in open industrial cells* (Robotics and Computer-Integrated Manufacturing) 61.

[15] Magrini E and De Luca A 2017 *Human-robot coexistence and contact handling with redundant robots* (IEEE/RSJ International Conference on Intelligent Robots and Systems (IROS), Vancouver, BC) pp. 4611-4617.

Design of a fuzzy controller for a membrane gas separation process control system

A I Koldaev[1], Zh M Koldaeva[1], A A Evdokimov[1], D V Boldyrev[1], I P Kuzmenko[2]

[1]North-Caucasus Federal University, Department Nevinnomyssk Institute of Technology, Gagarina Street, 1, Nevinnomissk, 357108, Russia

[2]Stavropol State Agrarian University, Zootekhnichesky Lane, 12, Stavropol, 355017, Russia

E-mail ventilator83@yandex.ru

Abstract. The production of nitrogen from the air through membrane gas separation processes is widely used in many industries. The problem of controlling the gas separation process, it is connected with multi-loop control using the control of several variables. This article proposes an intelligent controller for matching fluctuations in pressure and airflow of the gas separator. The intelligent controller is based on a fuzzy logic algorithm. Applying fuzzy logic in the proposed solution allows you to increase the speed of the control system. The effectiveness of the proposed controller was evaluated in comparison with the PID controller. The proposed fuzzy controller can improve the accuracy of the control system of gas separation, reduce the process transient duration and increase the service life of technological equipment.

Keywords: fuzzy logic; PID control; gas separation; membrane technology; nitrogen; control system

1. Introduction

Air separation for the production of enriched nitrogen and oxygen is of great importance to the chemical industry [1].

For practical applying, whether in laboratories or at large industrial enterprises, nitrogen is obtained in three main ways. All of these ways are based on the decomposition of atmospheric air: by the cryogenic decomposition of air, using short-cycle heat-free adsorption, and by the method of membrane diffusion [2].

The gas separation process based on membrane technology has a preferred application compared to other methods, since it is economical, compact in size, has a modular configuration and can provide low unit energy consumption [3].

Existing gas separation plants are a multi-stage multi-connected construction with non-linear connections between their constituent elements [1,3]. When designing multi-loop control using multi-variable control, it is important correctly to determine controlled variables and their combinations. Most control systems of gas separation plants are based on a traditional PID controller system. The presence of only three tuning parameters in the PID controller, which are configured only at the stage of system design, in some cases is insufficient to obtain the specified control quality, especially for systems with unstable properties, high delay, and systems that require high quality tracking at the same time setting and high quality of attenuation of external disturbances in conditions when there is the incompleteness of information about these disturbances and object properties.

An effective way towards overcoming the limitations in the management of complex objects is the applying of fuzzy controllers that allow shaping the effects on the object by changes in the state variables of the control object, for example, error and speed of error. The approach to constructing control systems based on fuzzy logic applies to many existing systems since in many cases the improvement of existing algorithms can be done at minimal cost using the existing hardware and software [4–6].

In this paper, for the plant of gas separation, the use of a fuzzy controller instead of a PID controller is proposed. The use of a fuzzy controller for the plant of gas separation will improve the accuracy and speed of the gas separation process.

2. Membrane technology for gas separation

Membrane processes are one of the main technologies for gas separation applications, the purpose of which is the separation gas mixtures using semipermeable membranes [7].

In the chemical industry, membrane methods are used to separate helium and hydrogen from natural gases, nitrogen, and oxygen from the air, etc [3,8].

Also, membrane technology is becoming increasingly important for medical purposes, especially in the process of enriching the level of oxygen concentration (O_2). Enrichment of O_2 level from nitrogen (N_2) streams using membrane processes is more economical than conventional cryogenic air distillation, mainly due to its low energy consumption, which subsequently leads to minimization of operating costs [8].

Purification of hydrogen, processing of natural gas, extraction of volatile organic compounds, and drying of gas are increasingly used in various industries due to membrane plants, but the membrane separation of gases remains the main market for the production of nitrogen from the air [9,10].

The membrane process is continuous (without a regeneration stage), it does not include chemicals, does not generate waste, offers the possibility of intensification, it is modular (easy to scale) and does not involve complex operations.

Membrane plants consist of membrane air separation modules, which are cylindrical tanks. Inside these tanks in parallel are many fibers consisting of special polymer materials [11,12].

Compressed air is supplied to the inlet of the membrane module, from where it is evenly distributed between all the individual fibers, arriving at their inner side (Figure 1). The fiber walls are membranes with an asymmetric pore arrangement through which water, hydrogen, and helium molecules diffuse quickly and easily to the outside of the fibers. Molecules of oxygen, as well as carbon dioxide CO_2, penetrate through the walls at an average speed. Principally the molecules of nitrogen, as well as argon contained in the air, remain on the inner side of the membranes [13,14].

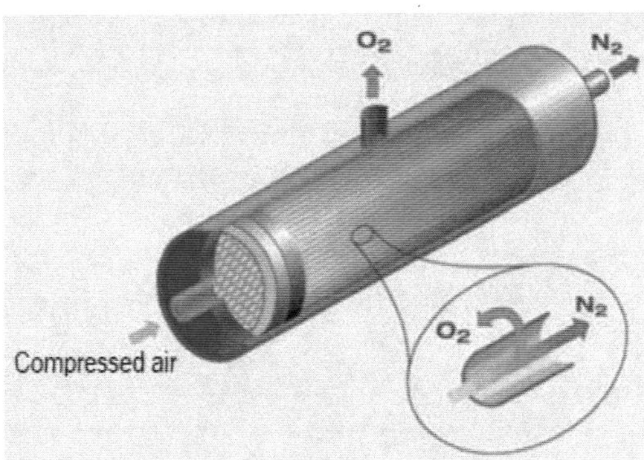

Figure 1. Membrane module

Depending on the requirements for nitrogen purity (usually from 90 to 99.9%), one-stage or multi-stage membrane process configurations are used.

The general structures of the synthesis of processes correspond to the structures shown in Fig. 2. The number of levels for various purity levels corresponds to the number and location of recirculation circuits for multi-stage processes.

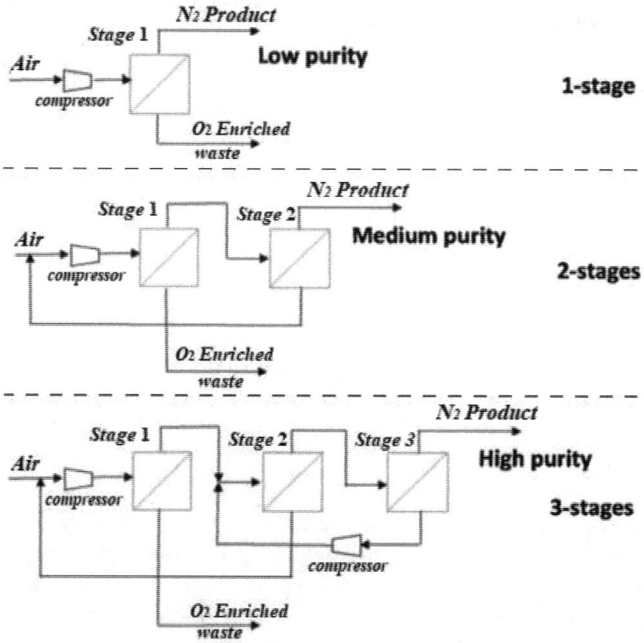

Figure 2. Structures of membrane processes with different nitrogen purities

As a rule, the "purity/performance" ratio can be adjusted as simply as possible with the help of a single flow regulator installed at the outlet of the membranes. At the same time, fluctuations in pressure or flow rate are more likely and may well lead to a drop in the purity of the gas. In this case, there is a need to improve the accuracy of regulation of pressure and gas flow [3,12].

3. Model of the gas separator

The gas dynamics in the gas separator model is based on the material balance equation and the ideal gas equation. The material balance equation determines the number of moles of total gas, component A (nitrogen) and component B (oxygen) in a gas separator. The ideal gas equation determines the total pressure and partial pressures of components A and B in a gas separator.

The diagram of a gas separator is shown in Fig. 3.

The equation of material balance, pressure and mole fraction are described as follows.

Changes in the number of moles of all gases in the material part per unit time can be expressed by the following equation.

$$\frac{dN_{tW}}{dt} = F - W - N_A \cdot Area - N_B \cdot Area \, , \qquad (1)$$

where tW is all gases in the material part,

N is the number of moles in the gas,

F is the amount of material entering the material part per unit time,

W is the amount of gas emitted from a portion of the material per unit time,

$Area$ is the membrane area.

Changes in the number of moles of gas component A in the material part per unit time can be expressed by the following equation

$$\frac{dN_{tW}}{dt} = F \cdot X_{AF} - W \cdot X_{AW} - N_A \cdot Area \,, \qquad (2)$$

where X_{AF} is the mole fraction of component A in F,
index AW denotes the gas component A in the material part.

Given the integral from time 0 to t, it can be expressed by the following equation. This is the number of moles in the material at time t

$$N_{tW}|_{t=t} = \int_0^t (F - W - N_A \cdot Area - N_B \cdot Area)dt + N_{tW}|_{t=0} \,. \qquad (3)$$

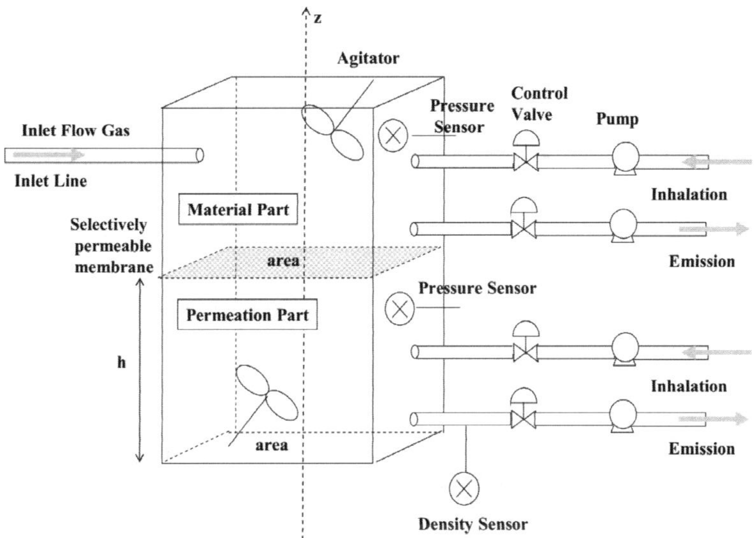

Figure 3. Diagram of the gas separator

Given the integral from time 0 to t, it can be expressed by the following equation. This is the number of moles of the gas component A in the material at time t

$$N_{AW}|_{t=t} = \int_0^t (F \cdot X_{AF} - W \cdot X_{AW} - N_A \cdot Area)dt + N_{AW}|_{t=0} \,. \qquad (4)$$

The pressure of all gases in the material is determined by the equation

$$P_{tW} = N_{tW} \cdot R \cdot T / V \,, \qquad (5)$$

where P is the pressure;
R is the gas constant;
V - volume in the material part (permeable part).

The mole fraction in the material part is determined by the equation

$$X_{AW} = \frac{X_{AW}}{N_W} \,. \qquad (6)$$

The change in the number of moles of all gases in the permeable part per unit time can be expressed by the following equation

$$\frac{dN_{tQ}}{dt} = N_A \cdot Area + N_B \cdot Area - Q \, . \tag{7}$$

Given the integral from time 0 to t, it can be expressed by the following equation. This is the number of moles in the permeable part at time t.

$$N_{tQ}\big|_{t=t} = \int_0^t (N_A \cdot Area + N_B \cdot Area - Q)dt + N_{tQ}\big|_{t=0} \, . \tag{8}$$

The equation of the material balance of the gas component A in the permeable part can be expressed by the following equation:

$$\frac{dN_{AQ}}{dt} = N_A \cdot Area - Q \cdot Y_{AQ} \, ; \tag{9}$$

$$N_{AQ}\big|_{t=t} = \int_0^t \left(N_A \cdot Area - Q \cdot Y_{AQ}\right)dt + N_{AQ}\big|_{t=0} \, . \tag{10}$$

The pressure of all gases in the permeation part

$$P_{tQ} = N_{tQ} \cdot R \cdot T/V \, . \tag{11}$$

The mole fraction in the permeation part

$$X_{AQ} = \frac{N_{AQ}}{N_W} \, . \tag{12}$$

4. Fuzzy control system

The structure of the fuzzy-logical part of the controller is shown in Fig. 4 [15]. It consists of fuzzification blocks, fuzzy rules (rule base), fuzzy inference and defuzzification.

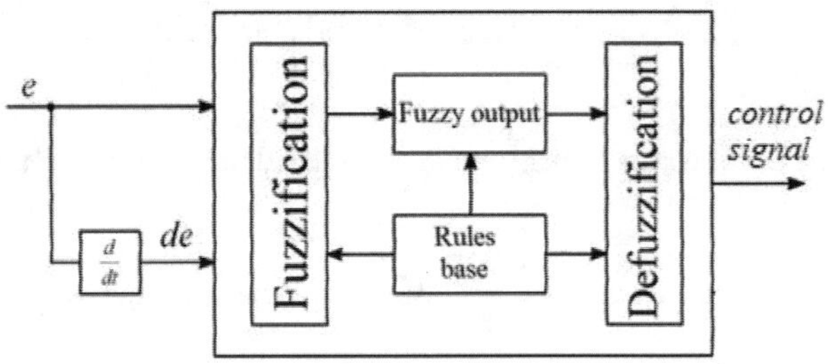

Figure 4. Fuzzy controller structure

The fuzzy model uses fuzzy rules, which are linguistic statements, including fuzzy sets, fuzzy logic, and fuzzy inference. A fuzzy rule plays a key role in representing expert control and experience in matching input variables of fuzzy controllers with output variables.

Figure 5 (left) shows the transitional characteristic of the object, divided into areas *A – I*.

The formation of linguistic rules can be demonstrated using the phase plane (Fig. 5, right), the horizontal axis of which represents the values of the derivative of the control error *de/dt*, and the vertical axis represents the values of the control error *e*.

The action starts from area A. Area A is characterized by a large error value and an almost zero value of the derivative by error. Therefore, in linguistic rules, this will be expressed as PB (positive big) and PZ (positive zero), respectively. When moving from area A to area B, the value of the error and the derivative of the error will change from PB to PM and PZ to PM, respectively.

Consider the period in which the output values increase from C to D areas, where the error and derivative by error change from NZ to NM and PB to PM, respectively. Considering transitions in all areas, it is possible to form a matrix of fuzzy statements presented in Table 1.

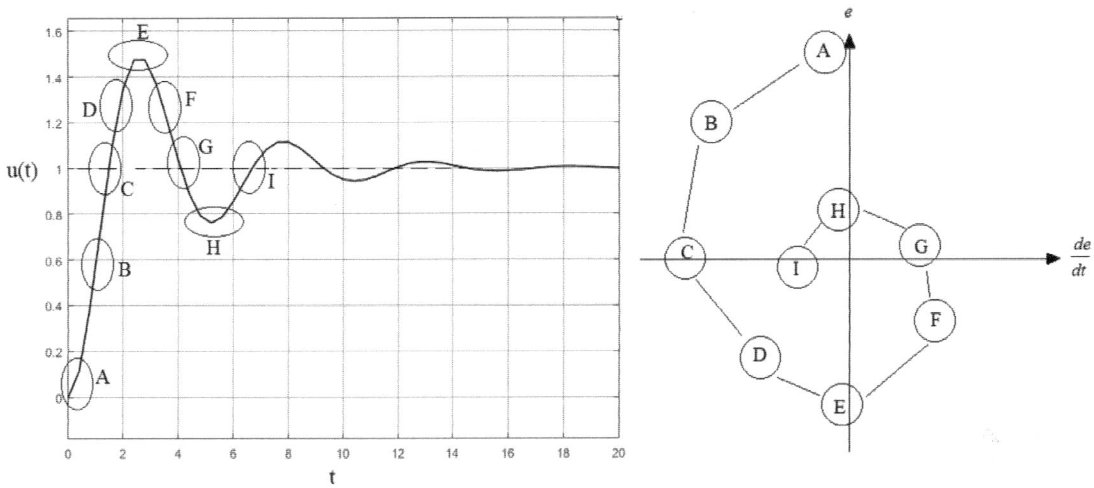

Figure 5. Transient response of control object (left) and phase plane (right)

Using the five membership functions, fuzzified input and output variables are normalized between +1 and –1. To get the fuzzy conclusion to a precise value, the centroid method is used. Fuzzy inference using five membership functions requires 25 rules.

An increase in the number of membership functions to seven leads to an increase in the number of rules to 49. However, this increases the cost of computing resources but does not lead to a significant improvement in control efficiency.

On the other hand, with three membership functions and 9 rules, calculations are faster. But compared to a system with five membership functions and 25 rules, the control efficiency is low in terms of overshooting and increasing the transition time. When creating the rule base, a table of linguistic terms is created (Table 1), based on data obtained during modeling under various operating conditions when the reference signal (target) and disturbances change.

Table 1. Linguistic terms.

e / *de*	**NB**	**NS**	**Z**	**PS**	**PB**
NB	NB	NB	NB	NS	Z
NS	NB	NB	NS	Z	PS
Z	NB	NS	Z	PS	PB
PS	NS	Z	PS	PB	PB
PB	Z	PS	PB	PB	PB

In table 1, the following notation is accepted: NB is negative big; NS is negative small; Z is zero; PS is positive small; PB is positive big.

5. Simulation Results

The object simulation was created based on a gas separator model using Matlab/Simulink. The separator model is a three-stage structure as shown in Figure 2.

As a control system for an object, PID control, and a fuzzy controller are compared. The simulation results are presented in figures 6 and 7.

IOP Conf. Series: Materials Science and Engineering **873** (2020) 012026 doi:10.1088/1757-899X/873/1/012026

Figure 6 shows that tracking the setpoint is best for a fuzzy controller in gas separator 3, followed by gas separators 2 and 1, respectively, which confirms the behavior of the model.

Figure 7, right, demonstrates that the process of gas separation with fuzzy control requires less time than using PID control. This provides lower energy consumption during the process of gas separation. Figure 7, left, shows that the control accuracy for fuzzy control is higher than with PID control.

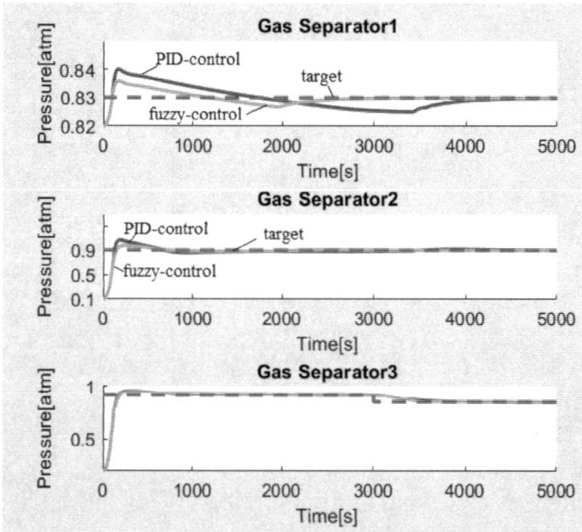

Figure 6. Gas separation simulation results

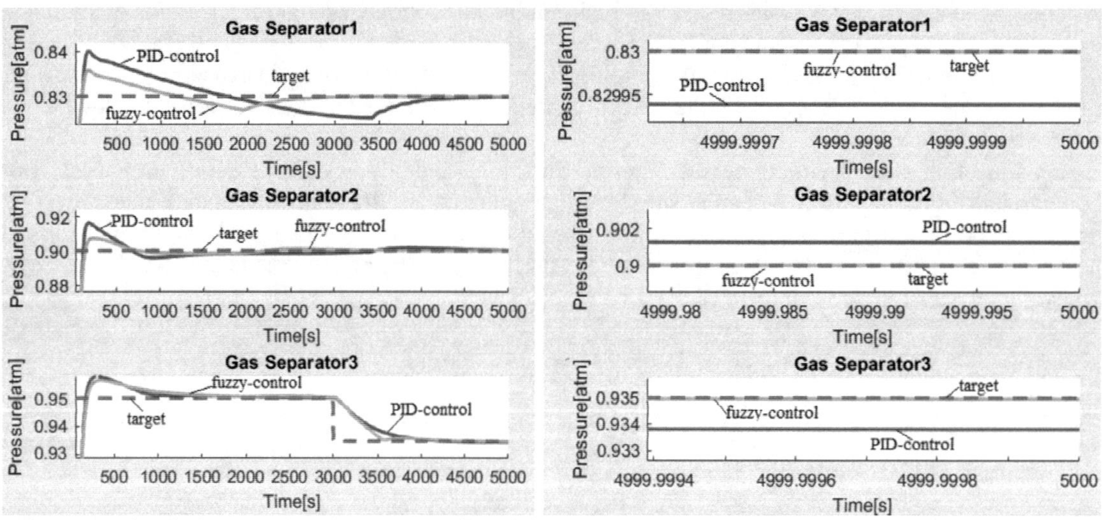

Figure 7. Simulation results of gas separation: left – increase along the pressure axis, right – increase along the time axis

Thus, the proposed gas separator control system with a fuzzy regulator allows you to obtain a given level of nitrogen concentration at a lower cost than when using PID control.

6. Conclusion

Today, the production of nitrogen-enriched air using membrane gas separation is the main method. In this work, the problem of constructing a control system for the process of obtaining nitrogen from the air in gas separator using a fuzzy controller is solved.

The object of control (a plant) is a gas separator. A simulation model is created based on a gas separator model using Matlab/Simulink. In the object control system, the PID controller and the fuzzy

IOP Conf. Series: Materials Science and Engineering **873** (2020) 012026 doi:10.1088/1757-899X/873/1/012026

controller are compared. The fuzzy control model in this example as a whole has better control characteristics compared to PID control and allows us to obtain a given level of nitrogen purity with a lower unit energy consumption.

References

[1] Murali R S, Sankarshana T and Sridhar S 2013 *Air Separation by Polymer-based Membrane Technology* Separation & Purification Reviews **42** pp 130–86.
[2] Chen X Y, Vinh-Thang H, Ramirez A A, Rodrigue D and Kaliaguine S 2015 *Membrane gas separation technologies for biogas upgrading* RSC Adv. **5** pp 24399–448.
[3] Bozorg M, Addis B, Piccialli V, Ramírez-Santos Á A, Castel C, Pinnau I and Favre E 2019 *Polymeric membrane materials for nitrogen production from air: A process synthesis study* Chemical Engineering Science **207** pp 1196–1213.
[4] Ghasem N M 2006 *Design of a Fuzzy Logic Controller for Regulating the Temperature in Industrial Polyethylene Fluidized Bed Reactor* Chemical Engineering Research and Design **84** pp 97–106.
[5] Lubentsova E V, Lubentsov V F, Koldayev A I, Evdokimov A A and Samoylenko D V 2015 *Fuzzy Temperature Controller Of The Exothermic Fermentation Processes* International Journal of Applied Engineering Research **10** pp 42753–42757.
[6] Chen L, Huang B and Liu F 2013 *Nonlinear system identification with multiple and correlated scheduling variables** IFAC Proceedings Volumes **46** pp 319–324.
[7] Belloni A 2008 *Industrial Gases Processing* ed H-W Häring (Weinheim: Wiley-VCH).
[8] Lin H, Zhou M, Ly J, Vu J, Wijmans J G, Merkel T C, Jin J, Haldeman A, Wagener E H and Rue D 2013 *Membrane-Based Oxygen-Enriched Combustion* Ind. Eng. Chem. Res. **52** pp 10820–10834.
[9] Sanders D F, Smith Z P, Guo R, Robeson L M, McGrath J E, Paul D R and Freeman B D 2013 *Energy-efficient polymeric gas separation membranes for a sustainable future: A review* Polymer **54** pp 4729–4761.
[10] Baker R W and Low B T 2014 *Gas Separation Membrane Materials: A Perspective* Macromolecules **47** pp 6999–7013.
[11] Schrier J 2012 *Carbon Dioxide Separation with a Two-Dimensional Polymer Membrane* ACS Appl. Mater. Interfaces **4** pp 3745–52.
[12] Hu T, Zhou H, Peng H and Jiang H 2018 *Nitrogen Production by Efficiently Removing Oxygen From Air Using a Perovskite Hollow-Fiber Membrane With Porous Catalytic Layer* Front. Chem. **6**.
[13] Sannomiya A, Fukui K, Nagaoka S, Suzuki Y, Iwaki M and Kawakami H 2007 *Gas transport properties of asymmetric polyimide membranes prepared by ion-beam irradiation* Journal of Polymer Science Part B: Polymer Physics **45** pp 262–269.
[14] Zhu X, Sun S, He Y, Cong Y and Yang W 2008 *New concept on air separation* Journal of Membrane Science **323** pp 221–224.
[15] Koldaev A I, Evdokimov A A, Samoylenko D V and Kocherov Y N 2018 *An Approach of the Wavelet-Fuzzy Controller of Vector Control System by the Induction Motor* 2018 International Multi-Conference on Industrial Engineering and Modern Technologies (FarEastCon) (FarEastCon) pp 1–6.

Development and testing of hardware-software complex for diagnostics of freight vehicles energy parameters

D V Kazakov[1, 3], I P Kuzmenko[2], A A Moskvitin[3], A S Abelyan[3]

[1]Nevinnomyssk Institute of Technology (branch of North-Caucasus Federal University), Gagarina Street, 1, Nevinnomyssk, 357108

[2]Stavropol State Agrarian University, Zootekhnichesky Lane, 12, Stavropol, 355017, Russia

[3]Armavir Institute of mechanics and technology (branch of Kuban State Technological University continues), Kirova street, 127, Armavir

E-mail: kazakov_nti@inbox.ru

Abstract Proposed method of energy parameters control of freight vehicles in a non-brake way is based on evaluation of their dynamic properties. An algorithm has been developed for calculating all the necessary energy parameters, on the basis of which analysis and correct technical and production operation of freight vehicles (FV) are carried out. The method of measuring speed and acceleration of FV was tested using a set of GPS and/or GLONASS sensors which showed its effectiveness.

Keywords: diagnostics, parameter control, power, transition process

1 Introduction

Freight transport is the most efficient kind of transport but it is also relatively expensive. The agro-industrial complex (AIC) is known as a complex system in which each element has its own unique function; in the case of failure or significant depletion of the resources of one of the elements the rest will not be able to function effectively [1]. With timely diagnosis and tracking of even minor losses in energy performance of machine and tractor stock and freight transport, serious problems in the operation of agricultural system can be prevented. Without making significant efforts and costs to control the parameters of the technical condition of freight vehicles engaged in cargo transportation in the AIC, the cost for repairing vehicles, thereby fuel costs as well as downtime of FV can be significantly reduced [2, 3, 4].

According to the results of a study on the example of an official DAF dealer in Russia, it was found that FV are operated with overload in 90% of cases. This type of operation leads to increased wear of parts and assemblies and as a result to a change in the energy parameters of the FV. In this case the frequency of maintenance does not correspond to the actual need, which can be established by constantly monitoring the parameters of FV using the developed hardware-software complex. Therefore, the task of developing and testing a hardware-software complex as well as software is an urgent task.

2. Materials and Methods

When introducing modern diagnostic methods, monitoring and data collection in the process of diagnostics and operation of FV, a non-brake method of diagnosis based on the assessment of dynamic properties of FV is promising and low-cost [5]. Engine power and torque are determined during its free acceleration without load [6, 7]. In this case, the angular acceleration of the engine crankshaft is measured at full fuel supply from the minimum stable idle speed to the maximum. A FV engine is loaded due to the inertia forces of moving and rotating parts of the engine and the car itself (the moment of inertia for each engine is a constant value) [8]. In our opinion, this type of diagnostics is faster than brake bench methods, in which the load applied to the engine crankshaft is created using a special brake. The data processing algorithm that implements the selected methods is shown in Figure 1.

Features of the hardware part of the complex:

- a set of GPS and / or GLONASS sensors, an accelerometer and a crankshaft sensor that provide a sufficient amount of data for calculations. During the development process, the prospect of adding additional sensors (pressure sensors before and after the oil filter, etc.) will be considered [11];
- a personal computer (PC) (at initial stages of development) that connects the sensors to a central processor with sufficient processing power to calculate the necessary parameters as well as an external storage device for storing them, during further development the possibility of adding a display to show the current parameters and performance indicators;
- bluetooth module and / or USB output, providing data output to various devices (personal computer, smartphone) for detailed display and analysis of parameters.

The software part of the complex will consist of the following modules:

- the main program for a PC (at the initial stages of development) and its own device (at the final stage of development) which will read data from sensors and calculate the necessary and additional parameters according to the developed algorithm, display basic information on an external storage device, then on the device's display;
- additional software for a PC or a smartphone, which allows to get detailed information, graphically to display the received data.

The operation of the hardware-software complex is provided by entering the initial parameters: external speed characteristics of the engine, transmission efficiency, gear ratios of the gearbox and final drive, wheel sizes.

The next step in the algorithm is obtaining data from sensors for speed, time and angular velocity of the crankshaft of the engine and then filling in the data amount, calculating the traction and dynamic characteristics of the FV and construction of acceleration and free acceleration curves. After that a specially developed algorithm compares and analyzes the reference characteristics of the FV and experimental ones, gives recommendations and a forecast of technical conditions. The algorithm will work on the basis of an iterative method.

3. Results

Assessment of the dynamic properties of FV can be performed using a mathematical model based on the well-known equation of motion [9, 10]. FV power balance is shown in Figure 2. The equation of the FV movement (power balance) has the form:

$$F_k - F_i - F_s - F_\alpha = 0 \quad (1)$$

where F_k is tangent force on the wheel, the driving force of FV, N;

F_i is force expended to overcome the inertia of translational and rotational masses of the car, N;

F_s is force of resistance to movement (wheel rolling, air resistance), N;

F_α is force expended to overcome ascent/descent, N.

$$F_k = \frac{M_d \cdot i_{kpp} \cdot i_{gp}}{r_d} \cdot \eta, \quad (2)$$

where M_d is torque developed by the engine, Nm;

i_{kpp} is gear ratio;

i_{gp} is gear ratio of the main transmission (differential);

$r_d$is actual rolling radius of the wheel, m;
η is transmission efficiency.

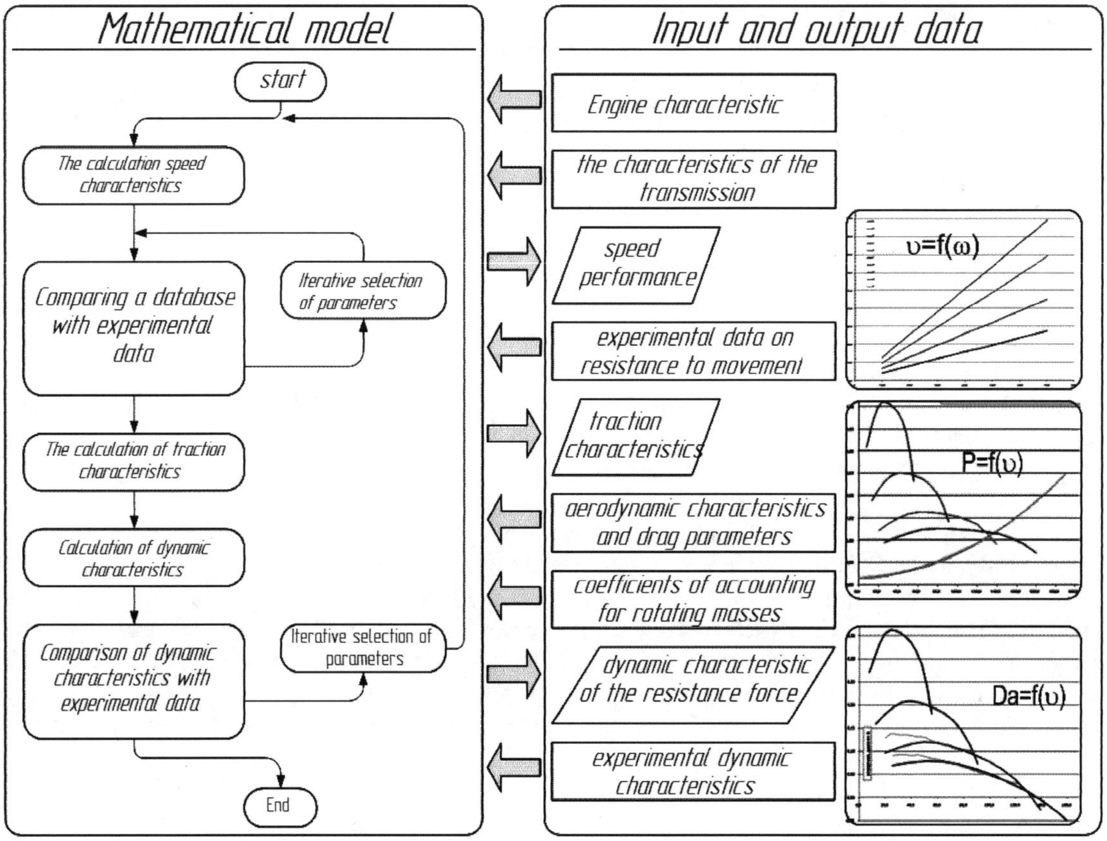

Figure 1. Algorithm of data processing by hardware-software complex

$$F_i=m_p \cdot a, \quad (3)$$

where m_p is reduced mass of CV, kg;
a is developed acceleration, m / s^2.

$$F_s = F_{sk} + F_{sv} \quad (4)$$

where F_{sk} is rolling resistance force of the car, these include friction force in bearings and rolling wheels, N;
F_{sv} is air resistance force, N.

$$F_\alpha = m \quad \cdot g \cdot \sin(\alpha), \quad (5)$$

where m is mass of FV, kg;
g is acceleration of gravity, m/s^2;
α is developed linear acceleration, m/s^2.

$$F_{sv} = C_x \cdot \frac{\vartheta^2 \cdot \rho}{2} \cdot S, \quad (6)$$

where S is tfrontal section of a vehicle, projection of body onto a plane perpendicular to the longitudinal axis, м^2;
ρ is air density, ρ = 1.29 kg/m^3;
C_x is drag coefficient of a vehicle, C_x= 0.25 ... 0.5;
ϑ is vehicle speed, m/s;

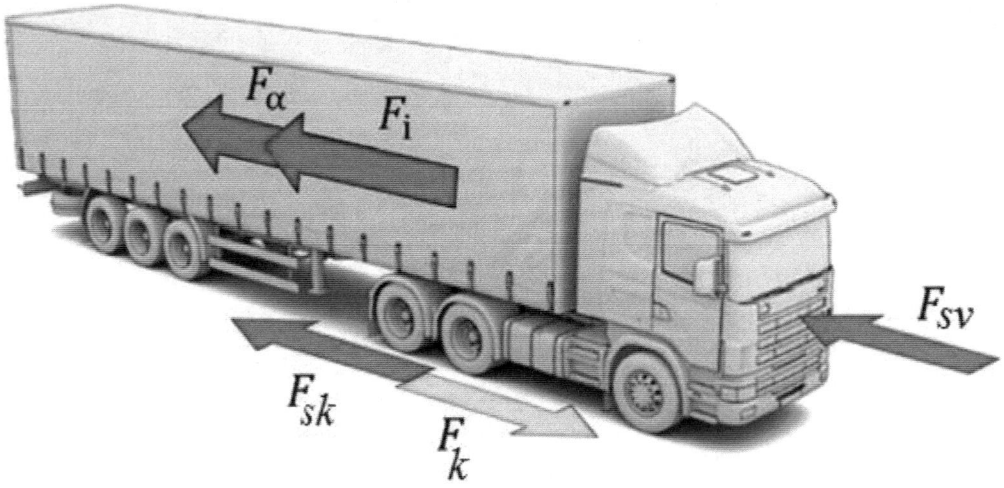

Figure 2. Power balance of FV

$$F_{sk} = m \cdot g \cdot k_1 + k_2 \cdot \vartheta + k_3 \cdot \vartheta^2 \quad (7)$$

where $k_1$is coefficient of rolling resistance which characterizes the constant component of rolling, is numerically equal to tabular coefficients of road resistance / reference to the method;

k_2 is coefficient characterizing a linear increase in rolling resistance versus vehicle speed;

k_3 is coefficient characterizing the quadratic component of resistance growth versus vehicle speed;

The reduced mass of the FV is determined from the expression

$$m_p = m \cdot \psi, \quad (8)$$

where ψ is coefficient of accounting for rotating masses;

m is operating mass of FV, kg.

The coefficient of rotating masses accounting is determined by the known method [12]. To implement the proposed mathematical model it is necessary to use the developed data processing algorithm, shown in Figure 1.

In order to test the proposed model and the hardware-software complex, preliminary experimental studies were carried out using a passenger car as an example. The car was accelerated, that means acceleration from the minimum engine speed (at which stable motion is possible) to the maximum possible (maximum power) from second to fifth gear under conditions of very fast maximum fuel supply. This is necessary to transfer data to the hardware-software complex in order to calculate the tangential traction force on the wheel that moves the car.

Acceleration is made in both directions on the selected section of the road in order to exclude the influence of slope and wind. In addition, the road section is selected so that at the end of the acceleration there was sufficient road section for the braking distance. A fragment of the experimental data on the acceleration of the car is shown in figure 3. Data recording began when the car was not yet moving.

The first stage is preparation for acceleration. The driver starts the engine and accelerates the car to 30 ... 40 km/h, after that he engages the necessary gear, the accelerator pedal is smoothly being released until the engine speed reaches 500...800 rpm. It is necessary to withstand this mode of movement for 3..7 seconds, then press the accelerator pedal until stop. After reaching the maximum engine crankshaft speed, the driver turns on the neutral gear and waits 3 ... 4 seconds (so that the fact of acceleration end is clearly visible from the speed graph data obtained after processing the GPS data), and the car begins to brake.

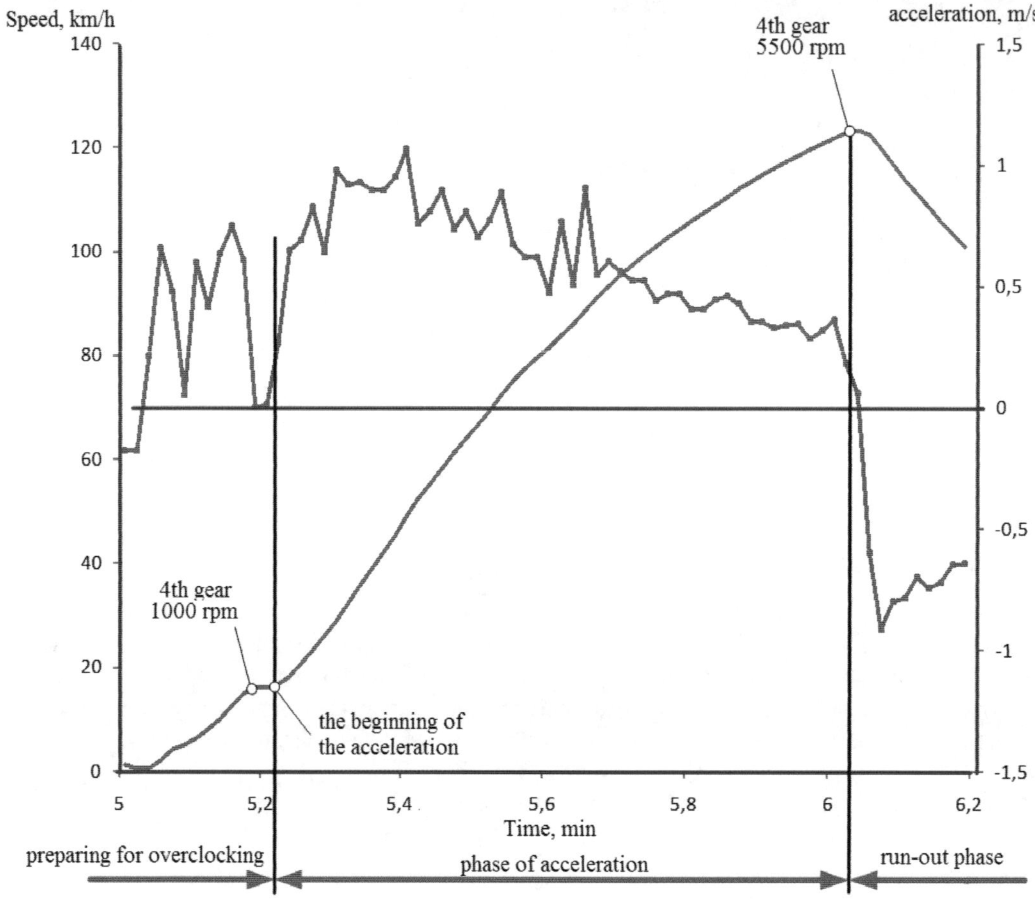

Figure 3. Acceleration of a car in fourth gear

Taking into account methodology features in order to obtain data that allow to assess more accurately characteristics of the vehicle it is advisable to accelerate it in all gears from minimum to maximum speed [12]. But in the first gear the acceleration is very short and as a result 3..4 points of the GPS graph are obtained, which does not allow to reliably determine the acceleration of the car. At higher gears: the fourth and the fifth, at maximum speed the car can accelerate to speeds of more than 140 km/h which increases both technical and administrative risks during the experiment.

Measurements of car run-out were made and experimental data fragment is presented in Figure 5. Free car run-out means the process of car rolling in neutral gear until it will stop completely. The higher the speed of accelerated car and accordingly the run-out path the more accurately rolling resistance forces can be determined. The car run-out is necessary in order to obtain the forces of resistance to movement: F_i, F_s, F_α. It is important to take into account that the maximum speed should be limited by a number of factors: length of chosen for research road section, traffic intensity, quality of coverage as well as other safety considerations. In this case a driver should try to drive the car strictly in a straight line and also avoid sharp maneuvers since they cause additional car braking. If the driver was forced to make a sharp maneuver the experiment should be repeated again. A fragment of experimental data on fixing car run-out is shown in Figure 4.

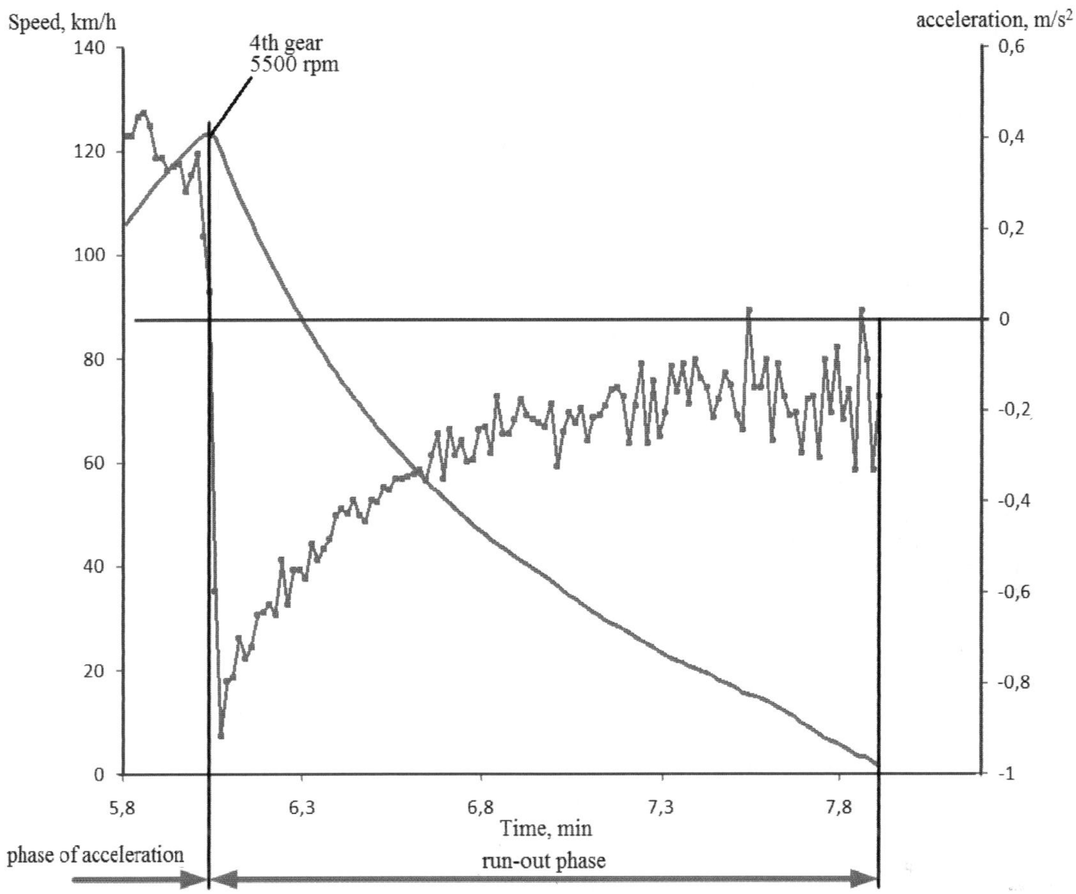

Figure 4. Car run-out

In specially worked out software that implements the developed algorithm, the resistance forces were calculated according to equation (1). The experimental data calculated by the iterative method and describing the resistance forces dependence from movement speed are shown in figure 5. Obviously, obtained dependence describes the studied regularities quite accurately.

4 Discussion

Figure 5 is of greatest practical interest with calculated data obtained by the developed algorithm. In this case, all the necessary parameters were calculated for making further decisions to improve efficiency of developed hardware-software complex. Having determined resistance strength to movement, the developed engine power and its fuel consumption indicators, which are especially accurately measured in diesel engines it is possible to build an effective system for technical and industrial operation of FT.

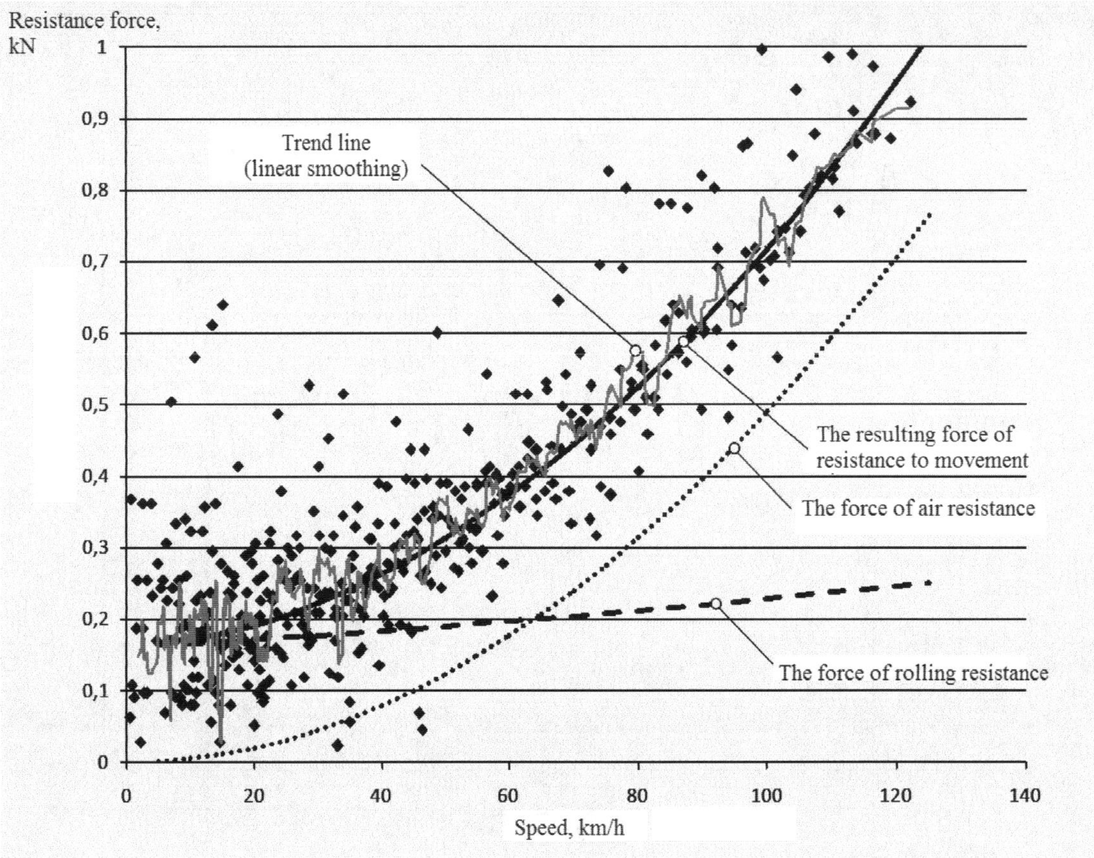

Figure 5. Experimental data of resistance forces to movement

The developed complex is intended for diagnosing vehicles, detecting a malfunction, partial loss or reduction of power characteristics and operating modes monitoring. Power is the main indicator of both operability and efficiency of FT. The consumers of the final product will be manufacturers involved in freight transportation as well as industries where it is important to monitor the technical condition of transport and analyze its performance in order to reduce costs, since one of the priority tasks is reducing the cost of provided services. In turn, fuel costs are reduced, downtime and inefficient use of FT due to breakdowns are also significantly reduced.

Conclusions

The presented mathematical model, the algorithm that implements it and preliminary experimental data confirm correctness of the chosen research direction and show the fundamental possibility of monitoring the technical condition of freight vehicles using GPS and / or GLONASS systems, a hardware-software complex and its software.

References

[1] Hu Y, Liu Y, Wang Z, Wen J, Li J and Lu J 2020 *A two-stage dynamic capacity planning approach for agricultural machinery maintenance service with demand uncertainty* (Biosystems Engineering), vol. 190, pp. 201–217, doi: 10.1016/j.biosystemseng.2019.12.005.

[2] Lebedev A and Arzhenovsky A, 2019 *Improving the efficiency of machine-tractor units* (Technical service of machines), no. 1 (134), pp. 46-52.

[3] Liu C, Lin H, Li Y, Gong L. and Miao Z., 2020 *Analysis on Status and Development Trend of Intelligent Control Technology for Agricultural Equipment* (Nongye Jixie Xuebao/Transactions of the Chinese Society for Agricultural Machinery) vol. 51, no. 1,pp. 1–18, doi: 10.6041/j.issn.1000-1298.2020.01.001.

[4] Pini M, Marucco G, Falco G, Nicola M and De Wilde W 2020 *Experimental Testbed and Methodology for the Assessment of RTK GNSS Receivers Used in Precision Agriculture* (IEEE Access), vol. 8, pp. 14690–14703, doi: 10.1109/ACCESS.2020.2965741

[5] Kazakov D, Kushcheva E, and Kugrysheva L 2015 *Method for monitoring the technical condition of mobile energy vehicles running on alternative fuels* (Autogas filling complex + alternative fuel), no. 8 (101), pp. 30-32.

[6] Shchetinin N, Arzhenovsky A, Maltsev D, Kazakov D, Morozov A 2005 *Method for determining the power of an internal combustion engine* (Patent of the Russian Federation), no. 2266527.

[7] Arzhenovsky A, Asaturyan S, Cheremisin Yu, Dolzhikov V and Chulanov A 2019 *Method for determining the power of an internal combustion engine* (Patent of the Russian Federation), no. 2682780.

[8] Arzhenovsky A, Kazakov D, Maltsev D, Asaturyan S and Mikryukov S 2008 *Device for measuring the acceleration of the crankshaft of an internal combustion engine over the entire speed range* (Patent of the Russian Federation), no. 2329510

[9] Kazakov D, Kushcheva E, and Kugrysheva L 2015 *Method for monitoring the technical condition of mobile energy vehicles running on alternative fuels* (Autogas filling complex + alternative fuel), no. 8 (101), pp. 30-32.

[10] Arzhenovskiy A, Kozlov D and Petrishchev N 2018 *Determining the traction characteristic of a tractor in operating conditions* (Agricultural machinery and technologies), 12(5). 25-30. DOI 10.22314/2073-7599-2018-12-5-25-30.

[11] Kazakov Dan Mikryukov Since 2010 *Determination of the coefficient of accounting for the rotating mass of a wheeled tractor* (Mechanization and electrification of agriculture), no. 7, pp. 26-28.

[12] Kazakov D and Kugrysheva L 2016 *Fixing and research of diagnostic parameters of cars running on alternative fuels* (Autogas filling complex + alternative fuel), no. 3 (108), pp. 14-16.

Anthropomorphic manipulator master-slave teleoperation using 3D depth sensor and color camera

V I Petrenko, F B Tebueva, M M Gurchinskiy, N Yu Svistunov, A K Nestulya

North-Caucasus Federal University, Stavropol, Russia

E-mail svistunovn4@gmail.com

Abstract. The article proposes a method for anthropomorphic manipulator teleoperation based on data processing using depth and color cameras of the Kinect controller. Processing data with Kinect allows to determine the coordinates of the nodal points of the operator's body, which, after recalculation, can be used to generate control signals for the manipulator. The article describes the algorithm and the mathematical apparatus for determining the angles of the operator's arm joints in real time. The proposed method can be used to control the joints of the manipulator of an anthropomorphic robot, and can also be adapted to control manipulators with a kinematic structure different from the structure of human arm, or to implement other control methods such as gesture control.

Keywords: Anthropomorphic manipulator, master-slave teleoperation, Kinect, depth sensor, joint angles.

1. Introduction

Recently, there has been a great interest in the topic of replacing a person when carrying out work in dangerous environments, emergency situations, and heavy routine work in industry. According to International Federation of Robotics, one of the directions in this field is the teleoperation of an anthropomorphic robot (AR) [1]. The teleoperation is based on the simultaneous motion trajectories formation for all degrees of freedom of AR manipulators using some kind of motion capturing device, e.g. an exoskeleton [2]. The use of AR teleoperation in undetermined and extreme environments is relevant because the human natural intelligence as a system for analyzing working conditions and decision-making still cannot be replaced with the artificial intelligence.

In addition to the growing demand for industrial robots in general [3], one of the trends in the development of robotics is the use of the motion tracking systems based on the depth sensors to capture operator's movements.

The aim of the work is to increase the convenience of the AR teleoperation through using motion capturing techniques. We develop an anthropomorphic manipulator teleoperation method using the Kinect controller to determine the coordinates of the operator's body nodal points. The calculations of the kinematics are based on the methods described in [4], [5].

The study [6] examined the use of sensory information for automation systems. The capabilities and applications of motion capture technologies in robotics are described in the review [7], including control systems based on depth and color cameras, which are used in agriculture for harvesting [8], in medicine [9] for rehabilitation, and in interactive control systems based on gestures [10], [11].

In [12], the Microsoft Kinect depth sensor is used to capture human movements: walking and turning. Unlike bulky motion capture suits, the sensor makes the system more convenient for the operator. Data on the position of the skeleton joints is extracted from Kinect and then processed to determine the joints angles of the robot.

2. Materials and methods

The Kinect controller (Figure 1), originally developed by Microsoft for the Xbox 360 game consoles family, is equipped with an RGB video camera, as well as an infrared emitter and receiver, which allow building a depth map and tracking object coordinates in three-dimensional space with a frequency of 30 Hz.

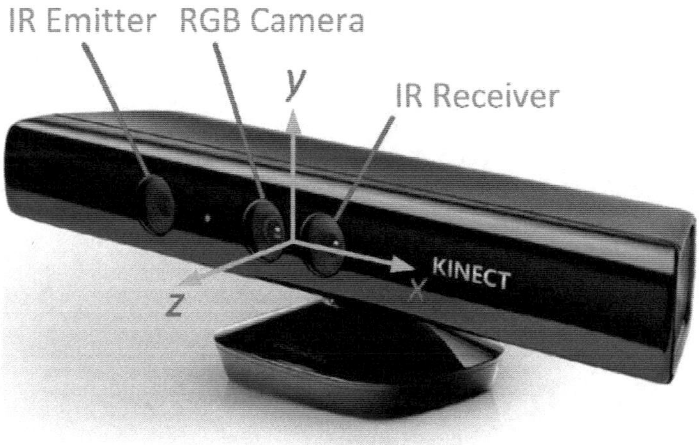

Figure 1. Sensors and coordinate system of the Kinect controller

Kinect software development kit allows to obtain the coordinates of the main nodal points of human body in real time (Figure 2). The coordinates of the right shoulder (8), left shoulder (4), as well as the elbow (5) and the wrist (6) of the operator's left arm are used to organize the teleoperation of the anthropomorphic manipulator.

Anthropomorphic manipulator teleoperation with Kinect involves the following steps:

1. Acquiring the coordinates of the shoulder, elbow and wrist of the operator.
2. Transition to the coordinate system associated with the shoulder of the left arm.
3. Operator's arm joints angles calculation.
4. Command formation for the manipulator.

The kinematic scheme of the operator's arm and the anthropomorphic manipulator is shown in Figure 3, where $B_1 - B_3$ are the humeral, ulnar, and wrist points; B_4 is the wrist center point; $A_1 - A_7$ – rotational kinematic pairs. The coordinate system $A_1x_0y_0z_0$ is considered as global. The location of the coordinate systems associated with the links is selected in accordance with the Denavit-Hartenberg (DH) representation, the DH parameters are listed in Table 1. L_1 and L_2 are lengths of B_1B_2 and B_2B_3 links respectively.

We use the following notation: ${}^{i}\mathbf{T}_j$ is homogeneous transformation matrix from the $\boldsymbol{j}$-th to the $\boldsymbol{i}$-th coordinate system, $\boldsymbol{i} < \boldsymbol{j}$, compiled in accordance with the DH representation;

$$ {}^{i}\boldsymbol{T}_j = \prod_{k=i+1}^{j} {}^{k-1}\boldsymbol{A}_k \,, \boldsymbol{i} < \boldsymbol{j} \tag{1}$$

$$ {}^{i-1}\boldsymbol{A}_i = \boldsymbol{T}_{z,\theta}(\theta_i)\boldsymbol{T}_{z,d}(d_i)\boldsymbol{T}_{x.a}(a_i)\boldsymbol{T}_{x,\alpha}(\alpha_i), \tag{2}$$

$$ \boldsymbol{T}_{z,\theta}(\theta) = \begin{bmatrix} \cos(\theta) & -\sin(\theta) & 0 & 0 \\ \sin(\theta) & \cos(\theta) & 0 & 0 \\ 0 & 0 & 1 & 0 \\ 0 & 0 & 0 & 1 \end{bmatrix}, \tag{3}$$

$$\boldsymbol{T}_{z,d}(d) = \begin{bmatrix} 1 & 0 & 0 & 0 \\ 0 & 1 & 0 & 0 \\ 0 & 0 & 1 & d \\ 0 & 0 & 0 & 1 \end{bmatrix}, \quad (4)$$

$$\boldsymbol{T}_{x.a}(a) = \begin{bmatrix} 1 & 0 & 0 & a \\ 0 & 1 & 0 & 0 \\ 0 & 0 & 1 & 0 \\ 0 & 0 & 0 & 1 \end{bmatrix}, \quad (5)$$

$$\boldsymbol{T}_{x,\alpha}(\alpha) = \begin{bmatrix} 1 & 0 & 0 & 0 \\ 0 & \cos(\alpha) & -\sin(\alpha) & 0 \\ 0 & \sin(\alpha) & \cos(\alpha) & 0 \\ 0 & 0 & 0 & 1 \end{bmatrix}, \quad (6)$$

where ${}^{i-1}\mathbf{A}_i$ is the homogeneous matrix of the complex transformation for adjacent coordinate systems; $\mathbf{T}_{z,\theta}(\theta)$ is homogeneous matrix of elementary rotation about the z axis by the angle θ; $\mathbf{T}_{z,d}(d)$ is homogeneous matrix of elementary shift along the z axis by the distance d; $\mathbf{T}_{x.a}(a)$ is homogeneous matrix of elementary shift along the x axis by the distance a; $\mathbf{T}_{x,\alpha}(\alpha)$ is homogeneous matrix of elementary rotation about the x axis by the angle α.

$\mathbf{T}_i$ is the matrix of transformation from the $\boldsymbol{i}$-th coordinate system into the global coordinate system. This matrix can be found by the following formula:

$$\boldsymbol{T}_i = {}^{0}\boldsymbol{T}_i, \boldsymbol{i} > 0. \quad (7)$$

In the text the first, the second and the third components of a certain vector $\boldsymbol{V}$ are denoted as $\boldsymbol{V_x}$, $\boldsymbol{V_y}$, $\boldsymbol{V_z}$, respectively.

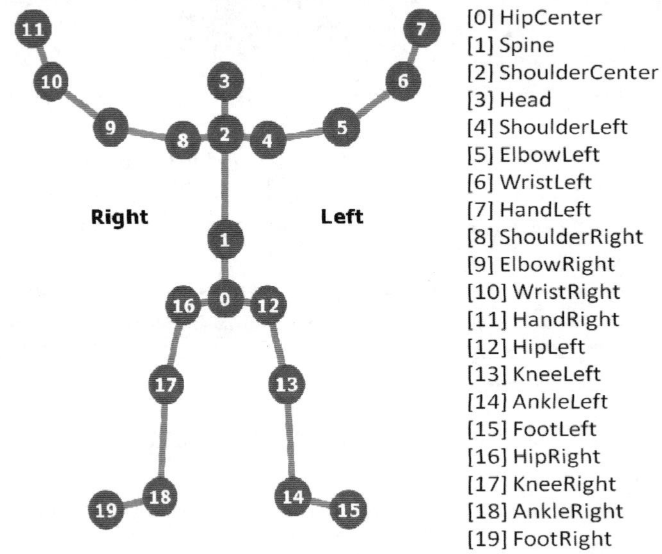

Figure 2. Skeleton joint points captured by the Kinect controller

Table 1. DH parameters of the manipulator and operator's arm

Link	$\boldsymbol{\theta}_i$, degrees	$\boldsymbol{d}_i$, cm	$\boldsymbol{a}_i$, cm	$\boldsymbol{\alpha}_i$, degrees	Range of $\boldsymbol{\theta}_i$, degrees
1	0	0	0	-90	-180…90
2	−90	0	0	90	-90…90
3	90	L_1	0	90	0…180
4	90	0	0	90	20…180
5	90	L_2	0	90	0…180

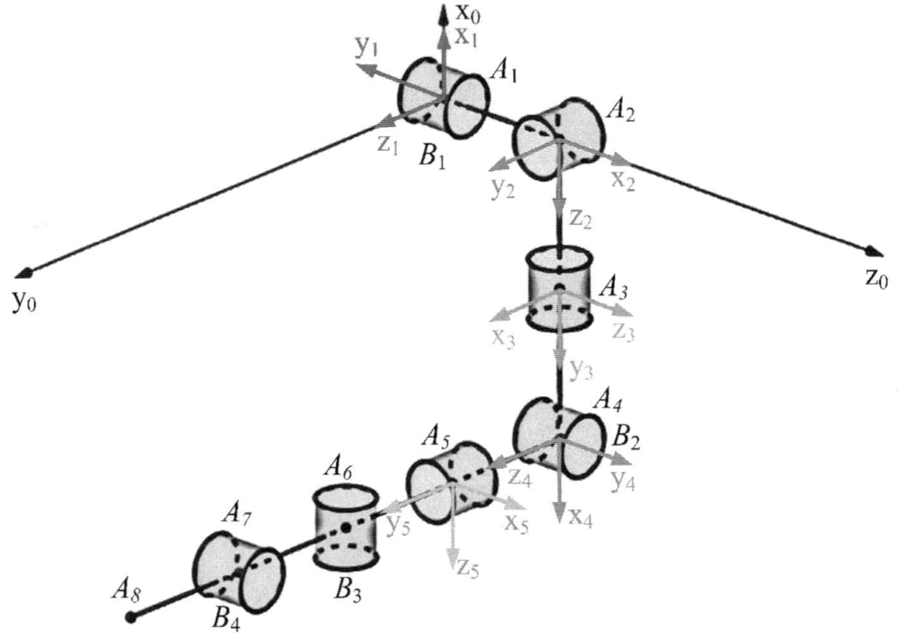

Figure 3. Kinematic scheme of the operator's arm and anthropomorphic manipulator

3. Operator's arm joint angles calculation

The radius vectors of points B_1, B_2, and B_3 in the coordinate system of the Kinect controller are denoted as $\boldsymbol{B}_1''$, $\boldsymbol{B}_2''$, $\boldsymbol{B}_3''$ respectively, and the radius vector of the right shoulder is denoted as $\boldsymbol{B}_0''$. It is assumed that the operator is facing the Kinect. In this case, the vector connecting the right and left shoulders may not be parallel to the x axis of the Kinect coordinate system (but it should be parallel to the xz plane). Therefore, to go to the global coordinate system $A_1x_0y_0z_0$, it is necessary to subtract the vector $\boldsymbol{B}_1''$ from $\boldsymbol{B}_2''$ and $\boldsymbol{B}_3''$, and rotate the resulting vectors around the y axis by the angle opposite to the angle φ between the projection of the vector $\boldsymbol{B}_1''\boldsymbol{B}_0''$ to the xz plane and the positive direction of the x axis:

$$\varphi = -\mathrm{atan2}(\boldsymbol{B}_{0x}'' - \boldsymbol{B}_{1x}'', \boldsymbol{B}_{0z}'' - \boldsymbol{B}_{1z}''), \tag{8}$$

where $\mathrm{atan2}(a, b)$ is the function that returns the angle between the radius vector of a point (a, b) and the positive direction of the abscissa axis, while the angle is counted in the direction from the abscissa axis to the ordinate axis and has a value in range from $-180°$ to $180°$.

The rotation matrix about the y axis by the angle α is

$$\boldsymbol{M}_y(\alpha) = \begin{pmatrix} \cos\alpha & 0 & \sin\alpha \\ 0 & 1 & 0 \\ -\sin\alpha & 0 & \cos\alpha \end{pmatrix}. \tag{9}$$

The transformed radius vectors $\boldsymbol{B}_1''$ and $\boldsymbol{B}_2''$ are denoted by $\boldsymbol{B}_2'$ and $\boldsymbol{B}_3'$, respectively:

$$\boldsymbol{B}_2' = \boldsymbol{M}_y(-\varphi)(\boldsymbol{B}_2'' - \boldsymbol{B}_1''), \tag{10}$$

$$\boldsymbol{B}_3' = \boldsymbol{M}_y(-\varphi)(\boldsymbol{B}_3'' - \boldsymbol{B}_1''). \tag{11}$$

The relative orientation of the Kinect coordinate system and the global coordinate system $A_1x_0y_0z_0$ after performing the offset and rotation transformations is shown in Figure 4. Thus, the homogeneous coordinates $\boldsymbol{B}_1^0$, $\boldsymbol{B}_2^0$, $\boldsymbol{B}_3^0$ of the radius vectors of points B_1, B_2 and B_3 in the global (zero) coordinate system are

$$\boldsymbol{B}_1^0 = (0 \quad 0 \quad 0 \quad 1)^T, \tag{12}$$

$$\boldsymbol{B}_2^0 = (\boldsymbol{B}_{2y}' \quad -\boldsymbol{B}_{2z}' \quad -\boldsymbol{B}_{2x}' \quad 1)^T, \tag{13}$$

$$\boldsymbol{B}_3^0 = (\boldsymbol{B}_{3y}' \quad -\boldsymbol{B}_{3z}' \quad -\boldsymbol{B}_{3x}' \quad 1)^T. \tag{14}$$

IOP Conf. Series: Materials Science and Engineering **873** (2020) 012028 doi:10.1088/1757-899X/873/1/012028

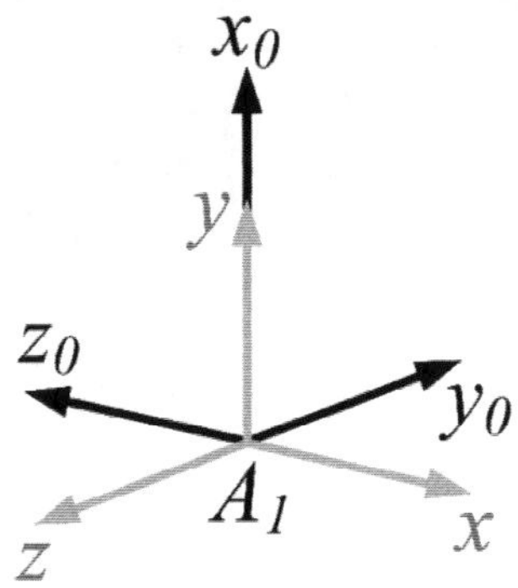

Figure 4. Kinect coordinate system relation to the global coordinate system

The angle θ_1 is calculated by the formula

$$\theta_1 = \text{atan2}(\boldsymbol{B}^0_{2x}, \boldsymbol{B}^0_{2y}) - 180°. \tag{15}$$

With known angle θ_1, the transition to the coordinate system associated with the first joint can be performed:

$$\boldsymbol{B}^1_2 = (\boldsymbol{T}_1)^{-1}\boldsymbol{B}_2, \tag{16}$$

The angle θ_2 is the angle between the positive directions of the axes x_1 and x_2 and can be found from the coordinates of the radius vector $\boldsymbol{B}^1_2$:

$$\theta_2 = \text{atan2}(\boldsymbol{B}^1_{2x}, \boldsymbol{B}^1_{2y}) + 90°. \tag{17}$$

The vector $\overline{B_2B_3}$ in the coordinate system associated with the second link is denoted as $\overline{\boldsymbol{B}^2_2\boldsymbol{B}^2_3}$ and can be found according to formula

$$\overline{\boldsymbol{B}^2_2\boldsymbol{B}^2_3} = (\boldsymbol{T}_2)^{-1}(\boldsymbol{B}^0_3 - \boldsymbol{B}^0_2). \tag{18}$$

The angle θ_3 can be calculated as the angle between the projection of the vector $\overline{\boldsymbol{B}^2_2\boldsymbol{B}^2_3}$ onto the x_2y_2 plane and the positive direction of x_2 axis:

$$\theta_3 = \text{atan2}\left(\left(\overline{\boldsymbol{B}^2_2\boldsymbol{B}^2_3}\right)_x, \left(\overline{\boldsymbol{B}^2_2\boldsymbol{B}^2_3}\right)_y\right). \tag{19}$$

The value of θ_4 is the angle between vectors $\overline{B_2B_3}$ and $\overline{B_2B_1}$:

$$\theta_4 = \arccos\left(\frac{(\boldsymbol{B}^0_3 - \boldsymbol{B}^0_2)\cdot(\boldsymbol{B}^0_1 - \boldsymbol{B}^0_2)}{|\boldsymbol{B}^0_3 - \boldsymbol{B}^0_2|\cdot|\boldsymbol{B}^0_1 - \boldsymbol{B}^0_2|}\right). \tag{20}$$

When calculating the angles θ_1, θ_2, θ_3 using formulas (15)–(19), the angle θ_3 can be outside the range $[0; 180°]$. This is due to the fact the angles θ_1 and θ_2 may be found based on B_2 coordinates in more than one way. An invalid value of θ_3 indicates that the wrong solution was chosen. In this case, the values of θ_1, θ_2 and θ_3 should be adjusted as follows:

$$\theta_1^* = \begin{cases} \theta_1 \; if \; \theta_3 > 0, \\ \theta_1 + 180° \; otherwise; \end{cases} \tag{21}$$

$$\theta_2^* = \begin{cases} \theta_2 \; if \; \theta_3 > 0, \\ -\theta_2 \; otherwise; \end{cases} \tag{22}$$

$$\theta_3^* = \begin{cases} \theta_3 \; if \; \theta_3 > 0, \\ \theta_3 + 180° \; otherwise, \end{cases} \tag{23}$$

where θ_1^*, θ_2^*, θ_3^* are corrected values of the angles θ_1, θ_2, θ_3.

After performing these calculations, a command for the manipulator is generated. It contains vector $\boldsymbol{\theta}$ of the joints angles:

$$\boldsymbol{\theta} = \{\theta_1^*, \theta_2^*, \theta_3^*, \theta_4\}. \tag{24}$$

Next, the calculation process is repeated for newly incoming input datasets from the Kinect controller.

4. Discussion

The proposed method for determining the angles of the operator's arm joints using the Kinect controller allows to perform real-time anthropomorphic manipulator teleoperation. However, this method does not involve control of the end effector of the manipulator due to the difficulty of achieving acceptable accuracy. This problem can be solved through the use of haptic gloves.

When the elbow angle value θ_4 is close to 180°, the accuracy of determining the angle θ_3 can be reduced significantly, since the length of the vector $\overline{\boldsymbol{B}_2^2\boldsymbol{B}_3^2}$ projection onto the plane x_2y_2 becomes close to zero. In this case, the angle θ_3 can be considered equal to the default value from Table 1, or fixed at a value calculated before the angle θ_4 crossed the threshold (for example, 165°). Also in this case θ_3 can be forecasted [13].

The depth and RGB cameras may be applied as an additional mean to the exoskeleton-based teleoperation systems [14] and during the human-robot cooperation in uncertain working environment [15-16].

5. Conclusion

In this paper the method for determining the joints angles of the operator's arm in real time using the Kinect controller was proposed. The method can be used to implement anthropomorphic manipulators teleoperation. It does not require expensive wearable equipment and makes the teleoperation more convenient for the operator. Also, it can be modified to control manipulators with a kinematic structure different from the structure of human arms, and to develop other control methods based on motion capture (e.g. using gestures).

References

[1] I. M. Kutlubaev, I. G. Zhydenko, and A. A. Bogdanov, "Basic concepts of power anthropomorphic grippers construction and calculation," in 2016 2nd International Conference on Industrial Engineering, Applications and Manufacturing, ICIEAM 2016 - Proceedings, 2016, doi: 10.1109/ICIEAM.2016.7910963.

[2] A. A. Bogdanov, A. F. Permyakov, and I. M. Kutlubaev, "Basic Principles of an Exoskeleton Construction for an Anthropomorphic Manipulator Control," in 2019 International Multi-Conference on Industrial Engineering and Modern Technologies, FarEastCon 2019, 2019, doi: 10.1109/FarEastCon.2019.8934739.

[3] "Executive Summary World Robotics 2019 Industrial Robots." [Online]. Available: https://ifr.org/downloads/press2018/Executive Summary WR 2019 Industrial Robots.pdf. [Accessed: 19-Mar-2020].

[4] V. I. Petrenko, F. B. Tebueva, M. M. Gyrchinsky, V. O. Antonov, and J. A. Shutova, "The method of forming a geometric solution of the inverse kinematics problem for chains with kinematic pairs of rotational type only," in IOP Conference Series: Materials Science and Engineering, 2018, vol. 450, no. 4, doi: 10.1088/1757-899X/450/4/042016.

[5] V. I. Petrenko, S. S. Ryabtsev, F. B. Tebyeva, V. O. Antonov, and A. S. Pavlov, "Determination of the spatial position and orientation of the links of the robot anthropomorphic grip by the solution of the direct and inverse kinematics problem," in CEUR Workshop Proceedings, 2018, vol. 2254, pp. 94–104.

[6] M. Pongratz and K. Mironov, "Accuracy of positioning spherical objects with a stereo camera system," in Proceedings of the IEEE International Conference on Industrial Technology, 2015, vol. 2015-June, no. June, pp. 1608–1612, doi: 10.1109/ICIT.2015.7125326.

[7] "Top 10 Robotic Kinect Hacks - IEEE Spectrum." [Online]. Available: https://spectrum.ieee.org/automaton/robotics/diy/top-10-robotic-kinect-hacks. [Accessed: 19-Mar-2020].

[8] R. K. Megalingam, G. V. Vivek, S. Bandyopadhyay, and M. J. Rahi, "Robotic arm design, development and control for agriculture applications," in 2017 4th International Conference on

Advanced Computing and Communication Systems, ICACCS 2017, 2017, doi: 10.1109/ICACCS.2017.8014623.

[9] Y. Bouteraa, I. Ben Abdallah, and A. M. Elmogy, "Training of Hand Rehabilitation Using Low Cost Exoskeleton and Vision-Based Game Interface," J. Intell. Robot. Syst. Theory Appl., vol. 96, no. 1, pp. 31–47, Oct. 2019, doi: 10.1007/s10846-018-0966-6.

[10] I. Ben Abdallah, Y. Bouteraa, and C. Rekik, "Kinect-Based Sliding Mode Control for Lynxmotion Robotic Arm," Adv. Human-Computer Interact., vol. 2016, 2016, doi: 10.1155/2016/7921295.

[11] S. Shirwalkar, A. Singh, K. Sharma, and N. Singh, "Telemanipulation of an industrial robotic arm using gesture recognition with Kinect," in CARE 2013 - 2013 IEEE International Conference on Control, Automation, Robotics and Embedded Systems, Proceedings, 2013, doi: 10.1109/CARE.2013.6733747.

[12] A. Sripada et al., "Teleoperation of a Humanoid Robot with Motion Imitation and Legged Locomotion," in ICARM 2018 – 2018 3rd International Conference on Advanced Robotics and Mechatronics, 2019, pp. 375–379, doi: 10.1109/ICARM.2018.8610719.

[13] V. I. Petrenko, F. B. Tebueva, M. M. Gurchinsky, V. O. Antonov, and A. S. Pavlov, "Predictive assessment of operator's hand trajectory with the copying type of control for solution of the inverse dynamic problem," SPIIRAS Proc., vol. 18, no. 1, pp. 123–147, 2019, doi: 10.15622/sp.18.1.123-147.

[14] A. A. Bogdanov, A. F. Permyakov and I. M. Kutlubaev, "Basic Principles of an Exoskeleton Construction for an Anthropomorphic Manipulator Control," 2019 International Multi-Conference on Industrial Engineering and Modern Technologies (FarEastCon), Vladivostok, Russia, 2019, pp. 1-4. doi: 10.1109/FarEastCon.2019.8934739

[15] Emanuele Magrini, Federica Ferraguti, Andrea Jacopo Ronga, Fabio Pini, Alessandro De Luca, Francesco Leali, Human-robot coexistence and interaction in open industrial cells, Robotics and Computer-Integrated Manufacturing, Volume 61, 2020, 101846, ISSN 0736-5845, https://doi.org/10.1016/j.rcim.2019.101846.

[16] E. Magrini and A. De Luca, "Human-robot coexistence and contact handling with redundant robots," 2017 IEEE/RSJ International Conference on Intelligent Robots and Systems (IROS), Vancouver, BC, 2017, pp. 4611-4617. doi: 10.1109/IROS.2017.8206331.

Research of wing in surface effect ship configuration and development of an experimental model of increased stability

I M Dantsevich, V N Prasolov, M N Lyutikova

Admiral F. F. Ushakov Maritime State University, Novorossiysk, Lenin Avenue 93, 353918, Russia

E-mail dantsevich.aumsu@gmail.com

Abstract. The article provides an analysis of the foundations of wing-in-surface-effect ship (WISE) development and describes the experiments conducted using the constructed three radiocontrolled and one virtual model of the WISE. The main problems identified during the experiments are described, as well as further prospects for the development of the project.

Keywords: wing in surface effect ship (WISE), ground effect, hydroaircenter, duplex tandem, wing ship, virtual modeling, inductive vortex

1. Introduction

Currently, existing modes of transport effectively use the conditions of the elements, but have the potential for development. The key to the development of existing transport is to find solutions to the main problems:

1. Limited movement speed of vessels in the water environment;
2. Large energy costs for overcoming gravity and preparing special areas for take-off, landing and maintenance of aircraft;
3. Limitations of the earth's surface and heterogeneity of cover when using land transport.

Actual experience in application of the laws of aero-and hydrodynamics has the potential for the development of a new type of transport with a synergistic use of air space, water and land surface. An example of this synergy is the movement of wing-in-surface-effect ship (WISE) in the air, but using the proximity of the earth or water surface:

1. The increase in lift capacity[1];
2. Reducing drag and increasing speed [1];
3. The ability to move over any smooth surface regardless of its composition [2, 3];
4. Safe height and possibility of smooth descent in case of emergency situations [2, 3].

At the moment, there are 4 theories of possible use of WISE:

1. Transportation of people and goods [5];
2. Rescue service [6];
3. Scientific and military surveillance [6];
4. Shock military systems [6].

In view of the existing potential for creating and implementing a new type of transport, 2 goals were set in the study. Firstly, the long-term goal is to create and implement WISE as a new type of transport that more effectively uses the potentials of water and air elements. Secondly, the goal of this stage of work is to study the existing layouts of WISE and create an experimental model of increased stability.

2. Methods for increasing the stability of various schemes of WISE configuration

To achieve the goals of the current stage of the research, the following methods were applied:

1. Collecting and summarizing information about existing WISE projects based on books, articles.
2. Development of prototypes of a new vehicle in the form of radio-controlled and 3D models.
3. Testing of prototypes for stability in the cruise mode during launches of radio-controlled models over a flat earth surface and virtual blowing of models in a virtual environment simulating the movement of the device above the surface 10 cm height.

In the long term, with the accumulation of sufficient information and experience, it is planned to create, test, refine and implement the effective WISE.

During the collection of information, it was found out that at the moment there are 3 lines of WISE development: "airplane", "back-deltoid", "tandem", each of which is developed in connection with the work of the corresponding designers: R. E. Alexeev, A.M. Lippish and G. V. Yorg. Prominent representatives of these lines are the machines: KM [2], [3], X-113 [7] and TAB 8 [8]. At the same time, the development of WISE is currently carried out mainly according to the first two types of layouts. For example, the aircraft scheme is developed in the domestic company RDC-Aqualines [9], and the second type of WISE is actively developed by the South Korean firm WingShipTechnology [10] and the Singapore firm WigetWorks [11].

Common features of the described devices are:

1. Movement at a height of 1 wing chord
2. Converting the incoming flow into a dynamic air cushion
3. Use of combinations of air and underwater means of stabilization and control.

For a visual introduction to the ground-effect, an experimental radio-controlled model of the hydroair-center HAC-1 was assembled, the design of which includes:

- the main fuselage, which is a flat-convex profile with a chord of 0.48 m and a span of 0.52 m, with a relative thickness of 10% and a maximum thickness at a distance of 25% from the leading edge.
- U-shaped tail stabilizer for yaw and pitch stabilization, as well as for the ability to mount the engines above the plane of the main fuselage.
- elevators and rudders,
- 2 brushless electric motors,
- additional wings with a positive V-shape to stabilize the machine on the roll and to install the ailerons at a distance from the center plane to create a sufficient moment of roll control.
- a rotary flap that allows you to transform the wing profile by changing the angle of deflection of the flow coming off the wing.
- catamaran floats-washers for limiting the minimum flight height of the carrier plane, reducing the area of contact between the hull and the underlying surface, and for creating an obstacle to the formation of inductive vortices on the carrier fuselage-wing.

Figure 1. HAC-1 model

During the HAC -1 tests, the calculated cruise mode was achieved, while low stability in pitch and roll was detected, which was manifested in leaving the calculated mode when the slightest disturbances in the form of wind gusts or changes in terrain occurred (Figure 1).

Additionally, the difficulty of maneuvering was revealed due to the deterioration of the ground-effect when trying to perform turns.

Taking into account the test experience, the experimental device HAC-2 was developed, which was based on the previous device, but the wing chord was doubled and a more forward alignment was installed (Figure 2).

Figure 2. HAC-2 model

HAC-2 after adjusting the S-shape by deflecting the flow flap to the 1-2 degrees top, the device showed greater stability in pitch, but it remained on the ground-effect only in case of active control. During uncontrolled flight, the device either descended to the ground, or went to a height of more than 1 chord. However, compared to the previous model, this device showed smoother fluctuations in pitch and roll, allowing the pilot to control the flight more accurately on the ground-effect.

The turns were performed only in the mode of increased engine power and in conditions of air flight in isolation from the ground-effect. However, compared to the previous model, this device failed to a lesser extent at the time of the roll and more easily moved to horizontal flight after performing the maneuver.

In parallel with the construction and testing of models it was revealed that the first models of designers who were engaged in WISE faced similar problems of stability in pitch.

For example, one of Alekseyev's experimental devices was named "the frog" [3] due to the appearance of oscillatory pitch changes that looked like jumps on a water surface. The same problems were encountered by the designer Troeng on the "Aerobot" machine [1].

It is known that the problem with pitch stabilization in WISE occurs due to the movement of the pressure center along the wing. The research given in the book "Airfoil boat" by Belavin clearly expressed the inverse relationship of the pressure center distance of the wing leading edge from the height of flight [1]. Thus, the higher the wing flies, the greater the moment of force that tends to increase the angle of attack and lift the device. This tendency manifests itself in a clear instability of the device in pitch and requires additional balancing forces.

In the WISE history a lot of techniques of pitch stabilization was applied, which represent:

1. Developed horizontal fins, raised high above the boat and maximally moved away from the plane of the wing forward or backward
2. S – shaped wing profile
3. Duplex tandem wing configuration
4. Reverse delta wing configuration
5. Means of contact with the surface (sledges, skis, hydrofoils, wheels, etc.)

To study of the proven methods of WISE stabilization by pitch, a radio-controlled model of the HAC-3 was built according to the scheme of the Lippisch WIG series with a reverse delta wing and a developed tail fins. At the same time, to speed up the work, the project was greatly simplified, which showed:

- Model span 0.5 m;
- Flat bent wing profile made of single-layer polystyrene foam plate;
- Placing controls and electronics directly on the upper plane of the devive.

At the same time, the main conceptual features of Lippisch models were preserved.

Figure 3. HAC-3 model

During the tests, the HAC-3 model showed greater pitch stability than the previous two models, which was manifested in an easier hold on the cruise mode (Figure 3). However, when interference occurred, the model also left the ground-effect and required intervention in the control to return to the cruise mode.

3. Development of virtual HAC model

As a result of testing 3 models we made a conclusion that in order to simplify pitch stabilization methods, you need a second pivot point on the ground-effect. This method was used by R. E. Alekseev when building his first SM1 and SM 2, as well as by Gunter V. Jorg, who built a whole series of devices with such component. We decided to design the fourth model according to the duplex tandem scheme, presented on the Figure 4.

During the calculations of the layout of the 4th model, it was decided to conduct initial design and testing in the virtual environment of the Solid Works program. It is assumed that in the virtual environment, even before the construction stage, it will be possible to design and test in more detail a layout that will hold the device on the ground-effect without interference with the control.

The layout of the HAC-4 apparatus was developed based on the ideas of Gunter V. Jorg about duplex-tandem with a convex-concave profile of the front wing and an asymmetric biconvex profile of the rear wing, as well as the ideas of R. E. Alekseev about a longitudinal cargo fuselage and enhanced engine blowing under the wing in the launch mode.

Initially, attention was focused on the study of the wings interaction in duplex-tandem, in this regard, many parts of the body were not installed on the layout. Taking into account the possibilities of virtual modeling, a full-scale size was chosen (span 7.87 m, length 16m, height 2m).

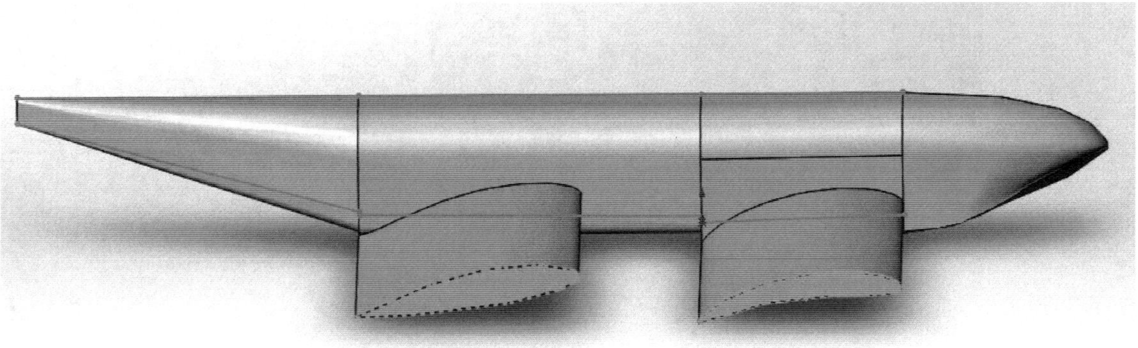

Figure 4. HAC-4 model

The model was blown by a stream of air at a speed of 120 m/s without attack angle at 0.1 m height above the surface.

A large inductive vortex has been identified that occurs at the ends of both wingtip and especially is demonstrated on the hind wing (Figure 5). The vortex disrupted the flow, shifting the air flow to the side and created a difference in lift between the front and rear wings. As a result of this flow, the model in real conditions would lower the rear wing and raise the front one, which would lead to destabilization of the model in pitch.

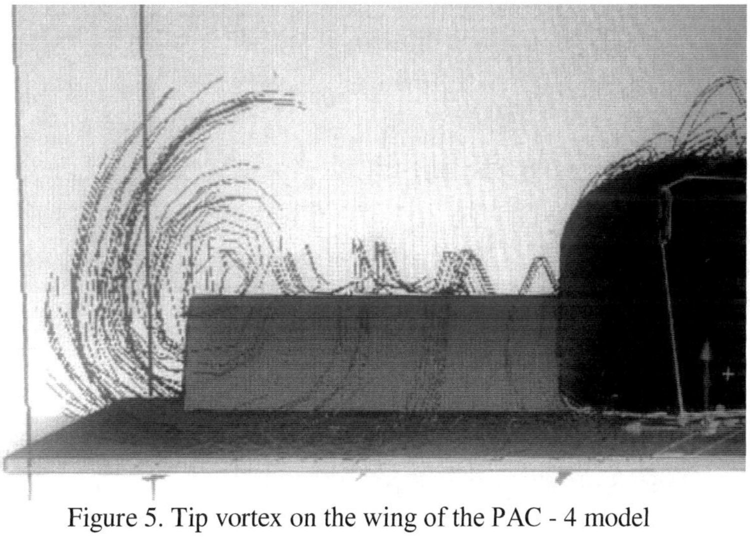

Figure 5. Tip vortex on the wing of the PAC - 4 model

To minimize inductive vortices, an endplate was designed that connects both wings, descends to the surface level and rises above the wing by an average of 0.25 m.

After installing the endplate, the end vortices decreased and the flow pattern of the wings became more longitudinal, providing sufficient and smooth flow around both the front and rear wings, as shown in Figures 6-7.

Figure 6. Flow around the upper wings of the HAC - 4 model

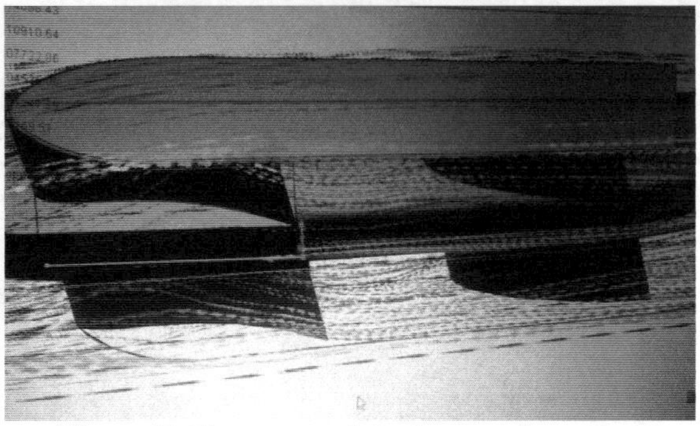
Figure 7. Flow around the lower wings planes of the HAC - 4 model

The phenomenon of inductive vortices is compensated by the installation of endplates of different shapes and length, as well as an increase in wing elongation. However, in this work the effect of end vortices clearly showed us the reality of virtual blowing methods, as they clearly demonstrated a phenomenon previously known to us only in theory.

The flow nature of the front and rear wings demonstrated the possibility of normal operation of two wings running behind each other at the same height, which again clearly proved the reality of using a tandem layout for the development of the future device. Project work HAC-4 gave us our first experience of developing and testing the model in a virtual environment, but the main part of the work on the model is still ahead.

4. Conclusion

All the phenomena and theoretical justifications discussed in this paper have already been described in the literature related to aircraft and ekranoplan development. However, it was important for us to get acquainted with the described phenomena in real experiment.

In this regard, we have made 4 important steps of our work:

1. reviewed the basics of WISE development;
2. experimentally tested the ground effect on three real and one virtual models;
3. in practice, we have identified problems of pitch stabilization;
4. initiated the research and use of methods for virtual design and testing of ekranoplan models.

Future research is planned to obtain:

– virtual tests with the identification of specific numerical values of the model's dive and nose up at different speed, height and angles of attack;

– adjusting the layout and working out the additional parts of the model;

– creation and research of a reduced radio-controlled prototype based on a machine developed in a virtual environment;

– testing of radio-controlled models with the revealing of specific digital data on the ground effect depending on the layout, speed, altitude, angle of attack, alignment and other factors.

References

[1] N I Belavin 1977 *Ekranoplany L.:* Sudostroenie 232

[2] Maskalik A Nagapetyan R Lukyanov A 2003 *Ekranoplans - transport vessels of the future* (Saint Petersburg: Sudostroenie) 352

[3] P Kachur 2008 Ekranoplans: past, present, future *Ekran. Past, Present. Futur. / / Equip. armament* **3** 29–36

[4] Yun L and Bliault A 2012 *High Performance Marine Vessels* 362

[5] R Ertekin 2016 Lawrence J. Doctors: Hydrodynamics of high-performance marine vessels *Journal of Ocean Engineering and Marine Energy* 2 **2** 247-248

[6] Akimoto H, Kubo S and Kanehira M 2010 Wing in surface effect ship with canard configuration *Int. J. Aerodyn.* 1 **1** 3

[7] Yun L, Bliault A and Doo J 2010 *WIG craft and ekranoplan: Ground effect craft technology* 1-450

[8] Wang H, Teo C J, Khoo B C and Goh C J 2013 Computational aerodynamics and flight stability of Wing-In-Ground (WIG) craft *Procedia Engineering* **67**15–24

[9] Tavakoli Dakhrabadi M and Seif M S 2018 Hydro-aerodynamic mathematical model and multi-objective optimization of wing-in-ground effect craft in take-off *Proc. Inst. Mech. Eng. Part M J. Eng. Marit. Environ.* 232 **4** 421-433

[10] Liang H, Zhou L, Zong Z and Sun L 2013 An analytical investigation of two-dimensional and three-dimensional biplanes operating in the vicinity of a free surface *J. Mar. Sci. Technol.* 18 **1** 12-31.

[11] Marshall D W, Newman S J and Williams C B 2010 Boundary layer effects on a wing in ground-effect *Aircr. Eng. Aerosp. Technol.* 82 **2** 99-106

The selection of parameters and control points in the geotechnical monitoring system

N V Dorofeev, A V Grecheneva, R V Romanov, E S Pankina

Vladimir State University, Faculty of Radio-electronics and Computer Systems, 23 Orlovskaya str., Murom, Vladimir Region, 602264, Russia

E-mail dorofeevnv@yandex.ru

Abstract. This article proposes an approach to the organization of the monitoring of geotechnical systems, which allows to increase the effectiveness of the automated control of destructive processes in geotechnical systems. The effectiveness of control is achieved by optimizing economic costs and technical constraints. The technical parameters of an automated geotechnical monitoring system are determined by the cost and importance of controlled objects. The criterion of optimality is the criterion of minimizing costs when implementing a geotechnical monitoring system with a constant value of possible damage. Possible damage in the geotechnical system during the development of destructive processes is expressed in monetary terms. The costs of implementing a geotechnical monitoring system are determined based on its cost, fixed costs for its maintenance, as well as technical parameters. The technical parameters in assessing the cost of implementing a geotechnical monitoring system are: the cost of measuring one controlled parameter, the number of measurement points, the cost of skipping the development of destructive geotechnical processes, the cost of a false alarm system. The proposed approach to optimizing the cost of implementing a geotechnical monitoring system is based on the monitoring method at key control points with a choice of bifurcation parameters. The proposed approach for optimizing the geotechnical monitoring system was tested in the city of Murom, Vladimir Region, and the Russian Federation while observing the development of suffusion processes of technogenic origin. The results of practical testing showed the possibility of reducing the cost of monitoring organization while maintaining control accuracy.

Keywords: registration, monitoring, geotechnical system, technical and economic analysis

1. Introduction

The development of the technosphere is accompanied by a change in the quantity and strength of the relationships between components of the natural and of the technical environment. Changes lead to accidents and disasters of various sizes when these changes negatively affect the stability of geotechnical systems [1-3]. Systems of the monitoring of the condition of geotechnical systems are built on two main approaches. One approach is to organize observations of changes in the geological environment, to isolate and predict the initial stage of the development of dangerous geodynamic processes. This approach takes into account hydrological, geological, climatic and other parameters. Another approach is to monitor technical and operational parameters of structures, identify and predict destructive processes, and evaluate the stability parameters of structures [4-6]. Different of technical, of organizational and of economic issues always arise in practice despite the advantages and disadvantages of both approaches. For example, what parameters of the geotechnical system need to be registered, what technical means should be used to register selected parameters, how to place technical means and how to collect data, how many sensors are needed and what places it is necessary to measure, what costs will be incurred, what will be the benefit, and etc.

The purpose of this work is to increase the efficiency of geotechnical monitoring systems in the conditions of the risk of the disturbance of the geodynamic stability and economic restrictions on the organization of monitoring work by developing the method of registering parameters of a geotechnical system based on data of the technical and economic analysis.

2. The method and approaches

The economic efficiency of the applying of the geotechnical monitoring system is achieved by reducing damage from the negative development of processes in the geotechnical system by reducing the measurement error and the probability of the false detection of precrisis situations in the geotechnical system, by increasing the likelihood of the correctly detecting of precrisis situations and predicting them at an earlier stage [7]. Thus, the economic efficiency of the introducing of the geotechnical monitoring system depends on the metrological, methodological and algorithmic support of the latter.

It should be noted that the economic efficiency from implementation will differ in the case of the organization of monitoring work on the basis of the monitoring system of the same type in various geotechnical systems. This difference can be several orders of magnitude. This is due to individual characteristics of the functioning of geotechnical systems and processes occurring in them, which (individual features) are not reflected or which are not fully reflected in the functioning of geotechnical monitoring systems. As a rule, geotechnical monitoring systems are developed for a specific object (geotechnical system) and have low adaptability to qualitative and quantitative changes in the geotechnical system [8].

At the beginning of the technical and economic analysis the mapping of economic damages is developed with the division into zones of significance and with the combining of components of the geotechnical system on objects of various levels (local, local, regional). For each zone, the possible economic damage and the significance of this zone are determined for the functioning of the geotechnical system taking into account the possible social criteria (1) and (2). Clustering of the map of zones is based on an expanding neural gas algorithm [9].

$$C_{GD} = \sum_{i=1}^{n} C_{iD} , \tag{1}$$

where C_{GD} is the total economic damage in the geotechnical system; C_{iD} is the economic damage of the separate area of the geotechnical system; n is number of zones.

$$C_{iD} = \sum_{j=1}^{m} (C_j(W_j)P_j r_j + S_j(r_j)), \tag{2}$$

where C_j is the significance of the j-th component in the i-th zone in the depending on its wear W_j, P_j is the cost of the j-th component in the i-th zone; r_j is the the risk (probability) of destruction, that characterize the possible degree of destruction, while 0 is the absence of damage, 1 is the complete destruction; S_j is the social damage expressed in cash.

In the event of a malfunction of each zone the calculated possible economic damage is the determining criterion for the implementation of the geotechnical monitoring system. The cost of implementing of the monitoring system should be lower than possible economic damage, while the costs of maintaining a geotechnical monitoring system for a certain time and it are set based on its technical parameters:

$$C_{DG} \geq C_I = C_S + \sum_{t=0}^{\Delta} (C_M + kC_S(1 - e^{-t/T_0})) + \sum_{p}\sum_{b} (p_{1\to 0}C_p + p_{0\to 1}C_{SM}), \tag{3}$$

where C_I is the cost of implementing of the geotechnical monitoring system; C_S is the cost of geotechnical monitoring system; t is time of the operation of the monitoring system, Δ is allowed period of the implementation of the geotechnical monitoring system; C_M is fixed costs of the maintaining of the geotechnical monitoring system; k is the correction factor in the range from 1.1 to 1.4; T_0 is the mean time between failures in the geotechnical monitoring system; p is unfavorable factor (analyzed parameters of the geotechnical system); b is bifurcation points for each parameter p; $p_{1\to 0}$ is the probability of the skipping of the transition of the parameter p in bifurcation point b by geodynamic monitoring system; C_P is the damage in case of missing an adverse event p; $p_{0\to 1}$ is the probability of false triggering of the geodynamic monitoring system when a transition of the parameter p to the bifurcation point is detected b; C_{SM} is the cost of the work done to eliminate the adverse event p in the event of the false positive.

Taking into account the individual characteristics of the geotechnical system, the geotechnical monitoring system is accepted for consideration of the issue about implementation in case if this system satisfies the following conditions:

$$\min(C_{\cos t}),\ C_{\cos t} = \{\forall C_I \mid (C_I \le C) \cap (\Delta \ge T)\}, \tag{4}$$

where C is eligible costs over the period of time T.

The complex nature of the functioning of geotechnical systems is described by a large number of heterogeneous parameters, some of which are interdependent. It is necessary to control the change in the values of these parameters over a large area. It is for assess the current state and predict of the stability of the geotechnical system. It is proposed to determine the parameters by the degree of dependence on each other, the degree of influence on the stability of the geotechnical system and the costs of measuring these parameters. It is for reduce of monitored parameters and, accordingly, reduce of the cost of implementing of the geotechnical monitoring system.

Based on the modular construction principle, the model of the geotechnical system or its individual areas is formed [10]. Based on this modular model, parameters p are determined for the assessment of the stability of the geotechnical system S_t [11]. Based on the analysis of variance, the degree of influence of the parameters p is estimated on the stability of the geotechnical system $S_t 1$. The result of the assessment is a vector

$$St_1 = \{f(p_1),..., f(p_n)\}, \tag{5}$$

where $f(p_i)$ is the degree of influence of the i-th parameter on the stability of the geotechnical system.

By analogy, for all parameters of the geotechnical system, the matrix Pij of the dependence of the i-th parameter on the j-th $f(p_i, p_j)$ is compiled, such that

$$P_{ij} = \begin{cases} f(p_i, p_j), i \ne j \\ 0, i = j \end{cases}. \tag{6}$$

Vectors of the stability of the geotechnical system St_i are compiled based on the matrix of mutual dependencies of the parameters P_{ij} by successively replacing the values of degrees of influence of the vector St of dependent parameters p on the set of corresponding parameters from the row of the vector P_{ij} for parameter p. Replacement is carried out until the values of the degrees of influence of the replacement parameters are greater than the set threshold ε when replacing the replaced ones.

Bifurcation points are determined for each vector parameter based on bifurcation diagrams for all parameters of the St vectors [12-14]. The degree of significance of the parameter is determined and the remoteness of the current state of the system is determined from these points. It is determining as well as bifurcation points. For the control, the parameters are selected from vector that satisfies the following conditions:

$$\min\left(\sum_p \left(\left(\left|b_p - \bar{p}\right| - D\right)\Delta SKC_{ME} St_i\{p\}/DS\right)\right), \tag{7}$$

where p are all elements (parameters) of the vector St_i; b_p is the the value of the bifurcation parameter, while $b_p = \min(\left|b_{pi} - p(t)\right|)$, b_{pi} i-th bifurcation point for parameter p; $\bar{p}$ is the current average value of the measured parameter p; D is the variance of the measured parameter p; ΔS is the area of the

IOP Conf. Series: Materials Science and Engineering **873** (2020) 012030 doi:10.1088/1757-899X/873/1/012030

influence of the parameter p; S is the total area of the analyzed area; K is the required number of parameter measurement points p on area S; C_{ME} is the cost of introducing one measurement point; $St_i\{p\}$ is the degree of influence of the parameter p on the stability of the geotechnical system.

In expression (7) the C_{ME} parameter determines the cost of measuring the parameter by the specific method, which is selected based on the technical and operational capabilities of the application at the particular measurement point. If it is possible to carry out measurements by several methods, the method is selected for the implementation that which (the cost of the installed equipment and work) requires less cost.

It is possible to reduce the cost of implementing of the geodynamic control system by applying the control method based on monitoring the stability of key points [15]. In accordance with this method, the control of hazardous processes is carried out at points that are most sensitive to the development of the initial phase of destructive processes and the manifestation of their precursors. Each key control current determines the general trend of parameter changes over a certain area. The increased sensitivity of these points allows you to record negative processes at earlier stages and changes in the control area.

It should be noted that posterior data and dependencies between the parameters of the geotechnical system may be absent at the initial time when monitoring the stability of the geotechnical system. In this case, the robust model of the analyzed sections of the geotechnical system is built and critical parameters and monitoring points are determined on its basis. As data is accumulated and dependencies are identified, the model is refined, and the functioning of the geotechnical monitoring system is adjusted.

Thus, the proposed method of the registration of parameters of the geotechnical system is based on the results of the technical and economic analysis. The method includes the following steps:

1. The assessment of risks and possible damages in case of disturbance of the stability of the geotechnical system and its individual sections, as well as the classification and allocation of the most significant monitoring zones.
2. The formation of a list of measured parameters.
3. The formation of a modular model of the geotechnical system or its analyzed area.
4. The assessment of the degree of mutual dependence of one parameter on another and the assessment of the degree of influence of each parameter on the stability of the geotechnical system.
5. The determination of bifurcation parameters and assessment of the significance of parameters in the analysis of the stability of the geotechnical system.
6. The founding of key control points and the definition of controlled parameters and methods for their registration.
7. The estimation of the cost of the geotechnical monitoring system with various configurations (number and types of measuring sensors)
8. The choice of the complete set of the geotechnical monitoring system according to economic and technical criteria.

3. Results

In order to verify the developed approaches, data of the detection of the suffusion process were processed during geotechnical monitoring, which was carried out under urban conditions in the city of Murom, Vladimir Region, and the Russian Federation (coordinates 55.557282, 42.056503) from August 2017 to May 2018. The monitoring area was 150x250 meters. As the result of data processing, key control zones were identified (yellow and green zones in Figure 1), which covered the most vulnerable sections of the geotechnical system, and the soil electromagnetic parameters (electrical resistance and permittivity of the medium) were selected as controlled parameters. According to the results of monitoring the suffusion process (red zone) developed in one of the identified zones, which provoked the collapse of the arch of the soil with the formation of the hole with the diameter of 4 meters (Figure 2).

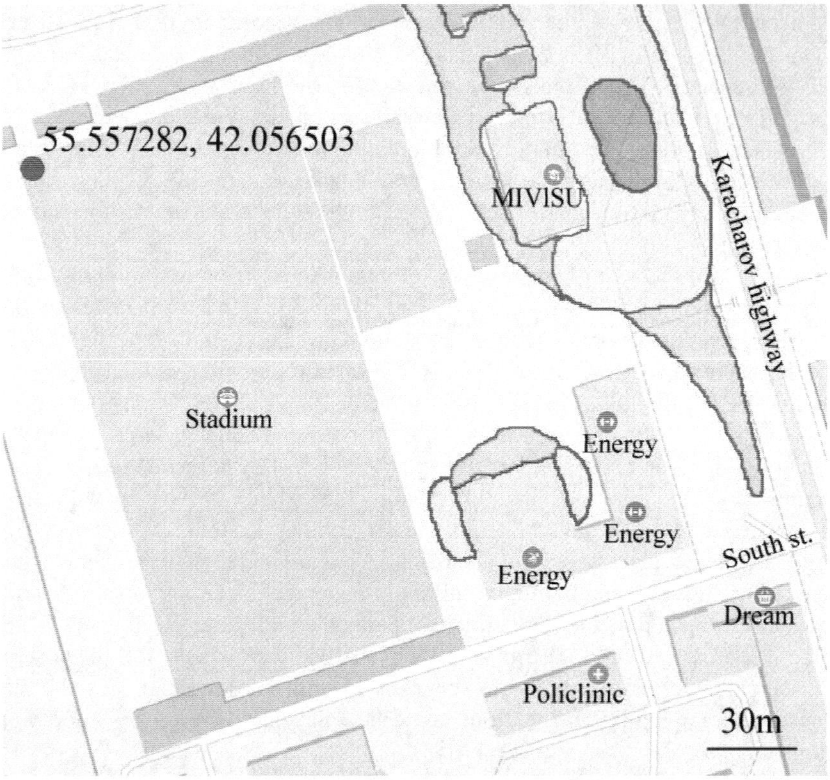

Figure 1. The identification of key control zones

Figure 2. The hole in the key control zone

The monitoring system was selected from the following systems: georadar of the type OKO-2, system based on constant current of type Meduza-48 and phase-measuring system of type S-1. Formed costs are shown in figure 3 taking into account their cost, operation, probability of detection and forecasting, as well as reliability. In the geotechnical system the total economic damage was estimated at 85.7 thousand dollars. Costs of monitoring work are shown in figure 4. These costs are without the application of the proposed method of registration of the parameters of the geotechnical system, and without the allocation of key zones and technical and economic analysis. These costs are for all analyzed area. The comparison of the costs is summarized in Table 1. It is incurred over 5 years.

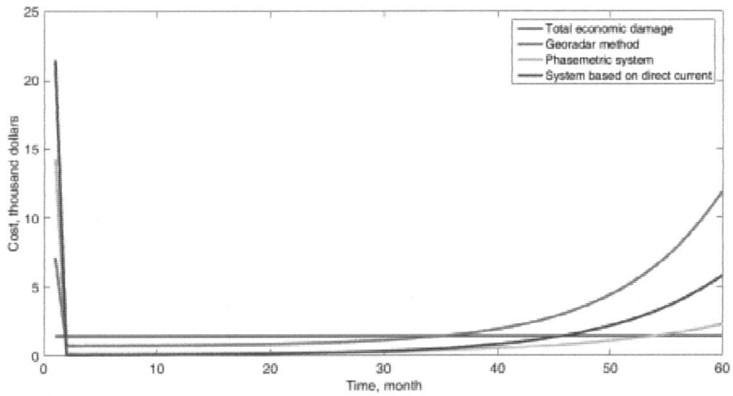

Figure 3. The cost comparison using the developed method.

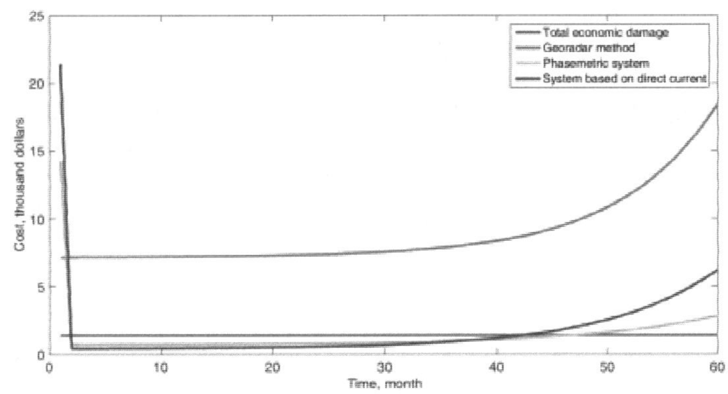

Figure 4. The cost comparison without a developed method.

Table 1. The cost comparison for various measurement methods over 5 years

Type of the monitoring system	Without the developed method, thousand dollars	With the developed method, thousand dollars
Georadar	535,2	155,8
System based on constant current	107,1	94,5
Phase-measuring system	82,8	49,0

4. Conclusion

The regular monitoring of the most important sections of the geotechnical system is governed by regulatory documents in the field of construction and operation of structures. However, in practice, for all buildings the constant conduct of such work is expensive and often ignored. As can be seen from the results described in this article, the developed method of the registration of parameters of the geotechnical system can reduce the cost of monitoring work, while maintaining the accuracy of forecast estimates, which will improve technological safety. To further reduce costs, it is possible to build a flexible monitoring schedule based on data on the period of development of destructive processes and a combination of various measurement methods. This approach is in good agreement with the objectives and goals of geotechnical monitoring.

5. Acknowledgment

This paper is an output of the science project executed with the support of the grant of the President of the Russian Federation No. MD-1800.2020.8.

References

[1] Inozemtsev V K and Redkov V I 2017 Geotechnical problems of the construction and operation of buildings in landslides with landslide processes *Bulletin of the Volga Regional Branch of the Rus. Academy of Architecture and Building Sci.* **20** 170-9

[2] Telichenko V I, Gutenev V V and Slesarev M Yu 2006 Approaches to the interpretation of environmental safety management systems in construction *Ecology of urbanized territories* **2** 6-11

[3] Sosunov I V 2010 Actual issues of emergency prevention, Scientific and methodological publication *EMERCOM of Rus., Federal State Institution Res. Institute of Civil Defense and Emergencies* 350

[4] Vasilyev G S, Kuzichkin O R, Romanov R V, Dorofeev N V and Grecheneva A V 2018 The practice of using a multi-pole electrical installation for monitoring the coastal zone of karst lakes *Int. Multidisci. Sci. GeoConf. Surveying Geology and Mining Ecology Management*, (Albena: Bulgaria) vol 18 (1.2) pp 727-34

[5] Kostarev S N, Sereda T G and Mikhailova M A 2013 Development of an automated monitoring and management system for natural-technical waste disposal systems *Fund. Res.* **6-2** 273-7

[6] Bondarik G K 2012 Geokibernetika - A tool for diagnosing and predicting the state of natural and natural-technical systems *Geoecol., Eng. geology, Hydrogeol., geocryology* **4** 364-70

[7] Leonov O A and Shkaruba N J 2012 Algorithm for selection of measurement means for quality control by technical and economic criteria *Bulletin of the Federal State Educ. Institution of Higher Prof. Educ. Moscow State Agroeng. Univer. named after V.P. Goryachkin* **2 (53)** 89-91

[8] Kuzichkin O R, Grecheneva A V, Gakhov R P, Dorofeev N V, Baknin M D and Gakhov B R 2018 Development and research of the geoelectric model of the local zone of geodynamic control *J. of Adv. Res. in Dynamical and Control Syst.* **10(13)** 620

[9] Sledge I J and Keller J M 2008 Growing neural gas for temporal clustering *19th Int. conf. on Pattern Reconition* (Tampa: Florida, USA) pp 1-8

[10] Grecheneva A V, Kuzichkin O R, Mikhaleva E S and Dorofeev N V 2018 Geotechnical monitoring of the buildings on the basis of analysis of transfer functions and cyclic vibrational

technogenic loads *J. of Adv Res, in Dynamical and Control Syst.* **10** (**02**) 1995-2003

[11] Kuzichkin O R, Grecheneva A V, Dorofeev N V and Mishunin V V 2018 Geotechnical monitoring of the objects based on the method of inclinometric control of own frequencies *J. of Adv Res, in Dynamical and Control Syst* **10**(13) 616-19

[12] Petrochenko V I and Petrochenko A V 2019 Optimization of design solutions for flood protection in river basins *Reclamation* **2** (**88**) 26-33

[13] Inozemtsev V K, Inozemtseva O V and Strelnikova K A 2010 The bifurcation criterion for stability of the "object-base" system based on the incremental base model *Construct. and Reconstruct* **1** (**27**) 16-22

[14] Nazarov D I 2015 Destruction of the structures of the mining building, energy and bifurcation analysis *Mining Inf. and Analytical Bulletin (Sci. and Tech. J.)* **7** 95-100

[15] Dorofeev N, Kuzichkin O and Eremenko V 2016 The method of selection of key objects and the construction of forecast function of the destructive geodynamic processes *Informatics, geoinformatics and remote sensing conference proceedings* (Albena: Bulgaria) vol 1 pp 883-90

Security methods for a group of mobile robots according to the requirements of Russian and foreign legislation

A S Basan [1], E S Basan[1], M A Lapina[2], V N Kormakova[3], V G Lapin[4]

[1] South Federal University, 2 Chekhov St., Taganrog, 347928, Russia

[2] North-Caucasus Federal University, 1 Pushkin St., Stavropol, 355017, Russia

[3] Belgorod State National Research University, 85 Pobeda St., Belgorod, 308015, Russia

[4] Stavropol regional clinical consulting and diagnostic center, 304 Lenina St., Stavropol, 355000, Russia

E-mail ele-barannik@yandex.ru

Abstract. This paper is devoted to the problem of creating security methods for mobile robotic systems. The urgency of the problem of ensuring the security of mobile robotic systems is associated with the presence of a contradiction between the growing popularity of the mobile robotic systems and the presence of many vulnerabilities. Development of a universal security system which should provide protection not only against an malicious attack, but also allow the system to cope with unintentional errors, interference or changes in the external environment, thereby improving the operation of group control algorithms for robots, and increasing the reliability and stability of the mobile robotic systems.

Keywords: Protection, mobile robots, standards, laws, requirements.

1. Introduction

Today, there is a trend associated with the growing popularity of mobile robotic systems, for example, in 2018, the global market for domestic robots reached $ 3.02 billion, according to the research company Research and Markets. According to experts asserts, by 2024 this revenue will be $ 9.13 billion, increasing by about 22.37% annually. The main purpose of robotic systems is to monitor and control the object. This research investigates the use of group control system which include distribution goals and tasks between robots. Of course, the use of a group control system to manage the operation of mobile robots is an effective solution that allows achieving a given goal in a shorter time, with the least loss of resources. It is critical in the case of robots with a limited energy supply [1]. The solution to the problems connected with the designer of the group control system is associated with the growth of interest, both in the scientific, and in the industrial and military spheres.

Nevertheless, a group of mobile robots equipped with a group control system is a rather complicated structure that is vulnerable enough to malicious attacks. The analysis of the group control systems from the point of view of information security revealed that such systems aren't satisfied security principles. However, the standard security methods which applied in information systems are not always applicable to mobile robots' groups. The main vulnerabilities of group control systems are: the presence of a wireless data transmission medium that allows to analyze the transmitted data by intercepting them; the ability to modify the transmitted data, as well as the ability to implement malicious data, causing a destructive effect on the network, due to wireless data transfer and insecure transmission channels; physical insecurity of mobile robotic tools, leading to the possibility of reverse engineering ; the location of the mobile robots outside of the controlled area, which can lead to interception of devices, as well as to the ability to influence environmental indicators, which can lead to disruption of the system.[2]

At the same time, there are the main stages of the group control system for the mobile robots: strategic, tactical and executive. The preparing attacks at the level of target distribution and group formation can lead to significant negative impact. Such as: a complete disruption of the group control system of robots, the incorrect distribution of goals, roles, the choice of incorrect ways to achieve goals. In this case, the following types of attacks are possible: spoofing, violation of the integrity of the transmitted data, compromise of trusted nodes, attacks aimed at disrupting the network clustering process — an attacker will be able to implement these types of attacks by penetrating the network or by conducting a man in the middle attack [3]. In addition, external attacker can affect this control level by conducting denial of service attacks and distributed attacks, as well as cyber-attacks involving physical impact. The following attacks are possible at the tactical control level: affecting on the performance of the robot's sensor system; carry out attacks on the position-tractor control, as well as attacks on the availability of nodes.

At the same time, despite the fact that today there are researches related to the development of security methods, the disadvantage of existing solutions are:

Incompleteness of threats and vulnerabilities databases for a mobile robotic system, as well as a list of possible attacks on the robotic systems.

Lack of an integrated approach to ensuring security in mobile robotic systems at the network, physical and application levels [4].

There is a problem associated with the development of a security system for the mobile robot groups: the difference between this type of network and the typical computer networks. This requires the development of special methods and approaches that must take into account the following factors: the limited computing and energy resources of mobile robots, which makes such solutions as artificial intelligence to detect an attacker difficult and resource-intensive. Thus, the development of a comprehensive security solution for a mobile robotic system is not a trivial task and requires an analysis of all possible threats and vulnerabilities, as well as building a model of the intruder, and also definition of a set of recommendations to increase the degree of system security [5].

2. The problem statement

The main aim of this research is to develop an architecture of protection system for a self-organizing group of mobile robots, taking into account the requirements of Russian and foreign legislation. To achieve this goal it is necessary to solve the following tasks:

- analysis of structural and functional characteristics of a group of mobile robots;
- analysis of types of group control systems for organizing operation a group of mobile robots;
- analysis of network structures for communication in a group of mobile robots;
- analysis of Russian and foreign standards in the field of information security of industrial control system and internet of things (as the most similar in structure and functionality to groups of mobile robots);
- implementation of a protection system in accordance with the developed methodology;
- analysis of the impact of protective equipment on the performance of the robotic system.

IOP Conf. Series: Materials Science and Engineering **873** (2020) 012031 doi:10.1088/1757-899X/873/1/012031

3. Development of a security system architecture for a mobile robotic system

3.1. Analysis of structural and functional characteristics of a mobile robotic system

The group control system for mobile robots consists of two main parts: a central station (CS) and on-board computers (BC) with a communication controller (CC).

The central station is installed stationary in the command center and solves the problem of planning the actions of the entire group of robots. In addition, the central station provides a human operator with each robot in the group in case of unforeseen situations, and is also used to set a target [6].

An on-board computer is designed to solve the problem of computational tasks such as distributing goals, determining trajectories, calculating the shortest path, etc. The communication controller is used to provide communication between the robot and the central station and with other robots of the group [7].

There are several ways to organize the process of controlling robots:

- operator control level;
- level of autonomous control level;
- group control level;
- centralized control level.

Let's consider a Flying Ubiquitous Sensor Networks (FUSNs) are one of the classes of WSN / USN wireless or all- pervasive sensor networks. The technology of these networks is based on the self-organizing association of many different sensors with low energy consumption in the network and their placement in hard-to-reach places [8].

Architecturally, they can be decomposed into:

- peer-to-peer networks,
- hierarchical (cluster) networks,
- centralized star topology network.

The architecture of the wireless sensor network, which represents the flying network, consists of three types of nodes:

The coordinator is the only stationary and "trusted" device in the network. It determines the routes for transmitting information, sets the network parameters, controls devices connected to the network, selects the necessary frequency channels, and also acts as a gateway to provide access to the external network. Depending on the application, it can perform various additional functions, for example, it can be a hub or a network manager that monitors the level of interference on a selected channel and can transfer the entire network to a channel of a different frequency;

Router – a device responsible for receiving, storing and transmitting information between nodes. It allows connecting end devices to the network and sends them packets with parameters from the coordinator. In hierarchical network end devices are connected to a node that acts as a router, which, in turn, has a connection to the coordinator. The coordinator can have a connection with several cluster groups;

Sensor nodes (end devices) - are collect data, such as temperature, pressure, dust, etc. Can control a remote object, if necessary. The touch device should have a small size, low power consumption and the ability to stay in sleep mode for a long time. Collectively form a sensory field, which is the place of monitoring[9].

The structural description of the analyzed UAV control system is presented in the table 1.

Table 1. Structural description of the UAV control system.

Control system elements	Structure
Operator	PC with any operating system, special software.
Data Transfer Standards	IEEE 802.15.4(ZigBee),IEEE 802.15.1(Bluetooth),IEEE 802.11 (WI- FI), IEEE 802.16-2004, BLE (Bluetooth Low Energy) [10], 6LoWPAN, LTE.
Protocols	TBRPF (Topology distribution base on reverse-path forwarding),DSR (Dynamic Source Routing), GPSR (Greedy Perimeter Stateless Routing),OLSR (Optimized Link State Routing Protocol), DOLSR (Directional Optimized Link State Routing Protocol), AODV (Ad-hoc On-Demand Distance Vector), HWMP (Hybrid Wireless Mesh Protocol), IEEE 802.11 n,s, ZigBee.
Navigation system	Omnidirectional transceiver antenna-feeder devices that monitor the location of an unmanned aerial vehicle
Hardware	Control system components: micro controller, GPS module, communication antenna, engine. Sensor system components: tactile sensors, optical sensors, sound sensors, position sensors, tilt sensors, infrared sensors, temperature sensors, gyro stabilizer.

3.2. Analysis of Russian and foreign legislation field of protection of industrial control systems

In the Russian Federation, one of the main government regulators, as well as special and control functions in the field of state information security, is the FSTEC (Federal Service for Technical and Export Control). The main regulatory document for ICS (industrial control systems) and ISPD (information systems of personal data), according to orders No. 17, 21 and 31, is the methodological document of the FSTEC "Measures for the protection of information in government information systems". The document describes the methodology for implementing organizational and technical measures to protect information in typical information systems (IS) [11].

This standards and recommendations are directed at realization of information security measures, starting with identification and authentication and ending with the informing and educating users. The information security process, according to the standards, should be structured as follows:

- definition of a basic set of protection measures according to the class of IS,
- adaptation of the basic set of protection measures ,
- refinement of the protection measures according to the threat model,
- addition, taking into account a different regulatory framework in terms of information protection.

In 2008, a safety standard for industrial control systems was issued. National Institute of Standards and Technology represents SP 800-82 Guide to Industrial Control Systems.

- In addition to this order, there are also foreign standards. For example is the NIST, the ISO, but these standards are mainly aimed at examining protection systems for the Internet of things. NIST has developed a Framework for Improving Critical Infrastructure Cybersecurity, where represented the necessary security subsystems that should be implemented in information systems (National Institute of Standards and Technology, 2018). For each of the proposed security subsystems, there are specified sections of the standards where the procedure for developing each subsystem. This Framework presents a large number of requirements, but it is also not clear how to select a specific requirement for the system, how to assess the need to protect one or another component of the system. At the first stage, it is proposed to conduct a risk identification, which includes the following stages:

- Asset Management -The data, personnel, devices, systems, and facilities that enable the organization to achieve business purposes are identified and managed consistent with their relative importance to organizational objectives and the organization's risk strategy.

- Business Environment - The organization's mission, objectives, stakeholders, and activities are understood and prioritized; this information is used to inform cybersecurity roles, responsibilities, and risk management decisions;

- Governance - The policies, procedures, and processes to manage and monitor the organization's regulatory, legal, risk, environmental, and operational requirements are understood and inform the management of cybersecurity risk.;
- Risk Assessment - The organization understands the cybersecurity risk to organizational operations (including mission, functions, image, or reputation), organizational assets, and individuals;
- Risk Management Strategy - The organization's priorities, constraints, risk tolerances, and assumptions are established and used to support operational risk decisions;
- Supply Chain Risk Management - The organization's priorities, constraints, risk tolerances, and assumptions are established and used to support risk decisions associated with managing supply chain risk.

The organization has established and implemented the processes to identify, assess and manage supply chain risks [12]. The following assets must be considered when assessing risks according to this document:

- Physical devices and systems within the organization.
- Software platforms and applications within the organization.
- Organizational communication and data flows.
- External information systems.
- Resources (e.g., hardware, devices, data, time, personnel, and software).
- Cybersecurity roles and responsibilities for the entire workforce and third-party stakeholders.

The document provides references to other standards that allow realized risk assessment: CIS CSC, COBIT 5, ISA 62443-2-1: 2009, ISO / IEC 27001: 201 NIST SP 800-53. Most of these documents are in the public domain, but are paid, which makes it difficult to study. Nevertheless, it can be concluded that when assessing risks, documents do not refer to methods for determining the current threats. Thus, the issues related to the identification of current threats are practically not worked out.

In 2008, the NIST Special Publication 800-82 standard was introduced [13]. This standard defines key components are:

- Control node. The control node consists of measurement sensors, a controller (includes equipment and actuators, such as PLC controllers, valves, switches, levers, motors) and variable systems.
- Human Machine Interface (HMI). Operators and engineers use the HMI to monitor, control and change set points, algorithms, control and set controller parameters.
- Remote diagnosis and support program. Remote diagnostic and support programs are used to prevent, recognize, and correct malfunctions.

This structure is fundamentally different from that presented in the FSTEC. In May 2015, the standard was released in the second version. The document describes the structure of ICS as follows. A typical ICS contains numerous control loops, human interfaces, and remote diagnostics and maintenance tools built using an array of network protocols on layered network architectures. Control loops utilize sensors, actuators, and controllers (e.g., PLCs) to manipulate some controlled process. A sensor is a device that produces a measurement of some physical property and then sends this information as controlled variables to the controller. The controller interprets the signals and generates corresponding manipulated variables, based on a control algorithm and target set points, which it transmits to the actuators. Actuators such as control valves, breakers, switches, and motors are used to directly manipulate the controlled process based on commands from the controller. As a whole, this document gives a clear understanding of what control systems are, a large number of examples of such systems, their architecture and description are given. This standard can serve as an example for creating a similar document for a robotic system. The document also provides a list of vulnerabilities specific to ICS. In addition, much attention is paid to network protection. Examples of firewall rules and network segmentation options are provided.

Another example is the Robot Security Framework (RSF) [14]. This article describes a security assessment system. Robotic system is divided into 4 components and evaluates the safety of each of them. At the same time, the assessment does not rely at all on the possible threats characteristic of each component of the system, and takes into account only the physical, network, firmware, and application. This Framework also lacks the ability to evaluate the intelligent control system of the robot, evaluate the robot if it is mobile, and the group control system. Authors hereby propose a framework based on four layers that are relevant divide them into aspects considered relevant to be covered. Also, they provide relevant criteria applicable for security assessment. For each of these criteria they identify what needs to be assessed (objective), why to address such (rationale) and how to systematize evaluation (method) [15].

In general, the standards provides for a wider set of requirements than the specified documents, as can be seen from Figure 1, where the number of requirements is presented vertically.

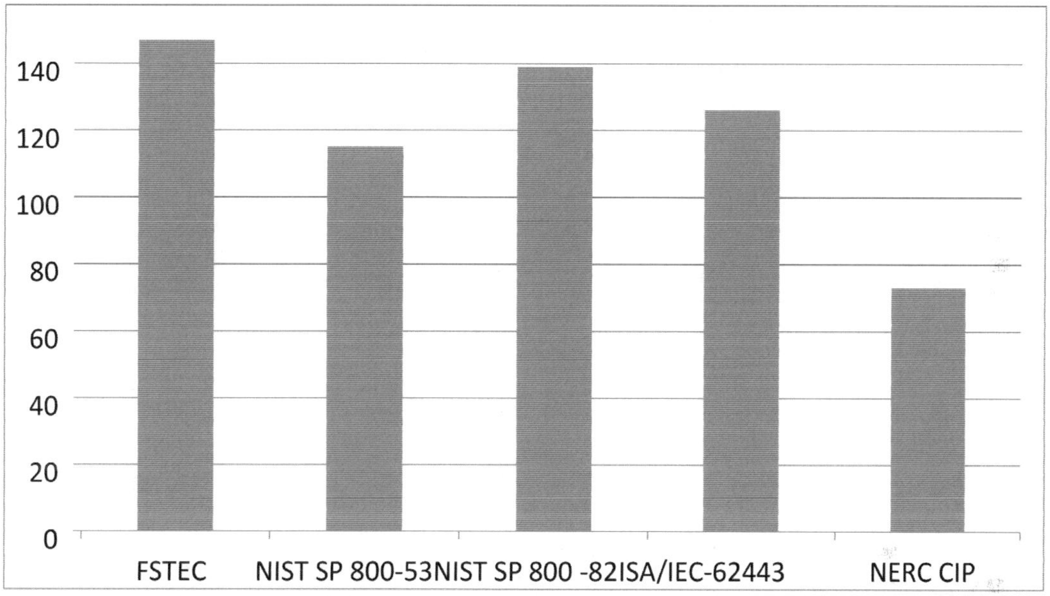

Figure 1. Comparison of the requirements of the FSTEC order with the requirements of International standards

3.3. Implementation of security requirements for a group of mobile robots

During the implementation of the security requirements, we installed and configured the firewall, Openvpn, USBGuard, wpasupplicant, denyhosts, eCryptfs [16].

The firewall provides security against network attacks. Openvpn provides security of transmitted information from interception and spoofing. Wpasupplicant provides settings to ensure network security with the use of security protocols such as WPA/WPA2. USBGuard and eCryptfs provide information security from physical downloads. Denyhosts limits the number of login attempts.

Before installing security systems, let's see how busy the system is. To display information about the system load, we will use the built-in htop program. Prior to installing security features, the system is loaded at 2.0%. After carrying out all the settings , the load of CPU was 18–23%. Based on this, we can conclude that with the installed means of protection, for the main activity of the device there is enough performance, about 80%.

Thus, we can conclude that the protection system is not fully implemented, the implemented protection measures provide protection against interference with the operation of the robotic system from the network, encrypted communication channels are installed, the stored information is encrypted, and the device is also protected from physical access to stored information. But the device is not completely protected from physical interference, an attacker can gain access to the components of the device and go unnoticed due to the lack of monitoring of connected components.

The following protection measures are shown in Table 2.

Table 2. implemented protection measures.

Protection measure	Implementation
Physical security	
Unused ports disable	Not implemented
When connecting to physical ports, require authentication	Implemented
Components and cables hidden under the case	Not implemented
Logging of disconnecting / connecting components to the event log	Not implemented
Logging device on / off events in the event log	Not implemented
Network security	
The network is password protected.	Implemented
Using complex passwords	Implemented
Using modern protocols	Implemented
Encryption and mutual authentication	Implemented
Disable unused network ports	Implemented
Operator receives real-time alerts for suspicious activity from the external network	Not implemented
Event logging attempts to connect to the security log	Implemented
Software security:	
Installed current OS versions	Implemented
No known vulnerabilities	Implemented
Verification of signatures before installation	Not implemented
Application Security:	
Programs do not have known vulnerabilities	Implemented
Confidential data is transmitted over an encrypted channel.	Implemented
Access to applications only from authorized users	Implemented

4. Conclusion

Groups of mobile robots are actively being introduced into many areas of activity, thereby questions about the security of robots remain open, including methods and means of protection for a group of mobile robots [17].

The main goal of the work was to research methods and develop protective equipment for a self-organizing group of mobile robots, taking into account the requirements of Russian and foreign legislation.

Description of the Groups of mobile robots helped to learn the control system of mobile robots, varieties of network typologies, components of mobile robots, data transfer protocols, and communication channels.

Comparison of standards allowed us to consider the advantages and disadvantages of Russian and foreign standards.

The use of protective equipment on an experimental bench made it possible to identify the effect of protective equipment on the performance of the entire system.

The result obtained, a comprehensive solution of protection measures for a group of mobile robots, can be used in the future to provide protection to a group of mobile robots, used for subsequent modernization in the development of security methods.

5. Acknowledgment
This work was supported by the RFBR grant 18-07-00212 "Development of a method and decision protocol for detecting anomalous node behavior in group control systems of autonomous mobile robots".

References

[1] Amullen E M, Shetty S, & Keel L H 2016 *Model-based resilient control for a multi-agent system against Denial of Service attacks* World Automation Congress (WAC) Rio Grande Puerto Rico pp. 1-6. doi: 10.1109/WAC.2016.7582963.
[2] Ani U P D, He H M, & Tiwari A 2016 *Review of cybersecurity issues in industrial critical infrastructure: manufacturing in perspective*. Journal of Cyber Security Technology 1(1) pp 32-74 doi: 10.1080/23742917.2016.1252211
[3] Basan E, Basan A, Grutsynin A 2019 *Overview of Information Issues for a Robotic System* In Proceedings of 19th Interantional Conference on Communication Technology (IEEE ICCT 2019) Xian, China pp 1275-1280
[4] Basan A, Basan E, Makarevich O 2019 *Analysis of ways to secure group control for autonomous mobile robots* In Proceedings of the 10th International Conference on Security of Information and Networks, Jaipur, India pp 134-139
[5] Basan A S, Basan E S, Stepenkin A A 2017 *Analysis and implementation of threats for mobile robot management systems* In Proceedings of the XIII Russian Scientific-practical Conference Mathematical Methods and Information Technology means, Krasnodar, Russia pp 20–23
[6] Faizal K & Palaniappan P LK 2014 *Risk Assessment and Management in Supply Chain* Global Journal of Researches in Engineering: G Industrial Engineering, 14(2), pp 19-30
[7] Hagele M 2016 *Robots conquer the world* IEEE Robotics & Automation Magazine 23(1), 118-120 doi:10.1109/MRA.2015.2512741
[8] Hoang T, Kirichek R, Paramonov A, Koucheryavy A 2016 *Supernodes-based solution for terrestrial segment of flying ubiquitous sensor network under intentional electromagnetic interference* In Proceedings of the ruSMART: Conference on Internet of Things Smart Spaces and NEW2AN: International Conference on Next Generation Wired/Wireless Networking, St. Petersburg, Russia pp 351-359
[9] Holm H, Karresand M, Vidström A & Westring E A 2015 *Virtual Industrial Control System Testbed* Swedish Defence Research Agency Stockholm, Sweden
[10] Kirichek, R V & Kucheryavy A E 2014 *Flying sensor networks* Electrosvyaz 11, pp 2-5
[11] Liang L, Ren W, Song J, Hu H, et al 2013 *The state of the art of risk assessment and management for information systems* In Proceedings of 9th International Conference on Information Assurance and Security (IAS) Gammarth, Tunisia pp 43-56 doi: 10.1109/ISIAS.2013.6947735
[12] Mitchell R, & Chen I R 2014 *Adaptive Intrusion Detection of Malicious Unmanned Air Vehicles Using Behavior Rule Specifications* IEEE transactions on systems, man, and cybernetics: systems, 44 (5) pp 2168-2216
[13] *National Institute of Standards and Technology* 2018 Framework for Improving Critical Infrastructure Cybersecurity version 1.1 pp 1-48
[14] Pshikhopov V, Medvedev M, Kolesnikov A, Fedorenko A, Gurenko R 2016 *Decentralized control of a group of homogeneous vehicles in obstructed environment* Journal of Control Science and Engineering (7192371) pp 1-9 doi: 10.1155/2016/7192371
[15] Pshikhopov V, Ali A 2011 *Hybrid motion control of a mobile robot in dynamic environments* In Proceedings of the IEEE International Conference on Mechatronics, Istanbul, Turkey pp 540–545 doi: 10.1109/ICMECH.2011.5971345
[16] Pshikhopov V K, Medvedev M Y, Gaiduk A R, Gurenko B V 2013 *Control system design for autonomous underwater vehicle* In Proceedings of the Robotics Symposium and Competition (LARS/LARC) Arequipa, Peru pp 77–82 doi: 10.1109/LARS.2013.61

[17] Phillips-Wren G 2012 *Ai Tools in Decision Making Support Systems*: a Review International Journal of Artificial Intelligence Tools 21(2) pp 1-13 doi: 10.1142/S0218213012400052.

IOP Conf. Series: Materials Science and Engineering **873** (2020) 012032 doi:10.1088/1757-899X/873/1/012032

Simulation of the stress-strain state of shells under internal pressure using the mixed finite element method, taking into account physical nonlinearity

D P Arkov, O V Kochetkova, N A Gureeva, A S Matveyev, E V Shiryaeva

Federal State Budgetary Educational Institution of Higher Education "Volgograd State Agricultural University", 26 Universitetsky Prospect, Volgograd, 400002, Russia

E-mail arkov-dmitriy@rambler.ru

Abstract. In this article, the finite element method (FEM) is used to model the stress-strain state (SSS) of shells experiencing internal pressure with elastic-plastic state being taken into account. Main geometric equations determining the location of an arbitrary point at load step are given. Mixed functionality used in implementation of FEM which allows to determine stresses and displacements simultaneously is obtained based on equality of possible and actual work of external and internal forces. Examples of stress-strain state of rotation shells with spherical and elliptical bottoms under internal pressure are provided. Static check of resultant external forces arising from applied pressure and resultant internal forces arising from stresses equality to zero is performed. Achieved results show correctness of the developed algorithm with elastic-plastic properties of the material being taken into account, ensuring that main requirements for the mathematical model are secured, namely adequacy, versatility and cost-effectiveness. Costs of resources computing are reduced by adopting a number of hypotheses and determination of stresses and strains at once ensuring cost-effectiveness and simplicity. Using developed scheme of the mixed functionality application in implementation of FEM the mathematical model acquires one of its main properties - "Potentiality".

Keywords: modeling, stress-strain state, finite element, displacements, stresses, functional, shell structures

1. Introduction

Different shells are among most common elements of engineering constructions and due to their curvilinear shape shells work as spatial elements. One of the most frequent numerical methods of stress-strain state of shells study is Finite Element Method (FEM). It allows to achieve fairly accurate solutions[1], therefore in [2] spherical and cylindrical geometry of the shell is studied for two different material configuration which are single-layer structures and multi-layer structures with an internal core embedded in them. Three-dimensional (3D) exact shell model and various two-dimensional (2D) computational models are compared by frequency and oscillation mode. Proposed numerical solutions are typical two-dimensional finite elements (FE) as well as classical and improved generalized two-dimensional differential quadrature (TDDQ) solutions. Problems regarding 2D Generalized Differential Quadrature solutions for FGM plates and shells analyses are discussed in [2-4]. 2D static analysis of FGM plates and shells is being proposed in [2-3].

In [5] the dynamic Koiter model is being viewed for hyperbolic parabolic shell, more precisely the saddle surface as its mean surface. Numerical calculation for this type of shell is the most complex due to its complex differential geometry.

Using a simplified model that reflects individual features of the object being studied allows us to see the relations of causes and effects, inputs and outputs more clearly and make necessary conclusions faster, and make correct decisions when modeling shells of different configurations [5-7, 10]. In [8] we consider a variant of Koiter shell model based on internal geometry methods by Michael Delfour and Jean-Paul Zolesio.

2. Informal statement of the problem

Analyzing the stress-strain state of complex elements and structures used in agro-industrial production and manufacturing processes in mechanical engineering and other industrial sectors its clear that usage of material modeling is difficult since it is necessary to operate with large number of criteria and restrictions that may be incompatible thus it's often impossible. Therefore, in [9], physicomathematical interpretation of alternative shear strain concept is proposed in order to get rid of intuitive aspect of its basic premise that total deviation w can be taken as sum of bending and transverse shear deviations. One of solutions to these problems is to use numerical solving procedures with a computational experiment [7-9, 11].

Among many schemes of FEM in calculations of the stress-strain state of shells special space is occupied by mixed methods, which proved to be effective here. Usage of methods based on Reissner variational principle for linear and geometrically nonlinear problems is repeatedly reported in literature, results of numerical calculations and their analysis are presented in [13, 14].

Browsing literature sources for calculation of plate and shell structures shows that modeling and analysis of the stress-strain state based on FE is of interest to researchers and engineers worldwide [2-9, 15-17]. There are practically no works devoted to the mixed formulation analysis of SSS with physical nonlinearity of material taken into account. This method allows us to determine displacements and stresses without additional operations.

3. Formalization and analysis of the model

In Cartesian coordinate system when rotation shell is deformed an arbitrary point M of median surface of the shell is determined by radius vector:

$$\vec{R} = x\vec{i} + r\sin\theta\vec{j} + r\cos\theta\vec{k}, \tag{1}$$

where x is axis coordinate; r is radius of rotation; $\vec{i}$, $\vec{j}$, $\vec{k}$ are unit vectors of Cartesian system; θ is angular coordinate being calculated counterclockwise from vertical.

Basis vectors of arbitrary point tangent to median surface are determined by differentiating the radius vector (1):

$$\vec{a}_1 = \vec{R},_s = \vec{i} + r,_s \sin\theta\vec{j} + r,_s \cos\theta\vec{k};$$

$$\vec{a}_2 = \vec{R},_\theta = r\cos\theta\vec{j} - r\sin\theta\vec{k}. \tag{2}$$

The normal to the median surface is defined by next equation:

$$\vec{a}_3 = \frac{\vec{A}}{r} = -\vec{i}\,r,_s + x,_s \sin\theta\vec{j} + x,_s \cos\theta\vec{k}. \tag{3}$$

Relations (2) and (3) can be written in matrix form:

$$\underset{3x1}{\{\vec{a}\}} = \underset{3x3}{[M]}\underset{3x1}{\{\vec{i}\}}, \tag{4}$$

Derivatives of basis vectors of point M of the median surface are defined by differentiation (2), (3), expressed in terms of orts of Cartesian coordinate system in matrix form it will look like following:

$$\{\vec{a},_s\} = [N_1]\{\vec{i}\};\ \{\vec{a},_\theta\} = [N]\{\vec{i}\}, \tag{5}$$

IOP Conf. Series: Materials Science and Engineering **873** (2020) 012032 doi:10.1088/1757-899X/873/1/012032

Arbitrary point M^t receives displacement $\vec{V}$ when shell is deformed [18]. This vector can be represented by components in basis of corresponding point M of the median surface:

$$\vec{V} = v^1\vec{a}_1 + v^2\vec{a}_2 + v_3\vec{a}_3 = \{\vec{a}\}^T\{v\}, \tag{6}$$

where $\{v\}^T = \{v^1\ v^2\ v^3\}$.

At (j+1)-th loading step displacement of the point M^t is determined by a vector, which in terms of vectors of local basis of the point M of the median surface and the corresponding components is expressed as

$$\vec{w} = w^1\vec{a}_1 + w^2\vec{a}_2 + w^3\vec{a}_3 = \{\vec{a}\}^T\{w\}, \tag{7}$$

where $\{w\}^T = \{w^1\ w^2\ w^3\}$.

Movement of the point M after the (j+1)-th loading step is determined by the total vector $\vec{V} + \vec{w}$. Position of the point M^{T*} in deformed state of the shell is determined by radius vector

$$\vec{R}^{t*} = \vec{R}^t + \vec{V}. \tag{8}$$

Deformations at the point M^t are defined as differences between the components of metric tensors of deformed and initial states and determined by relations that can be represented in matrix form:

$$\underset{6x1}{\{\varepsilon\}} = \underset{6x3}{[L]}\underset{3x1}{\{v\}}, \tag{9}$$

where $\{\Delta\varepsilon\}^T = \{\Delta\varepsilon_{11}\ \Delta\varepsilon_{22}\ \Delta\varepsilon_{33}\ 2\Delta\varepsilon_{12}\ 2\Delta\varepsilon_{13}\ 2\Delta\varepsilon_{23}\}$ is vector-string of strain tensor components at the point M^t; $[L]$ is matrix of algebraic and differential operators.

In calculations for geometrically linear deformation between the increments of deformations and the increments of displacements at the (j+1)-th step of loading relations could be written in matrix form [18]:

$$\{\Delta\varepsilon\} = [C_\sigma^P]\{\Delta\sigma\}, \tag{10}$$

Arbitrary eight-node finite element with nodes i, j, k, l, m, n, p, h is accepted as finite element. Nodal unknowns take stress increments at nodal points and movements of nodal points at the loading step. To perform numerical integration, an arbitrary octagon is mapped to a cube whose local coordinates change within $-1 \le \xi, \eta, \zeta \le 1$.

Displacement vector components of the finite element inner point are expressed in terms of the nodal values of the displacement components using the matrix relation

$$\underset{3x1}{\{w\}} = \underset{3x24}{[A]}\underset{24x1}{\{w_y\}}, \tag{11}$$

where
$$\underset{3x24}{[A]} = \begin{bmatrix} \underset{1x8}{\{\varphi(\xi,\eta,\zeta)\}^T} & \{0\}^T & \{0\}^T \\ \{0\}^T & \underset{1x8}{\{\varphi(\xi,\eta,\zeta)\}^T} & \{0\}^T \\ \{0\}^T & \{0\}^T & \underset{1x8}{\{\varphi(\xi,\eta,\zeta)\}^T} \end{bmatrix}.$$

Stress increment tensor components of the finite element inner point are expressed in terms of the nodal values of the stress increment tensor components by following matrix relation

$$\underset{6x1}{\{\Delta\sigma\}} = \underset{6x48}{[G]}\underset{48x1}{\{\Delta\sigma_y\}}, \tag{12}$$

Replacing the internal forces increments actual workload at the loading step with the difference between the possible and additional internal forces workload at the loading step and taking into account the matrix relations (10), (11) and (12), the function at the loading step is written as:

$$\Pi_{LNS} \equiv \underset{1x48}{\{\Delta\sigma_y\}^T} \int_V \underset{48x6}{[G]^T} \underset{6x24}{[B]} dV \underset{24x1}{\{w_y\}} - \frac{1}{2} \underset{1x48}{\{\Delta\sigma_y\}^T} \int_V \underset{48x6}{[G]^T} \underset{6x6}{[D]} \underset{6x48}{[G]} dV \underset{48x1}{\{\Delta\sigma_y\}} -$$

$$-\frac{1}{2} \underset{1x24}{\{w_y\}^T} \int_S \underset{24x3}{[A]^T} \underset{3x1}{\{\Delta p\}} dS - \underset{1x24}{\{w_y\}^T} \int_S \underset{24x3}{[A]^T} \underset{3x1}{\{p\}} dS + \underset{1x24}{\{w_y\}^T} \int_V \underset{24x6}{[B]^T} \underset{6x24}{\{\sigma\}} dV = 0. \tag{13}$$

Minimizing the function (13) by nodal unknowns $\{\Delta\sigma_y\}^T$ and $\{w_y\}^T$ we get system of equations:

$$\frac{\partial \Pi_R}{\partial \{\Delta\sigma_y\}^T} \equiv - \underset{48x48}{[H]} \underset{48x1}{\{\Delta\sigma_y\}} + \underset{48x24}{[Q]} \underset{24x1}{\{w_y\}} = 0;$$

$$\frac{\partial \Pi_R}{\partial \{\Delta w_y\}^T} \equiv \underset{24x48}{[Q]^T} \underset{48x1}{\{\Delta\sigma_y\}} - \underset{24x1}{\{f_p\}} - \underset{24x1}{\{\Delta f_p\}} + \underset{24x1}{\{f_\sigma\}} = 0. \tag{14}$$

The system of equations (14) can be represented in traditional form for the finite element method

$$\underset{72x72}{[k]} \underset{72x1}{\{z_y\}} = \underset{72x1}{\{F\}}, \tag{15}$$

where

$$\underset{72x72}{[k]} = \begin{bmatrix} -\underset{48x48}{[H]} & \underset{48x24}{[Q]} \\ \underset{24x48}{[Q]^T} & \underset{24x24}{[0]} \end{bmatrix}$$ is matrix of deformation of FE at the loading step;

$$\underset{1x72}{\{z_y\}^T} = \left\{ \underset{1x48}{\{\Delta\sigma_y\}^T} \; \underset{1x24}{\{w_y\}^T} \right\}$$ is vector of nodal unknowns of FE;

$$\underset{1x72}{\{F\}^T} = \left\{ \underset{1x64}{\{0\}^T} \left(\underset{1x8}{\{\Delta f_p\}^T} + \underset{1x8}{\{R\}^T} \right) \right\}$$ is nodal forces vector of FE;

$\{R\}^T = \{f_p\} - \{f_\sigma\}$ is discrepancy.

4. Results

This paragraph shows the results of solving test problems, in which adequacy of the results obtained through described algorithm basis was checked.

Results obtained with numerical experiments and results obtained with equations of structural mechanics solution basis have been compared. Purpose of these tests was to check both the capabilities of developed algorithm and the convergence of results.

Test example 1. The stress-strain state of a rotation shell with spherical bottom under internal pressure is considered (the diagram is shown in figure 1). The following initial data are accepted: q=2,39 MPa, E=2×10^4 MPa; μ=0,3. The shell has been divided along Meridian into 75 elements, with a thickness of is 3, 4, 5, 8, 10 and 11 elements. Results of calculations at step 30 are shown in sheet 1. First line shows the number of nodes along thickness of the shell, second one shows normal stresses in the inner fiber of section 2-2 and the third one shows the stress intensity σi in the same fiber of section 2-2.

It can be seen that even with 5 elements in thickness of the shell sufficient computational process convergence has been achieved.

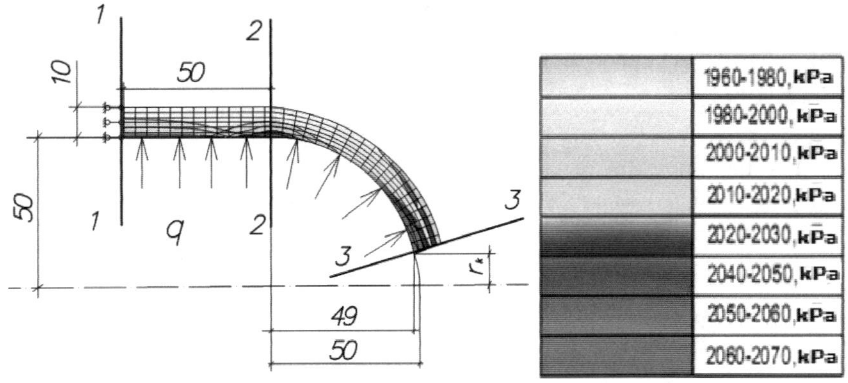

Figure 1. Load distribution diagram for cross-section thickness

Figure 1 shows zones of plastic deformations in the longitudinal section of the shell. When loading the first plastic deformations occur in the section 3-3 of shell's inner surface.

Table 1. Calculation results.

Number of elements	3	4	5	8	10	11
σ_{xx}, normal stress	1062,34	1074,64	1126,23	1126,84	1125,35	1126,74
σ_i stress intensity	2195,21	2205,44	2214,62	2214,113	2213,73	2214,51

The statics equation of zero equal sum of external and internal forces projections on axis of the shell has the form

$$\sum_{i=1}^{n} \sigma_{xx} \Delta h_i \cdot 2\pi r_i = q\pi\left(R^2 - r_K^2\right),$$

$N_q^{\cdot} = q\pi\left(R^2 - r_K^2\right)$ is resultant of external forces arising from applied pressure;

$N^{in} = \sum_{i=1}^{n} \sigma_{xx} \Delta h_i \cdot 2\pi r_i$ is resultant of internal forces arising from stresses in the cross-section 2-2;

$N_q^{\cdot} = 3680,16кH$; $N^{in.} = 3614,02\ кH$ the difference is $\delta = 1,87\%$.

Result shows the correctness of the developed algorithm for the elastic–plastic properties of material taken into account.

Test example 2. The stress-strain state of the rotation shell with an elliptical bottom under internal pressure is considered (scheme is shown in figure 2). The following input data is accepted: q=1,74 MPa, E=2×10^4 MPa; μ=0,3.

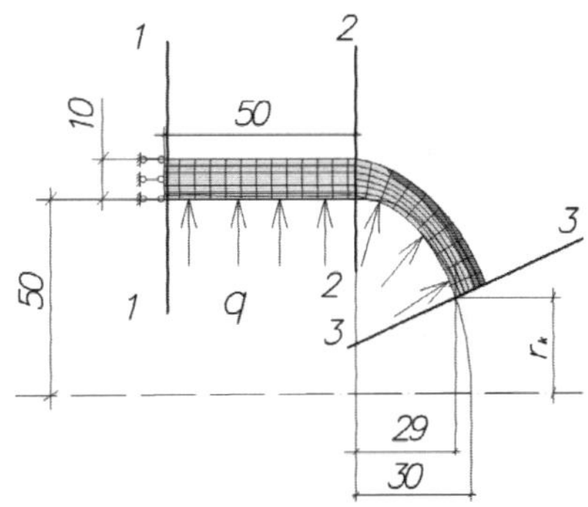

Figure 2. Scheme of load distribution through the thickness of the cross section

Static check: sum of the projections of external and internal forces on the shell axis is zero, was performed for section 2-2.

$N_q^{\cdot} = 2822,17\,кН$; $N^{in.} = 2884,24\,кН$ the difference is δ = 2,24%.

Figure 2 shows the zones of plastic deformations in longitudinal section of the shell. When loading first plastic deformations occur in the section 3-3 of the shell's outer surface.

Test example 3. The stress-strain state of the rotation shell with an elliptical bottom under internal pressure is considered (scheme is shown in figure 3). The following input data is accepted:

q=2,91 MPa, E =2×10^4 MPa; μ =0,3.

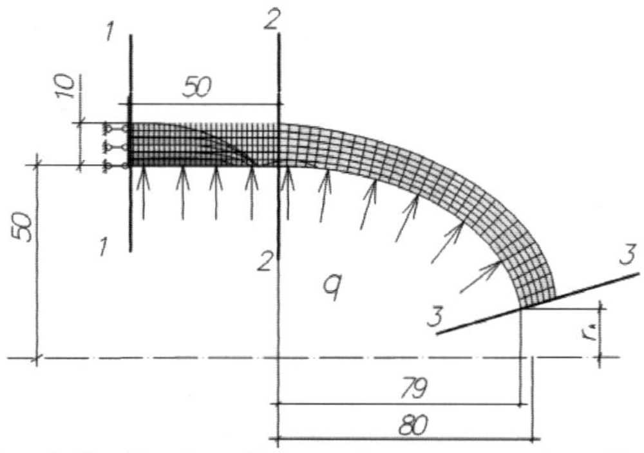

Figure 3. Load distribution diagram for cross-section thickness

In this example the resultant in section 2-2 of internal stresses N^{in}=3782,37 kN and resultant from external pressure Nq=3841,17 kN. The statics equation of sum of external and internal forces projections on the axis of the shell in section 2-2 equality to zero was performed with accuracy δ = 1,56%.

In all cases the convergence of numerical calculation results was checked depending on the number of elements in the thickness of the shell. Results showed that a sufficient number of elements in the thickness is equal to 5.

5. Conclusion

Results show correctness of the developed algorithm with elastic–plastic properties of the material being taken into account.

It is known that the main requirements for mathematical models are those of adequacy, universality and economy. Developed finite element allows us to meet mentioned above requirements. By adopting a number of hypotheses and determining both stresses and deformations, resources computing is reduced ensuring cost-effectiveness and simplicity. Using developed scheme of mixed functional application in implementation of FEM the mathematical model acquires one of its main properties called "Potentiality".

References

[1] Arkov D P 2012 *Application of the mixed functional finite element method to the calculation of plates and shells taking into account physical nonlinearity* (Cand. techn. Sci. diss. Volgograd, p 156)

[2] Alibeigloo A and Nouri V 2010 *Static analysis of functionally graded cylindrical shell with piezoelectric layers using differential quadrature method* (Composite Structures) 92(8), pp 1775-1785

[3] Akbari R and Alashti M, Khorsand M H, Tarahhomi 2013 *Thermo-elastic analysis of a functionally graded spherical shell with piezoelectric layers by differential quadrature method* (Scientia Iranica) Volume 20(1), pp 109-119

[4] Asanjarani A and Satouri S, Alizadeh A, Kargarnovin M H 2015 *Free vibration analysis of 2D-FGM truncated conical shell resting on Winkler-Pasternak foundations based on FSDT* (Proceedings of the Institution of Mechanical Engineers) Part C: Journal of Mechanical Engineering Science 229(5), pp 818-839

[5] Xiaoqin Shen and Qian Yang, Linjin Li, Zhiming Gao, Tiantian Wang 2020 *Numerical approximation of the dynamic Koiter's model for the hyperbolic parabolic shell* (Applied Numerical Mathematics) vol 150, pp 194-205

[6] Hernandez E and NaranjoJ C 2020 *Velloj in Modelling of thin viscoelastic shell structures under Reissner–Mindlin kinematic assumption* (Applied Mathematical Modelling) vol 79, pp 180-199

[7] Semenov A 2019 *Mathematical model of deformation of orthotropic shell structures under dynamic loading with transverse shears* (Computers & Structures) vol 221,pp 65-73

[8] Lebiedzik C 2007 *Exact boundary controllability of a shallow intrinsic shell model* (Journal of Mathematical Analysis and Applications) 335(1), pp 584-614

[9] Mitsuru E 2016 *An alternative first-order shear deformation concept and its application to beam, plate and cylindrical shell models* (Composite Structures) 146(20), pp 50-61

[10] He X and Du H, Ying Z, Wang L, Melnik R. 2019 *Modeling static microstructure of shape memory alloy via* (Legendre wavelets collocation method Journal of Physics: Conference Series) 1419(1), 012003

[11] Cheng Q and Sun W, Cheng Y, Chen L, Li T 2019 *Nonlinear Control Strategy Based on Lyapunov Function for MMC Under Grid Voltage Unbalance Condition* | [电网电压不平衡条件下MMC的基于Lyapunov函数非线性控制策略] Gaodianya Jishu/High Voltage Engineering 45(12), pp 3984-3992

[12] Errico F and Tufano G, Robin O, Ichchou M, Atalla N 2019 *Simulating the sound transmission loss of complex curved panels with attached noise control materials using periodic cell wavemodes* (Applied Acoustics) 156, pp 21-28

[13] Tyukalov Y Y 2019 *Calculation of circular plates with assuming shear deformations* (Conference Series: Materials Science and Engineering) 687(3), 033004

[14] Kress G and Filipovic D 2019 *Exact model for the response of moderately thick laminates to transverse forces* (Composite Structures) 227, 111261

[15] Morse L and Sharif Khodaei Z, Aliabadi M H 2019 *A dual boundary element based implicit differentiation method for determining stress intensity factor sensitivities for plate bending problems* (Engineering Analysis with Boundary Elements) 106, pp 412-426

[16] Klochkov Yu V and Nikolaev A P, Ishchanov T R, Andreev A S, Klochkov M Yu 2020 *Consideration of geometric nonlinearity in finite element strength calculations of thin-walled structures of the shell type* (Construction mechanics of engineering structures and structures) vol 16. (1), pp 31-37

[17] Klochkov Yu V and Nikolaev A P, Vakhnina O V, Klochkov M Yu 2019 *Variants of defining relations of the deformation theory of plasticity in the calculation of the rotation shell based on the finite element method* (Construction mechanics of engineering structures and structures) vol 15(4) pp 315-322

[18] Arkov D P and Gureeva N A 2011 *Application of the mixed finite element method for strength calculations of silos intended for grain storage* (Proceedings of the lower Volga agrodiversity complex: Science and higher professional education) vol 1 (21), pp 189-197

Measurement of Strain Induced on a Deep Placement Fertilizer Applicator by Static load and Finite Element Simulation

I Patuk[1], H Hasegawa[2], I Borodin[3], R Pavlyuk[4]

[1]Graduate School of Science and Technology, Niigata University, 8050 Ikarashi, 2 no-cho, Nishi-ku, Niigata, 950-2181, Japan

[2]Institute of Science and Technology, Niigata University, 8050 Ikarashi, 2 no-cho, Nishi-ku, Niigata, 950-2181, Japan

[3]Engineering and Technology Institute, Primorskaya State Academy of Agriculture, Bluhera st., 44, Ussuriisk 692510, Russia

[4]Stavropol State Agrarian University, Zootekhnichesky Lane, 12, Stavropol, 355017, Russia

E-mail: yaroslav_w93@mail.ru

Abstract. This paper briefly reveals study results performed on a deep placement fertilizer applicator (DPFA) in order to measure strain gage resistances and calibrate a strain gage piece of equipment in laboratory conditions. The finite element method (FEM) simulation was performed in order to identify stress concentrated areas by strain gage attachments, and to determine optimum material properties for a DPFA prototype fabrication. New laboratory arrangements and theories of resistance force determination of a DPFA were developed in order to simulate different soil depths (0.15, 0.20, 0.25m) and apply a static load (5000N) in laboratory conditions. Unique practical knowledge was obtained in terms of different strain gage applications on the DPFA in its rigid form to obtain strain gage resistances. It was found that the strain sensitivity of the strain gages depended on the work dimensions of strain gage characteristics.

Keywords: Deep placement fertilizer applicator; static load; FEM; strain gage; soybean cultivation

1. Introduction

The development of new machinery and technologies in agriculture is one of the key functions in sustainable production. Nowadays, agricultural machinery manufacturers produce machinery and tools (e.g. conventional plows, harrows, chisel plows, etc.) with significant quantity and excessive design (e.g. over mass, size, and safety factors) [1]. Therefore, improvement by way of streamline design would significantly reduce the amount of used material and the final cost.

Deep placement fertilizer applicators (DPFA) are still one of the most important issues in developing agricultural tools in new agricultural machinery. Currently, the development of a DPFA has become a key point in utilizing a deep placement fertilizer technology (DPFT) in crop production. Since the DPFT is an advanced and cost-effective method, which serves not only in decreasing the wastage and excessive use of fertilizer, but also in the even supply of nutrition in the soil and the mitigation of the negative effects caused by fertilizer applications on the environment [2]. In Addition, the subsoil tillage by the DPFA may significantly enhance crop growth through the tillage of compacted soil layers which provides free air and water exchange. Additionally, it increases labor-savings by the effect of slow-release fertilizers [3-4].

To apply the DPFT during crop planting, new DPFAs need to be developed for subsoil tillage with simultaneous application of slow-release fertilizer. Although several studies have been effectively carried out on deep placement fertilizer machinery and subsoiler tools [5-8], there is limited practical knowledge on the usage of DPFT and design optimization of DPFA. Thus, this study was carried out in order to obtain practical knowledge of a DPFT for soybean cultivation in Primorsky Krai (PK), Russia. Moreover, the results of this study will be used for future machinery development in terms of shape, streamline design and the reduction of excessive design.

The main purpose of this study was to measure strain resistances by applying a static load on a DPFA with attached strain gages in laboratory conditions. These measurements are needed to obtain data on strain gage resistance and calibrate strain gage equipment before field experiments to calculate traction resistance and soil resistance. Another purpose was to perform a simulation on the DPFA to evaluate the mechanical behavior by stress and strain analysis of the static simulation through the Finite Element Method (FEM). This simulation was needed to acquire the locations of stress concentrations on the DPFA where strain gages can be attached for practical experiments.

2. Materials and Methods

2.1 Designed DPFA

This study was based on a chisel plow model, which was made by Sandakov et al. [8]. The model was excessively designed and used for subsoil cultivation. However, we have designed a new DPFA plow, which is not only satisfactory with our technical issues for soybean cultivation but meets the issues of conserving resources due to its streamline design (Figure 1).

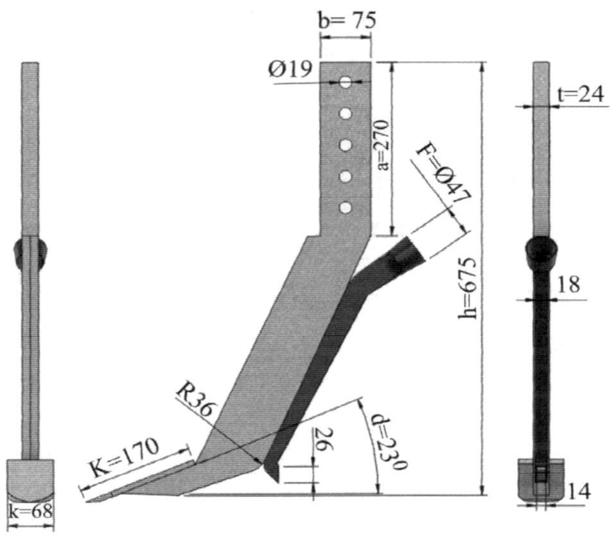

Figure 1. Designed DPFA, (mm)
a – frame connector height; b – shank width; h – shank height; K – knife length; k – knife width; R – curvature radius of the shank; d – penetration angle; F – fertilizer pipe inlet; t – shank thickness

The main issues were to achieve deep tillage and simultaneously apply fertilizers into the soil at a depth of 0.15, 0.20 and 0.25m. In addition to this, the fertilizer pipe attached to the chisel shank should not increase the tractive resistance made by the chisel plow. In the case of the tractive resistance, we decided to make a fertilizer pipe with a round inlet and square outlet, attaching it on the back of the plow to avoid resistance from the front.

2.2 FEM

To evaluate the mechanical behavior of the DPFA and acquire the stress locations for strain gage attachments, several types of FEM analysis were used, including structural and stress analysis depending on the variable loads applied. This FEM simulation software (Fusion 360 by Autodesk, Inc.) was used in collaboration with an advanced computer aided design (CAD), which is commonly implemented in the agricultural engineering industry for design and manufacturing processes [9].

According to the previous study, the static force induced on a chisel plow meets the dynamic force at targeted soil depths of 0.3, 0.35 and 0.4m and at operating speeds of 4, 4.5 and 5km/h, respectively [8]. Therefore, due to the heavy soil property in PK (clay soils with 25-40% of clay), and our experience, 6000N has been considered as a maximum static load for the FEM simulation.

2.3 Laboratory setup

A laboratory setup was developed to simulate and apply the static load to the DPFA in laboratory conditions. Figure 2 shows the main features of the laboratory setup.

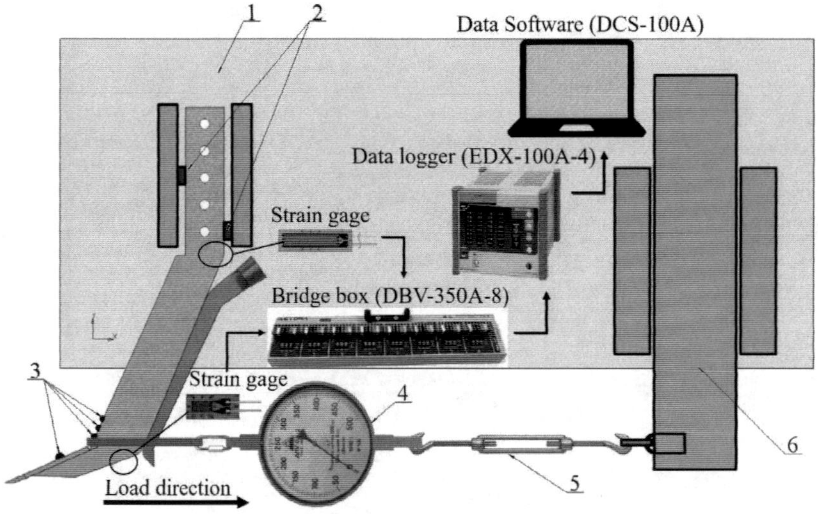

Figure 2. Laboratory setup

1 – metal worktop; 2 – adjustable boundary conditions of bolt connectors; 3 – applied force centers; 4 – dynamometer (5000N); 5 – adjustable load turnbuckle; 6 – adjustable bracket depth

The main idea of the laboratory arrangement was to fix the DPFA on a worktop and apply an interval static load. This configuration was made on a metal worktop, and all parts were fixed by an arc welding. In order to simulate the different frame fixations of the DPFA, we used two adjustable cylinders (Figure 2 (2)) with a diameter of 19mm. Furthermore, to simulate different soil depths (0.15, 0.20 and 0.25m), we determined force (loads) centers (Figure 2 (3)).

2.4 Strain gage equipment

The strain gages attached to the DPFA, data logger amplifier with data acquisition software and the bridge box were manufactured by Kyowa Electronic Instruments Co., Ltd. Japan.

Two types of foil strain gages (KFGS-30-120-C1-11 L5M2R and KFGS-20-120-C1-11 L5M3R) were used with the 2 and 3-wire systems, as well as different gage widths and lengths in order to capture and identify different strain resistances. The EDX-100A-4 data logger amplifier with the DCS-100A data acquisition software was used to record data manually and periodically with different time intervals. The one-touch type bridge box (DBV-350A-8) was used to connect the strain gages with the data logger amplifier.

3. Results and Discussion

3.1 FEM Simulation

Once the design was completed, a 3D model was generated. It performed a static analysis using the software AutoCAD Fusion 360. This simulation showed the specified areas with a different range of colors to visualize the material's behavior of the DPFA under different types of loads and boundary conditions. Thus, the optimum material property for the DPFA fabrication, and maximum stress locations where the strain gages were attached were determined. Figure 3 shows the boundary conditions and the areas of applied stress.

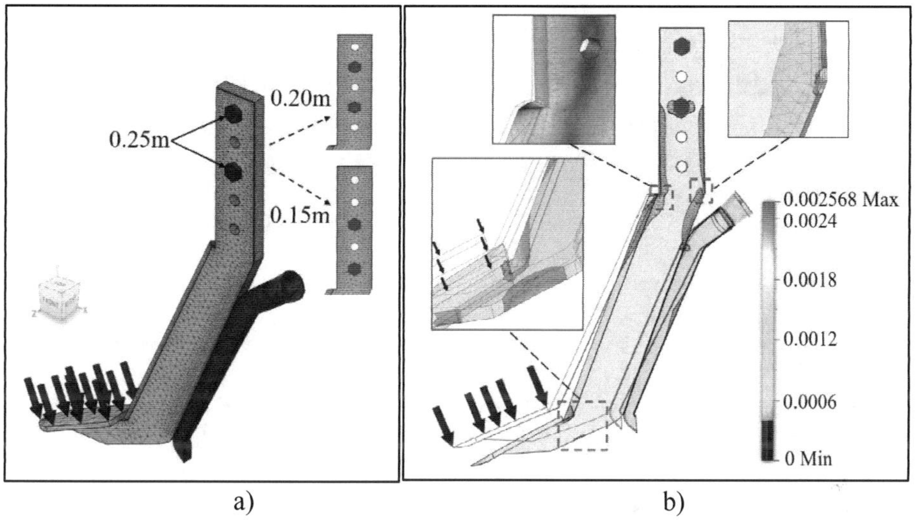

Figure 3. FEM simulation
a) – boundary conditions; b) – stress and strain analysis

In real subsoil conditions, a soil load will be evenly distributed throughout the subsoil area of the DPFA, thus in the FEM simulation, different loads were applied on the maximum load surface areas of the knife. Therefore, we simulated a different DPFA fixation depending on different soil depths as shown in figure 3 (a). Moreover, the FEM simulation showed the most stressed areas (Figure 3 (b)), where we attached the strain gages to obtain strain gage resistances and calibrate all strain gage equipment. In addition, by multiple experiments, we have determined the material properties of the DPFA to have optimum yield and ultimate strength. The determined optimum material properties of the DPFA are presented in Table 1.

Table 1. Material properties of the DPFA.

DPFA parts	**Shank and Knife**	**Fertilizer pipe**
Material	High strength steel	Common steel
Young's modulus (MPa)	207×10^{-3}	205×10^{-3}
Poisson's ratio	0.33	0.29
Yield strength (MPa)	780	207
Ultimate strength (MPa)	950	345
Maximum stress (MPa)	379.9	150.1
Maximum strain (mm/mm)	25.6×10^{-4}	11.3×10^{-4}
Mass (kg)	11.2	0.58

Next, based on the results of the simulation and material properties, a DPFA prototype was manufactured and strain gages were attached for laboratory experiments. Figure 4 illustrates a manufactured DPFA with attached strain gages fixed to the laboratory setup.

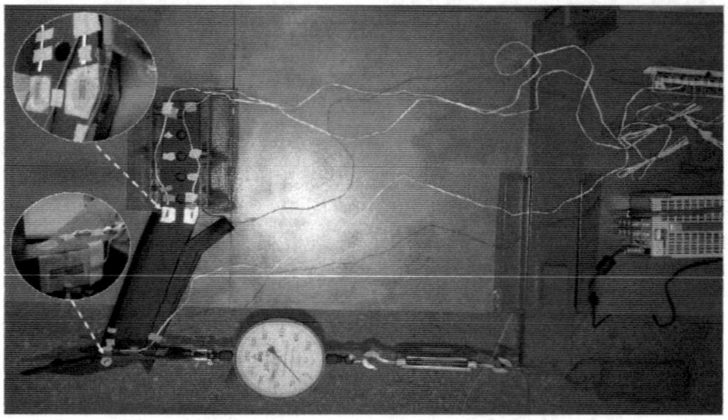

Figure 4. DPFA fixed on the laboratory setup

3.2 Determination of resistance force of DPFA

By the FEM simulation in the CAD Fusion software, the results were obtained, however, to simulate the static load and determine the strains induced on the DPFA in laboratory conditions, it was necessary to determine the centers of application of a resultant force depending on the different soil depths. A resultant force is a force acquired by combining a system of forces on a rigid body. The defining feature of a resultant force is that the resultant force has the same action on the rigid body as the initial system of forces [10].

Accepting the assumption, that the soil resistance force applied to the DPFA is proportional to the front area of the DPFA. In this case, we used the method of determining the centers of parallel forces on the DPFA, such as the method of determining the gravity center of the rigid bodies.

We divide the front area of the DPFA (the front area of the DPFA is perpendicular to the axis of soil movement) into three geometric figures, symmetrical to the vertical axis. Figure 5 illustrates the scheme of the determined centers of a resultant force application depending on the different soil depths with all dimensions for calculation.

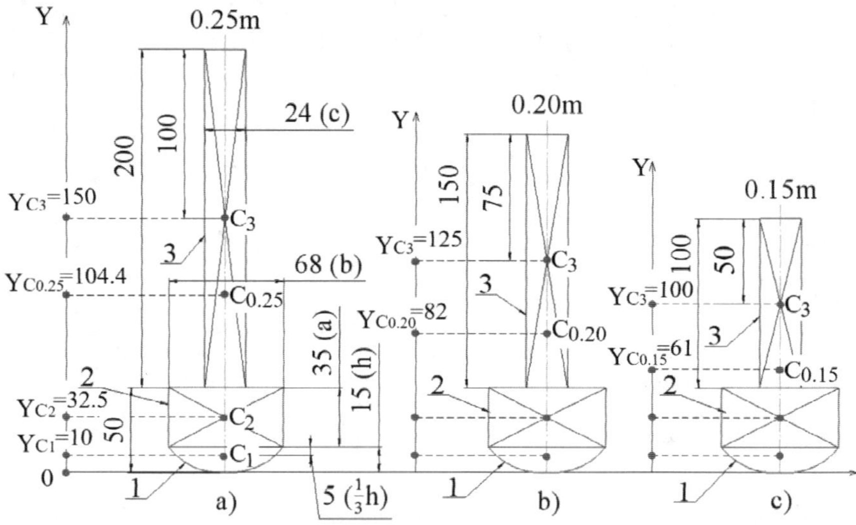

Figure 5. Scheme of a resultant force application depending on the different soil depths, (mm)
1 – triangle of the bottom knife part; 2 – rectangle of the top knife part; 3 – rectangle of the shank

According to the scheme (Figure 5), we determined the coordinates of the centers C1, C2, C3 of the application of resistance forces of individual sections (1,2,3) of the DPFA, along the Oy axis as follows: C1 = 10; C2 = 32.5; C3 = 150 for soil depth of 0.25m (C1 and C2 have unchanged values, C3 has a variable value, depending on the soil depth).

To determine the center of application for the resistance force on the DPFA, we used the formula for determining the center of parallel forces (center of gravity) [11], as follows:

$$Y_C = \frac{1}{S_{Total}} \sum (Y_{Ci} \cdot S_i) \tag{1}$$

where: Y_C is the general center of application of the resistance force on the DPFA; S_{Total} is the total area of the DPFA (front area applied in the soil), mm^2; Yc_i are coordinates of the resistance force center of each DPFA areas, mm; Si is the area of each section of the DPFA, mm^2.

A brief example of the calculation for 0.25m applied force center is given below:

$$Y_{C_{0.25}} = \frac{Y_{C_1} \cdot S_1 + Y_{C_2} \cdot S_2 + Y_{C_3} \cdot S_3}{S_1 + S_2 + S_3} \tag{2}$$

Where: S_1=510 and S_2=2380 are areas of individual geometric shapes of the knife (Figure 5 (1, 2)), mm^2; S_3=4800 is the area of the shank (Figure 5 (3)), mm^2. However, according to the figure 5 (b, c), the coordinate of the center resistance force will be changed, and the area S_3 (which is the rectangle of the shank) will be decreased.

$$Y_{C_{0.25}} = \frac{10 \cdot 510 + 32.5 \cdot 2380 + 150 \cdot 4800}{510 + 2380 + 4800} = 104.4\text{mm}$$

By the calculation, three applied force centers were obtained for the shank, and one center for the knife of the DPFA depending on the soil depth. Thus, for the laboratory experiment, we applied a static load on the force centers (Figure 2 (3)) at a distance from the lower part of the DPFA ($Y_{C0.25}$=104.4mm; $Y_{C0.20}$=82mm; $Y_{C0.15}$=61mm) for the shank, and (Y_k=28.5mm) for the knife.

3. 3 Strain gages resistance result

By using the laboratory configuration (Figure 2), the static load 5000N was applied with an interval of 500N stepwise, and data recording was done manually. Three unique strain gage resistances were obtained depending on the simulation of different soil depths. Figures 6 and 7 show the results of obtained strain gage resistance by two different strain gages attached to the upper part of the shank. The negative (–) direction of the strain gage's resistance shows that a foil strain gage deforms in the reverse direction.

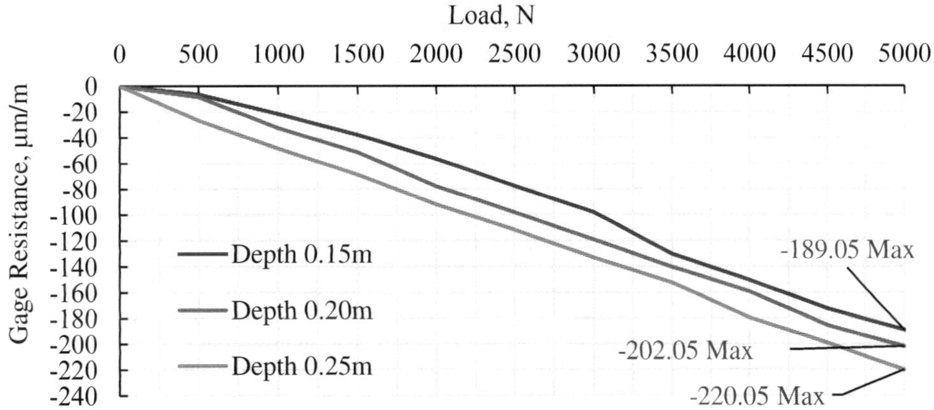

Figure 6. Strain gage resistance of KFGS-30-120-C1-11 L5M2R

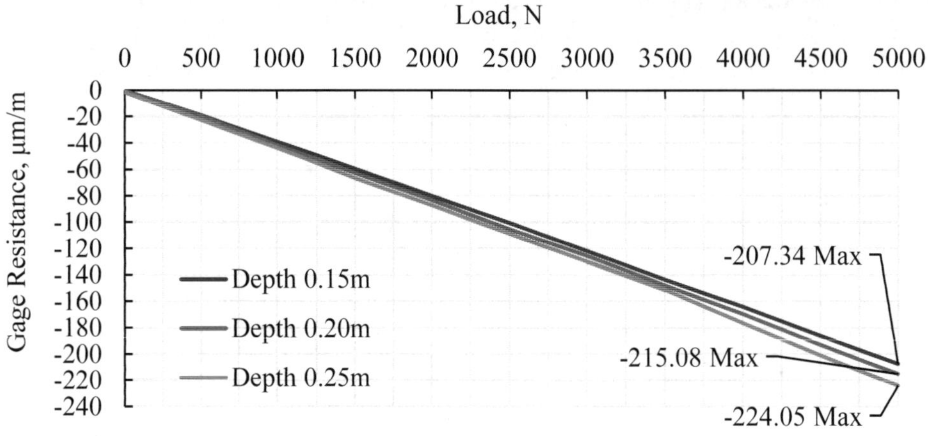

Figure 7. Strain gage resistance of KFGS-20-120-C1-11 L5M3R

Due to the different strain gage characteristics, especially gage width and length, we can notice that there is an entirely different resistance sensitivity under the same experimental conditions. Moreover, the strain gage with wider gage width (KFGS-20-120-C1-11 L5M3R) shows not a significant difference in the resistance between different simulation conditions. However, during the measurements, it was noticed that the KFGS-20-120-C1-11 L5M3R had fewer noises in comparison with the KFGS-30-120-C1-11 L5M2R which showed more stability in terms of data recording. These effects of noises can be explained by the different wire systems and connections between the strain gage and bridge box. Figure 8 shows the difference between the two strain gages (gage characteristics (Grid)) attached to the upper part of the shank of the DPFA.

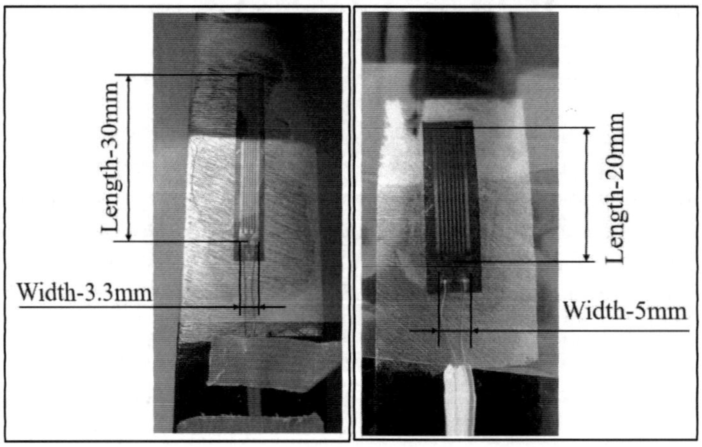

a) b)

Figure 8. The work dimensions of used strain gages
a) – KFGS-30-120-C1-11 L5M2R (2-wire system); b) – KFGS-20-120-C1-11 L5M3R (3-wire system)

We can conclude that in order to capture more accurate strains induced by a load in a rigid body, it is better to use the strain gages with a longer gage length since it leads to obtaining accurate strains. Plus, the 3-wire gage connection system will help to avoid unnecessary noises and measurement errors. For further field experiments to calculate traction resistance of the DPFA and soil resistance induced on a subsoil area, we consider using the strain gages (KFGS-30-120-C1-11), however, with the 3-wire system.

4. Conclusion

This study was focused on the DPFA to evaluate the mechanical behavior through stress and strain analysis of the static simulation by FEM, and perform a laboratory experiment to obtain data on strain gage resistances and calibrate strain gage equipment before field experiments.

In the FEM simulations were found the most stressed areas from where we attached the strain gages for laboratory experiments. Plus, the FEM determined the material properties of the DPFA with optimum yield and ultimate strength. The laboratory experiments and theory of resistance force determinations were developed to obtain strain gage resistance, and calibrate strain gage equipment. Practical knowledge was obtained in terms of different strain gage characteristic types. It was found that the strain sensitivity depends on the work dimensions of a strain gage. Additionally, the wire gage connection system affects the accuracy of the data recorded by the reduction of noises.

5. Acknowledgments

We would like to thank the Ministry of Education, Culture, Sports, Science and Technology (MEXT), Japan, for the opportunity to make this study under providing financial support within a Ph.D. scholarship at Niigata University, Japan. Furthermore, we gratefully thank Primorskaya State Academy of Agriculture, Russia, for technical support and providing a laboratory for the experiments. We also thank Mr. Eli Joel Pappas, JET Programme Participant and Linguistician, Niigata, Japan, for editing a draft of this manuscript.

References

[1] Yurdem H, Degirmencioglu A, Cakir E and Gulsoylu E 2019 Measurement of strains induced on a three-bottom moldboard plough under load and comparisons with finite element simulations *Measurement* **136** pp 594-602

[2] Yao Y, Zhang M, Tian Y, Zhao M, Zhang B, Zhao M and Yin B 2018 Urea deep placement for minimizing NH3 loss in an intensive rice cropping system *Field Crops Research* **218** pp 254-266

[3] Topakci M, Celik H K, Canakci M, Rennie A E, Akinci I and Karayel D 2010 Deep tillage tool optimization by means of finite element method: Case study for a subsoiler tine *Journal of Food, Agriculture & Environment* **8**(2) pp 531-536

[4] Fujii T, Hasegawa H, Ohyama T and Sinegovskaya V T 2015 Evaluation of tillage efficiency and power requirements for a deep-placement fertilizer applicator with reverse rotational rotary *Russian agricultural sciences* **41**(6) pp 498-503

[5] Ahamed M S, Ziauddin A T M and Sarker R I 2014 Design of improved urea super granule applicator. *Int. J. Appl. Science and Engineering Research* **3**(1) pp 98-104

[6] Hang C, Gao X, Yuan M, Huang Y and Zhu R 2018 Discrete element simulations and experiments of soil disturbance as affected by the tine spacing of subsoiler *Biosystems Engineering* **168** pp 73-82

[7] Ebrahimi R, Mirdamadi H R and Ziaei-Rad S 2018 Operational modal analysis and fatigue life estimation of a chisel plow arm under soil-induced random excitations *Measurement* **116** pp 451-457

[8] Sandakov T, Sandakova N, Chang L, Hasegawa H and Radnaev D 2019 Optimum design of a chisel plow for grain production in the Republic of Buryatia, Russian Federation *AMA, Agricultural Mechanization in Asia, Africa and Latin America* **50**(1) pp 73-78

[9] Celik H K, Caglayan N, Topakci M, Rennie A E and Akinci I 2020 Strength-based design analysis of a Para-Plow tillage tool *Computers and Electronics in Agriculture* **169** 105168

[10] McCarthy J M and Soh G S 2010 *Geometric design of linkages* Vol 11 Springer Science & Business Media

[11] Lachuga Y F and Ksendzov V A 2010 *Theoretical mechanics* (Moscow: Kolos) p 576

Improvement of the process of dosing sugar beet seeds using a cone-shaped sowing apparatus

E V Kulaev, M V Danilov , L I Vysochkina, D S Kalugin and V K Maliev

Stavropol State Agrarian University, Zootekhnichesky Lane, 12, Stavropol, 355017, Russia

E-mail lubasha_vis_67@list.ru

Abstract. To improve the quality of sowing seeds of sugar beets, the constructive and regime parameters of the high-speed cone-shaped device, which ensures the transportation of seeds with a cone-shaped dispenser, the increase in the grasping capacity of cells, the process of sowing seeds into cells and laying seeds in cells. As a result of analytical studies, regularities in the change in the supply of sugar beet seeds by a cone-shaped sowing device are revealed depending on the linear velocity, the angle of inclination of the cells and the thickness of the sowing ring. The high-speed seeder provides high accuracy of seed dosing, high uniformity of seeding and seed distribution without damage in the seed bed. This contributes to the survival and stability of the culture in the first days after sowing, which ultimately will ensure a reduction in the cost of growing sugar beet and increase yield. The analytical dependencies obtained and the calculation procedure can be used in the design of mechanical precision seeding machines. Empirical regression models are used to adjust the sowing apparatus to optimal operating conditions.

Keywords: sugar beet, cone-shaped seeder, sowing ring, sowing devices.

1. Introduction

Sugar beet is a technical culture, giving carbohydrate-rich root vegetables. With high agrotechnics, the yield of beet can be 8 t/ha or more. The most important factor in increasing the yield of sugar beet is the sowing of seeds in a compressed agrotechnical time with the optimal placement of plants in the area of nutrition. At present, seeders with mechanical and pneumatic sowing devices are used for sowing sugar beet. World practice proves that seeders with a mechanical sowing apparatus are better in seeding than pneumatic seeders [1-3]. Mechanical sowing apparatuses by design and principle of operation are divided into reel, cup, screw, vibratory, centrifugal, spoon, cellular and apparatus for laying a seed moisture-soluble tape. Apparatuses of cellular-tape type are used on precision seeders «Mammen» firm «Giraud», «Dalman», «Hillsheg» firm «Overum», «Stenhei» [4]. Disk sowing machines are installed on seeders SKNK-6, STH-4A, STSN-6A, SST-12V, McCormick, John Deer, Monodrill, Tuz, Ebra, Prodacts, Webb, «Gloucester», «Exaclete» and others [4]. The cellular disk unit is used on the seed drills of «Webb» and «Gloucester», «Monocenter», «Exaclete», «Fendt», «Unicorn-2», «Sembner», «Monocenter-SP», «A-695», «A-765», «Tume-Mono», «Tank», «Massey Ferguson», «Palm» [5, 6, 7]. The sowing units of the drum type are used on the seeders such as «Palm Agromatic», «Unisem», «Monosem», SPP-12, SGU, STG-18, «Unikorn» and others. Seeding apparatuses of disk-and-spoon type are installed on seeders «Smat», «Nibex», «John Deere» and other.

The remaining devices for seed dosing have not found wide application, some due to the complexity and high cost of construction, others due to poor quality of sowing [8-9].

From the analysis of the operation of mechanical sowing machines, it is revealed that they have a number of disadvantages: the damage of seeds, which is obtained by removing excess seeds with the help of ejectors; the need for thorough seed calibration; the need for seed discs with different sizes and number of cells.

Also, in seeders with a mechanical seeder, the speed of movement is limited by the permissible rotational speed of the sowing disk [10-11]. The aim of the study is to improve the quality of sowing seeds of sugar beet, by improving the design of the sowing apparatus, ensuring the accuracy of dosing and the absence of damage to the seeds.

2. Materials and methods

To overcome the above-mentioned drawbacks, a model of a mechanical cone-shaped sowing device for precise seed sowing has been proposed. A mechanical cone-shaped precision seeding machine (Figure 1) includes a hopper 1 with an outlet window 2 located on the surface of the cylindrical body 3 on which the conical dispenser 5 is supported, connected to the shaft 10 by means of a drive bolt 11.

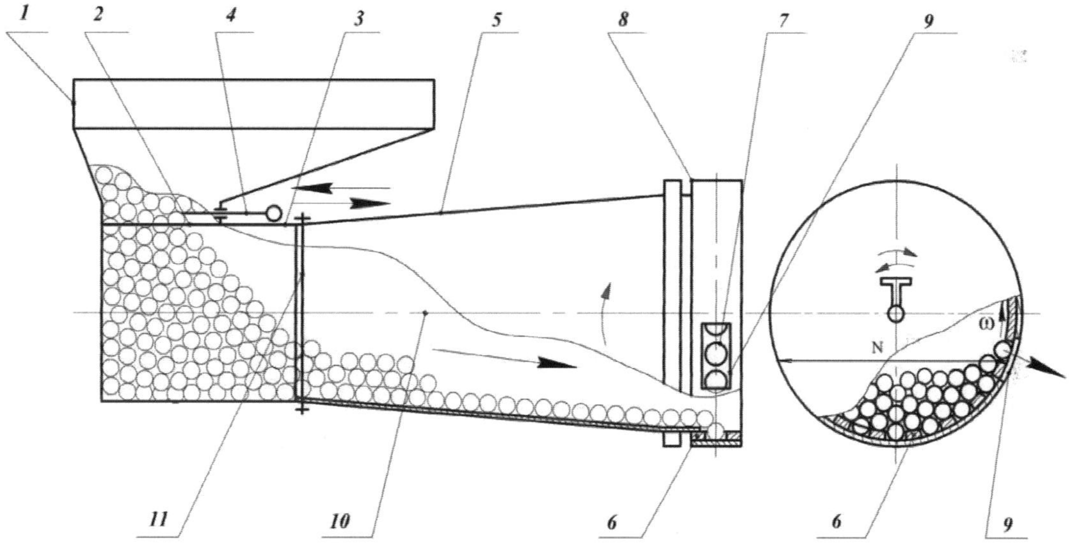

Figure 1. Scheme of a cone-shaped seeder: 1 - hopper; 2 - outlet window; 3 - cylindrical body; 4 - damper; 5 - conical dispenser; 6 - sowing ring; 7 - through cells; 8 - rim's circumference; 9 - groove; 10 - drive shaft; 11 - bolt drive.

The amount of seed to be poured is controlled by the damper 4. On the side of the larger diameter, a sowing ring 6 is attached to the cone metering device with uniformly distributed through cells 7. With the cap on the end part 8 of the rim's circumference which is made a groove 9. The lid is rotatable around its axis changing the angle of the slot arrangement, and recorded. The drive cone dispenser through a chain transmission from the dispenser shaft drills. The operation of the seeding machine is carried out as follows, the seeds from the hopper through the exit window under the action of gravity fall into the cavity of the cylindrical body. Crumbling at the angle of the natural slope, the seeds fall into the inner cavity of the rotating cone dispenser. Due to the frictional force and tilting of the conical wall of the dispenser during rotation of the seeds move along its axis and reach a certain layer seeder ring with cells.

The main factors influencing the exact seeding process by mechanical sowing machines are the type, shape and size of the sowing disk cells, the rotation speed of the sowing disk, the presence of input and output facets on the cells [12-16].

Seeds from the hopper through the exit window enter the cavity of the cylindrical body, crumble at the angle of the natural slope λ and form a layer of h_c height on the surface of the rotating cone metering device. The height of the layer of h_c seeds entering the inner surface of the cone metering device depends on the area of the hopper exit window, the diameter and length of the cylindrical body. The value of this layer is regulated by the damper. The cross-sectional area of the seed layer S_c captured by a rotating cone dosimeter is determined from the expression:

$$S_c = \frac{1}{2}[l_d \cdot R_k - N(R_k - h_c)], \tag{1}$$

where l_d is the arc length, m; R_k is the radius of cylindrical body, m; N is the chord, m.

The height of the seed layer h_c, entrained rotating conical dispenser is determined by the formula:

$$h_c = 2R_k - \frac{L \ \ - L_o}{ctg\lambda}, \tag{2}$$

where L is the length of cylindrical body, m; L_o is the length of the hopper's outlet window, m; λ is the angle of natural slope of seeds, deg.

Seeds with a layer of h_c are fed to the inner surface of the cone dispenser from the side of smaller diameter. Feeding of grain material is continuous. When the cone dispenser rotates with the angular velocity ω, the seeds rise together with the inner surface of the dispenser to the point determined by the elevation angle β, after which they begin to slide down. The seed rolls out at an angle of ε, which ensures the seeds move along the axis of rotation due to the frictional force and the slope of the cone forming the dispenser, at an angle $\alpha_к$ (Figure 2).

To obtain the angle of the largest slope at point M, draw through MD a vertical plane and rotate it around the line MD as around the axis until this plane becomes perpendicular to the edge of the angle of the dihedral angle, so that the angle CFD = 90 °.

Then the angle MFD = γ is the angle of the largest tilt at point M, since the edge of the angular plane of the dihedral angle will be perpendicular to all lines drawn in the vertical plane MFD.

We define the angles: δ, γ and E, assuming the slope angle of the generator of the dispenser $\alpha_к$ and the angle of seed raising β. By the known rules of trigonometry after the transformation, we get:

$$sin\delta = \frac{tg\delta}{\sqrt{1 + tg^2\delta}} = \frac{sin\alpha_k}{\sqrt{tg^2\beta + sin^2\alpha_k}} \tag{3}$$

$$cos\gamma = \frac{1}{\sqrt{cos^2\alpha_k + sin^2\alpha_k + tg^2\beta}} \cdot cos\alpha_k \cdot cos\beta \tag{4}$$

$$tg\varepsilon = \frac{cos\alpha_k \cdot sin^2\delta}{sin\alpha_k} = \frac{sin^2\alpha_k}{2(tg^2\beta + sin^2\alpha_k)} \tag{5}$$

From formula (4) we conclude that the maximum angle of tilt γ is greater than $\alpha_к$ and more β; for $\alpha_к$ = 0, we obtain $\gamma = \beta$, and for $\beta = 0$ - $\gamma = \alpha_к$; at $\beta = 90°$ or at $\alpha_к = 90°$, we obtain $\gamma = 90°$.

From formula (5) we conclude that the deflection angle is $\varepsilon = 0$, both at $\alpha_к = 0$ and at $a_к = 90°$. The formula (5) is reduced to the form:

$$tg\varepsilon = tg\alpha_k \left[\frac{cos^2\alpha_k}{tg^2\beta + sin^2\alpha_k}\right], \tag{6}$$

from which it can be seen that at small tilt angles $\alpha_к$ and at $\beta \cong 30...35°$

$$tg\varepsilon = tg\alpha_k; \quad \varepsilon = \alpha_k. \tag{7}$$

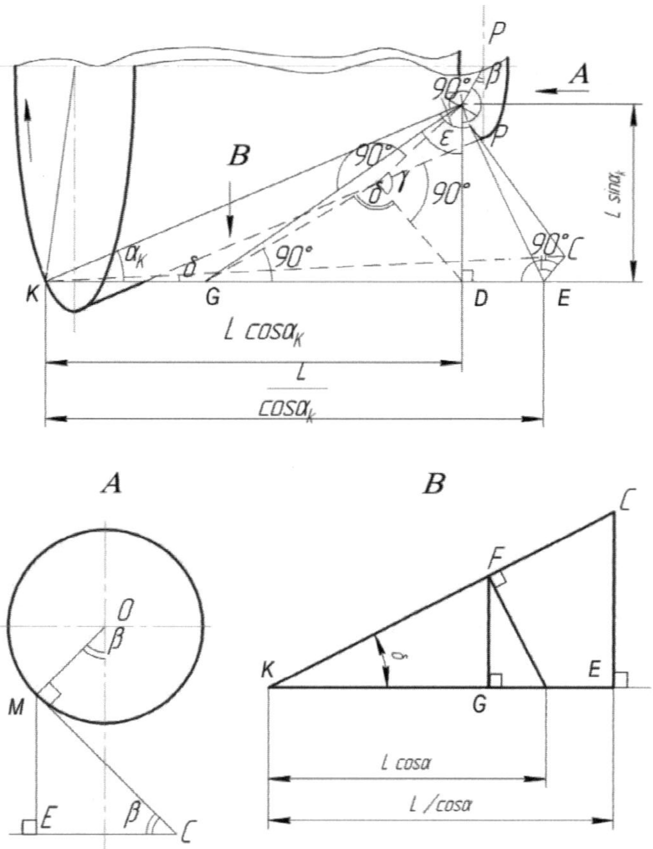

Figure 2. Scheme for determining the angle of the seed slope ε

We take, with sufficient accuracy for technical calculations, that the relative trajectory of the seed movement over the surface of the cone metering device is the correct helix with a lifting angle W, which is determined from the expression:

$$W = \varepsilon + \alpha_k = 2\alpha_k = const. \quad (8)$$

Thus, the relative trajectory of seed movement on the dispenser's tapered surface unfolded in the plane is shown as a straight line running at an angle W of the unfolded circle so that the seeds have passed the entire length of the dispenser L, determined by the dependence:

$$L = V_\Pi \cdot tgW = V_\Pi \cdot tg2\alpha_k, \quad (9)$$

Feed rate is determined:

$$V_\Pi = V_0 \cdot tg2\alpha_k = \frac{\pi \cdot n \cdot R}{30} \cdot tg2\alpha_k \quad (10)$$

where n is the frequency of rotation of the conical dispenser, min^{-1}; R is the radius of the conical dispenser, m.

Consequently, the feed rate of the conical dispenser is directly proportional to the ratio of its diameters, angular velocity and tangent of the double tilt angle of the cone generator. From the length of the conical dispenser feeder *L*, the feed rate at steady state is completely independent. To increase the exciting capacity of the cells, cells are proposed that are located at an angle α_c (Figure 3) to the radius of the sowing ring. The length of the cell L_c in this case is determined from the expression:

$$L_c = \frac{d_c}{cos\alpha_c}. \quad (11)$$

where L_c is the cell length, m; d_c is the cell diameter, m; α_c is the angle of inclination of the cell to the radius of the sowing ring, deg.

As a result, leaving the diameter of the cell d_c unchanged, we increase the path L_c (cell length) that the seed passes through in the slip phase above the cell.

The process of seed sinking into the cells in the phase of sliding along the surface of the sowing ring depends, in the main, on the magnitude of the absolute angular velocity of the seeds, which varies from the rotational speed of the sowing ring. Exceeding the limiting value of this speed will result in the seeds not sinking into the cells of the sowing ring. Thus, in the process of sinking seeds into the cells, the rotation frequency of the sowing ring, the ratio of the mean diameter of the seed to the length of the cell, is closely related.

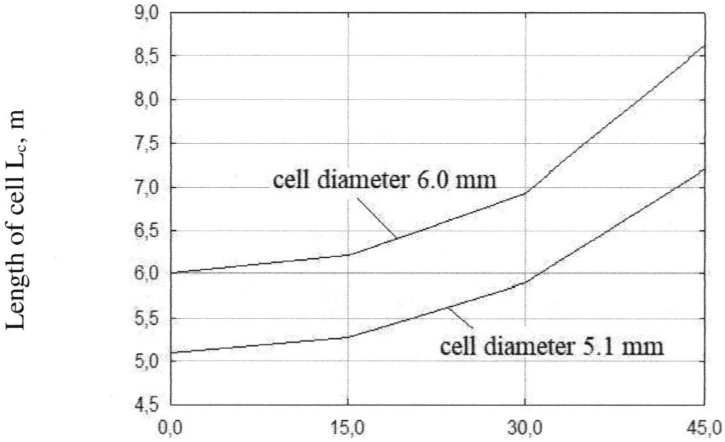

The slope angle of the cell α_c.
Figure 3. Change in cell length L_c depending on the cell inclination angle α_c

In the graphical form, the expression (11) is represented in Figure 3. As can be seen from the obtained expression and the presented graph (Figure 4), the angle of inclination of cells α_c has the greatest influence on the seed yield angle χ from the cells of the sowing ring.

The number of seeds supplied to the cell depends on their diameter and the thickness of the seed ring. To place only one seed in the cell, it is necessary that the diameter of the cell is equal to or exceeds the diameter of the seed by some Δd, depending on the ratio of the speed of movement of the seed and the cell.

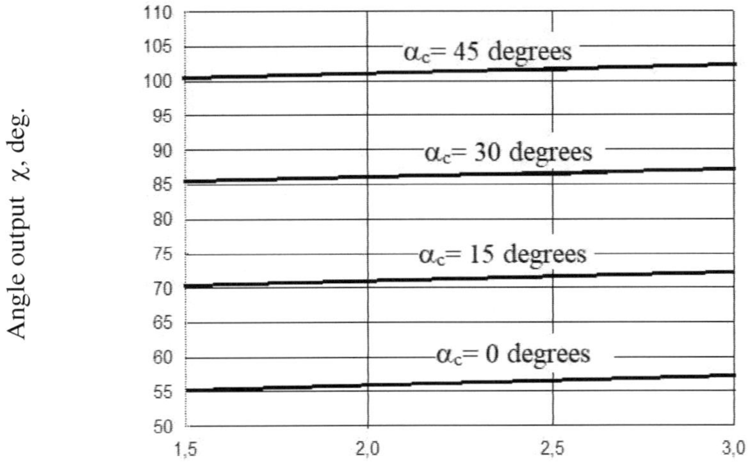

Figure 4. Graph of changes in the angle output χ seeds depending on the angular speed variation and sowing ring ω angle cell α_c

Let us establish the relationship between the diameter D_c of the cells, the excess of the cell depth over the seed Δt located in it and the immersion δ of the upper seed in the cell.

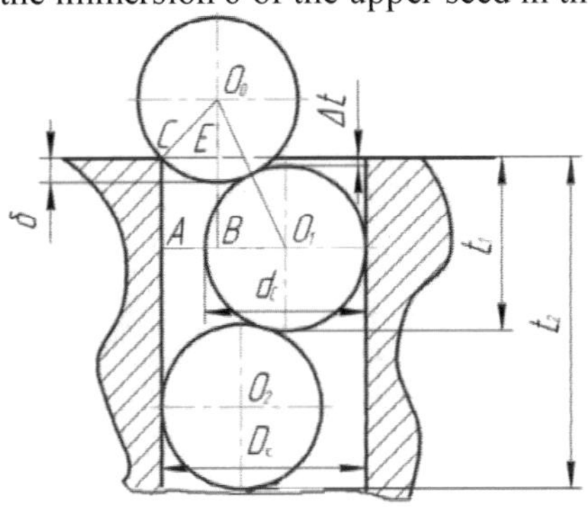

Figure 5. The scheme of laying seeds in a cell with different disk thickness

We get the expression:

$$D_c = \frac{d_c}{2} + [\delta \cdot (d_c - \delta)]^{\frac{1}{2}} + [d_c^2 - (d_c + \Delta t - \delta)^2]^{\frac{1}{2}} \quad (12)$$

From formula (12) we find $\delta = f(d, \Delta t, D)$, having previously adopted the notation

$$k = D_c - \frac{d_c}{2}; \; d_{c1} = d_c + 2\Delta t, \qquad \beta = k^2 + \Delta t^2 + 2d\Delta t$$

After mathematical transformations we get:

$$\delta = \frac{(2k^2 d_c + d \cdot \beta) \pm \sqrt{(2k^2 \cdot d + d \cdot \beta)^2 - \beta^2(d^2 + 4k^2)}}{d^2 + 4k^2} \quad (13)$$

The seed supply by the cells of the seed ring depends on the length of the cells L_c (Figure 6) and the thickness of the seed ring t_{BK}. Let us establish the relationship between the cell length L_c, the seed diameter d_c, the excess of the cell depth over the Δh seed located in it, and the immersion δ of the upper seed in the cell.

The length and angle of the cell are found from the equation:

$$L_c = \frac{d_c}{2}(cos\alpha_c - sin\alpha_c) + \Delta h \cdot sin\alpha_c + \sqrt{d_c^2 - (d_c - \delta + \Delta h)^2} + \frac{d_c}{2cos\alpha_c} \quad (14)$$

The immersion δ of the upper seed in the cell is determined from the equation:

$$\delta = d_c + \Delta h - \sqrt{d_c^2 - \left[L_c - \left(\frac{d_c}{2}cos\alpha_c - \frac{d_c}{2}sin\alpha_c + \Delta h \cdot sin\alpha_c + \frac{d_c}{2cos\alpha_c}\right)\right]^2} \quad (15)$$

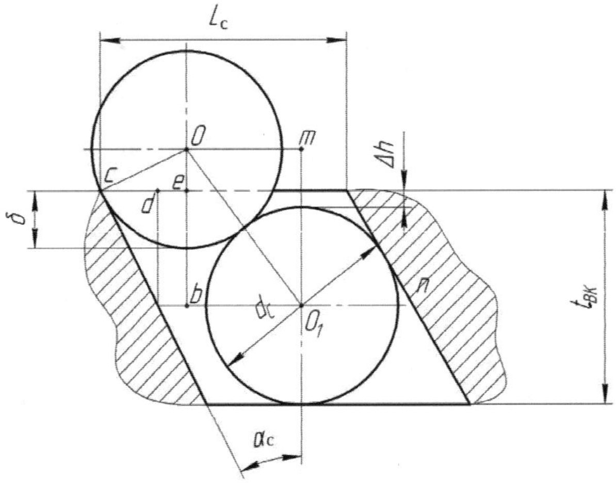

Figure 6. Scheme of seed placement in inclined cells

The thickness of the seed ring t_{BK} is determined from the equation:

$$t_{BK} = \delta + \sqrt{d_c^2 - \left[L_c - \frac{d_c}{2}\left(cos\alpha_c - sin\alpha_c + \frac{2\Delta h \cdot sin\alpha_c}{d_c} + \frac{1}{cos\alpha_c}\right)\right]^2} \quad (16)$$

3. Results and discussion

In accordance with the method of investigating the properties defined seeds needed for experimental studies: the size of sugar beet seeds are normally distributed, the length of the test seed varieties ranging from 3.46 to 5.59 mm, the width varies from 3.26 to 5.58 mm , the thickness of the seeds ranges from 2.31 to 4.37 mm. Volume-weight properties of sugar beet seeds are concluded in the following intervals of variation: the mass of 1000 seeds is 19.25 ... 22.58 g, the bulk density of seeds is 252.51 ... 261.21 grams per liter, the angle of natural tilt is 38°18 '... 44°08'.

On the basis of theoretical studies, it was established that the following factors have the greatest influence on the process of single-grain seed supply by the apparatus: linear velocity of the cell of the sowing ring V_c, $m \cdot s^{-1}$; angle of slope of cells α_c, deg.; thickness of the sowing ring t_{BK}, mm.

Obtained statistical model establishes a mathematical relationship between the parameter optimization - the number of passes made by the cells - Y_{P0}, the number of twins Y_{P2}, secondary meshes serving Y_M, and linear velocity meshes X_1, X_2 angle meshes and a thickness of the seed ring X_3.

$$Y_{PO} = -10{,}95 + 67{,}49V_c + 0{,}065\alpha_c + 2{,}05t_{BK} - 0{,}34V_c\alpha_c - 9{,}66V_ct_{BK} - 0{,}0116\alpha_ct_{BK} \quad (17)$$

$$Y_{P2} = -7{,}92 + 20{,}86V_c + 0{,}155\alpha_c + 2{,}365t_{BK} - 0{,}355V_c\alpha_c - 7{,}399V_ct_{BK} - 0{,}0136\alpha_ct_{BK} \quad (18)$$

$$Y_M = 0{,}885 - 0{,}749X_1 + 4{,}25 \cdot 10^{-3}X_2 + 5{,}75 \cdot 10^{-2}X_3 \quad (19)$$

As a result of determination of rational constructive and technological parameters of experimental mechanical cone-shaped sowing apparatus by constructing sections response surfaces, it was found that the high quality of single-grain seed dosing ($Y_{P0} \leq 2\%$, $Y_{P2} \leq 2\%$, $M \geq 1$) is provided at an angle cell seeding tilt ring $\alpha_c = 30$ degrees, the thickness of the sowing ring is 2.6 mm and the linear velocity of the sowing ring is 0.20 $m \cdot s^{-1}$.

4. Conclusions

Sowing machines for sowing sugar beet seeds do not provide high accuracy of seed dosing, and the presence of an ejector and a reflector in the sowing apparatus results in damage to the seeds. To eliminate these drawbacks, a mechanical cone-shaped precision seeding machine is proposed, in the design of which there are no ejecting and reflecting devices, as a result of which there is no damage to the seeds. Increased the gripping capacity of cells by using obliquely located cells. The regularities of the change in the supply of sugar beet seeds by a cone-shaped sowing device are determined, depending on the linear velocity, the angle of inclination of the cells and the thickness of the sowing ring. As a result of experimental studies, a statistical mathematical model of the process of supplying sugar beet seeds with a cone-shaped sowing device was obtained. Optimal design-technological parameters cone-shaped mechanical sowing device, which complied with the requirements for agronomic crop of sugar beet, are: the angle of inclination of the seed cell rings 30 degrees, the thickness of the seed ring 2.6 mm at a linear velocity of the seed ring 0.2 $m \cdot s^{-1}$.

References

[1] Turan J, Visacki V, Mehandžić S, Findura P, Burg P and Sedlar A 2014 Sowing Quality Indicators for a Seed Drill with Overpressure *Acta Univ. Agric. Silvic. Mendelianae Brun. 62.* pp 1487-92
[2] Mursec B, Vindis P, Janzekovic M, Cus F and Brus M 2007 Analysis of quality of sowing by pneumatic sowing machines for sugar beet *Achievements Mater. Manuf. Eng.* 22/1 pp 85-88
[3] Zubrilina E, Vysochkina L, Danilov M and Maliev V 2017 Design modification of the seed distributor of the pneumatic seeder for corn sowing *Eng. for Rural Development Proceedings* pp 772-778
[4] Merkes R 2000 Technology of cultivation of sugar beet in Germany *Sugar beet* №10 p 18
[5] Sharaby N, Kalsov A and Butovchenko A 2019 A comparative analysis of precision seed planters pp 244-250
[6] Melo R et al. 2013 Qualidade na deposição de sementes de milho em semeadoras em um solo cearense *Revista Ciência Agronômica*, Fortaleza, v. 44, n. 1, pp 94-101
[7] Bottega E, Rosolem D, Neto A, Piazzetta H and Guerra N 2014 Qualidade da semeadura do milho em função do sistema dosador de semente e velocidade de operação *Global Science and Techn*, Rio Verde, v.7, n.1, pp 107-114
[8] Tuğrul K Buzluk S and Boyacioğlu A 2012 Seed drilling distance applications in sugar beet cultivation. *Afr J of Agr Res* Vol. 7(28), pp 4110-17
[9] Nadykto V 2019 Modern Development Paths of Agricultural Production Trends and Innovations: Trends and Innovations p 814
[10] Ovchinnikov V 2017 Results of the study of damage to small seeds from the disk sowing apparatus *Vestnik Mordovskogo universiteta*, Volume 27, pp 190-197
[11] Turan J, Visacki V, Sedlar A, Pantelić S, Findura P, Máchal P and Mareček J 2015 Seeder with Different Seeding Apparatus in Maize Sowing. *Acta Univ. Agric. Silvic. Mendelianae Brun.* 63. pp 137-141
[12] Jia H, Chen Y, Zhao J, Guo M, Huang D and Zhuang J 2018 Design and key parameter optimization of an agitated soybean seed metering device with horizontal seed filling. *Int. J of Agr. and Biol. Eng.*, 11(2) pp 76-87
[13] Manisha S and Ajay V 2018 Neural Network Prediction of Performance Parameters of an Inclined Plate Seed Metering Mechanism and its Reverse Mapping for Rice. *Int. J. Curr. Microbiol. App. Sci.* 7(10) pp 3494-15
[14] Tiago P. da S. Correia, Saulo F. G. de Sousa, Paulo R. A. Silva, et al 2016 Sowing performance by a metering mechanism of continuous flow in different slope conditions. *Engenharia Agrícola*, 36(5) pp 839-45
[15] Mursec B, Vindis P, Janzekovic M, Cus F and Brus M 2008 Testing of quality of sowing by pneumatic sowing machine. *J of Achs in Materials and Manuf Eng.* vol 26 pp 81-88
[16] Bakhtiari, M.R. & Ahmad, Desa. (2017). Design of a vacuum seed metering system for kenaf planting. *Agr Eng Int: CIGR Journal.* 19. pp 23-31.

Analysis of a nonlinear system dynamics in the Morlet wavelet basis

I M Dantsevich, M N Lyutikova, AY Novikov, S A Osmukha

Admiral F. F. Ushakov Maritime State University, Novorossiysk, Lenin Avenue 93, 353918, Russia

E-mail dantsevich.aumsu@gmail.com

Abstract. The article illustrates the methodological study basis of the multi-frequency (packet) spectrum of sea waves in the influence on the associated system "ship-outboard complex of geophysical equipment". The selection of the main harmonic of vibrations in the spectrum allows us to implement a compensator of the outboard system according to the stability conditions determined by the Morlet polynomials obtained from the decomposition of the observed dynamics of the system.

Keywords: Morlet wavelet, dynamics of outboard equipment, remotely operated underwater vehicle, outboard geo-physical system, the heave compensator, packet structure of sea waves, suspension system, statistical characteristics of unrest.

1. Introduction

Research of underwater robots Remotely operated underwater vehicle (ROV) is carried out in order to develop technologies for conducting underwater technical work. The power of modern ROVs working class is 300 kW or more. Technologically, ROVs have almost zero buoyancy, so the goal of the research is to develop control technologies that allow to compensate the dynamics of ROV when working with a set of working tools and manipulators.

The main task of the current study is to develop methods for "fast" calculation of sets of hydrodynamic coefficients based on field observations of the system dynamics for providing descent and lifting operations on carrier vessel board and ROV.

The results of studies of the observed dynamics of the system in the Morlet wavelet basis have shown that mathematical methods of decomposition, analysis and synthesis are effective with a smaller amount of calculations. The subsequent implementation of "fast" calculations in Morlet banks is performed on the basis of spectral density invariants of the decomposition power.

Frequency identification methods of control systems for marine technological complexes allow us to accurately determine quantitative indicators of forces and moments of dynamics based on the parametric description of the object's dynamics and its systems.

The paper [1] considers a systematic approach to modeling the dynamics of marine engineering systems related to nonlinear systems. The linearization of model parameters is achieved by converting forces and moments registered directly in the dynamics from the frequency domain to the time domain. The dynamics of elementary objects that together make up a more complex nonlinear system is considered.

In [1-2], the researchers Tristan Pe'rez, Thor I. Fossen and Kristin I. Pettersen described in detail mathematical models of underwater robots. The research results are implemented in many approaches to the design of navigation and ROV control systems. As follows from earlier works by Antonelli and

Gianluca [3-4], computational approaches to hydrodynamics are widely used with the statistically averaged perturbation characteristics and identification of the control system by matrices of the attached masses of the control object.

The commonality of approaches unites the estimation of the perturbations spectrum of hydrodynamic forces and moments acting on the control object by means of Fourier analysis.

The second approach involves decomposing of the observed dynamics in the bases of orthogonal functions, which makes it easier to obtain numerical models in the course of working out the hydrodynamics of the complex.

2. Materials and Methods

The dynamics of the phase components of the wave spectrum with a set of orthogonal functions can be described by compressing and mapping the Morlet basis [5], which is named the Morlet wavelet.

Modeling of a complex two-component model of the ship's mass distribution and a tethered geophysical system on a developed sea wave is performed in order to compensate the dynamics of the geophysical system.

Let us determine that the scheme of a two-link system with a special pitching compensator allows compensation of vertical movements of the suspension of the ROV lowered over the ship's side and the cable-rope, as shown in Figure 1.

Three coordinate systems are considered when modeling the carrier-ship system and the complex of outboard equipment: geocentric oxyz; linked o'x'y'z'; movable OXYZ [5, 6].

The natural experiment studied the dynamics of the outboard system consisting of ROV and power box with power equipment – garage-digger outboard system. Usually, garage-digger contains converter electronics and a winch with a cable rope with ROV at the free end. The cable-rope transmits electrical power that provides ROV supply, as well as provides information about the control system and on-board video cameras.

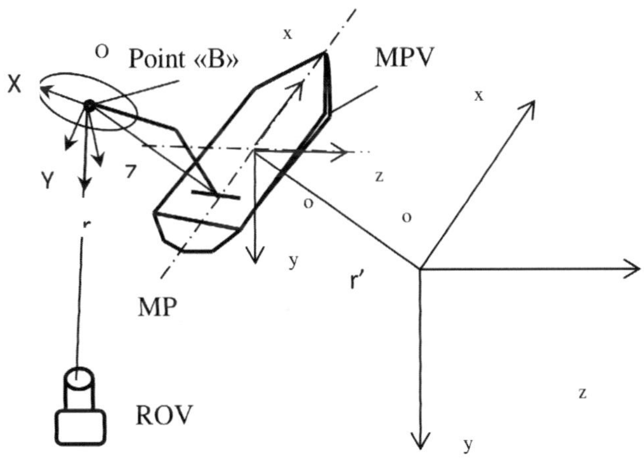

Figure 1. Dynamics conditions of the carrier ship and the ROV system in conditions of longitudinal and transverse rolling

The center of system gravity (point B) oscillates along an elliptical trajectory (Figure 1). MP is midplane (diametral plane "DP"), MPV is main plane of the vessel.

The working length of the communication cable-rope (CR) ship's carrier and the ROV garage-digger (GD) binary system is up to 6000 meters. The operating voltage transmitted to the plug-in converter is usually 2500 volts, which means that the cable weight can be reduced.

Figure 2 shows experimental measurements of system dynamics with the following parameters: the point of descent of the cable-rope (CR) from the last roller of the ship's fodder hoisting device (FHD)

IOP Conf. Series: Materials Science and Engineering **873** (2020) 012035 doi:10.1088/1757-899X/873/1/012035

and the underwater vehicle. The coordinates are: X = -18.6 m (aft of the ship from amidships), Y = -0.5 m (down from the MVP), Z = 0 (from DP).

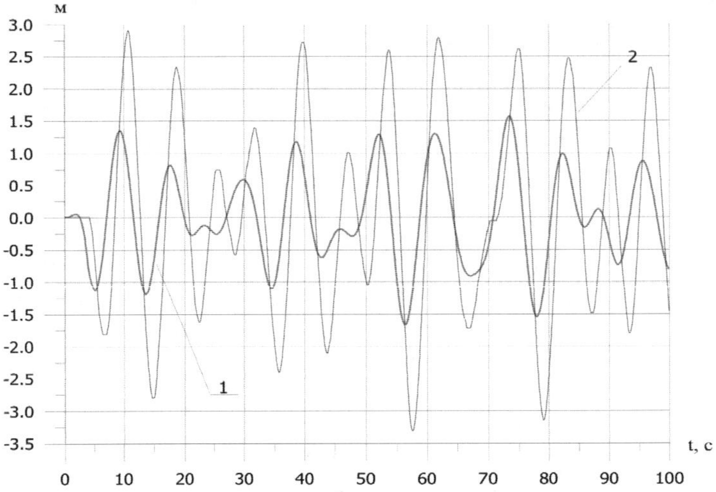

Figure 2. Results of registration during the natural experiment of vertical vibrations of the equipment complex with the inactive compensator

The natural experiment of vertical vibrations of the equipment complex with the inactive compensator is carried out with the indexes: cable length 6000 m; fading sea waves 5 points; course angle of movement (CAM)=90 degrees; 1 is vertical movement Y_{sp} of the point B (red); 2 is vertical movement Y_{gd} of the complex (blue).

When calculating the behavior of ROV while the ship is pitching, we used a spatial mathematic model of the system as the device itself and the power supply unit. The dynamics of CR was described by a system of nonlinear partial differential equations [6].

For a related coordinate system, the position of the center of gravity of the suspension point is defined as:

$$m\frac{d^2r}{dt^2} = (m+\lambda_i)g + F_i, \quad (1)$$

where r is the position vector of the mass center; $m+\lambda$ is the mass system with added masses of water; F_i is the reaction force of the suspension and Archimedes force outboard system.

Equations of moments of forces of the connected coordinate system are:

$$\begin{aligned} J_{x'}\frac{d^2\varphi}{dt^2} &= -\left[\sum_i F_i \times r\right]_{x'} \\ J_{y'}\frac{d^2\psi}{dt^2} &= -\left[\sum_i F_i \times r\right]_{y'}, \\ J_{z'}\frac{d^2\xi}{dt^2} &= -\left[\sum_i F_i \times r\right]_{z'} \end{aligned} \quad (2)$$

where r is the position vector of the gravity center of the suspension system masses; F_i is the reaction forces acting on the suspension system (Archimedean force and the suspension reaction force) in the system of connected coordinates x', y', z' and angles φ, ψ, ξ [6].

Equations (1-2) for a complex of outboard equipment consisting of a garage-deepener and an underwater vehicle, it is necessary to take into account the attached water masses. The complex represents shapes of a complex streamlined shape.

Two cases of modeling the dynamics of the outboard equipment complex can be determined: at depth and on the surface at the time of docking of the complex and onboard equipment on the carrier vessel deck.

Figure 2 shows the results of calculating the vertical movement of the ROV when the GD winch is not working. The results refer to a 5-point attenuating wave, CAM = 900, and the length of the CR of GD is 500 m. The length of the CR of the ROV connecting the ROV and GD is 8 m.

It should be noted that due to the irregularity of GD oscillations, the relative speed of the ROV and GD at their junction is also a random value.

To estimate the possible values of the relative vertical velocity GD and the ROV and jerks in CR during docking we performed calculations of the GD and the ROV movement with regular oscillations of GD at an average frequency and an amplitude of 3% exceedance for the state intensity of 4 and 5 points, CAM = 90^0 and CR = 500 m and GD equal to 6000 m.

To reduce the relative velocities of GD, ROV and the intensity of jerks in the CR, it is necessary to take measures to reduce the fluctuations of GD.

Such measures include holding the vessel at the junction of the ROV and GD in the position of CAM = 1800, reducing the intensity of the wave at which the connection is made, and reducing the depth of immersion at which the GD is located at the junction. However, the most effective way to reduce the intensity of GD vibrations is to include a vertical GD vibration compensator in the FHD.

The compensator can be either a passive type in which movement of the piston is proportional to change of the cable wire tension or active type, the moving piston of which is controlled by a dedicated system.

The work [7] shows that for effective operation, a passive compensator must have a very low stiffness, which is practically impossible for implementation. The feasibility of using an active compensator requires additional evaluation.

The efficiency of an active compensator depends on its design parameters, mainly on the size of the piston stroke and the speed of its movement, as well as on the characteristics of the piston movement control system. It is obvious that the movement of the piston leads to a proportional change in the released length of the CR on which the ROV is lowered (and raised).

3. Results

3.1. Mathematical model of a low-stiffness compensator

Studies of pitching compensators are carried mostly to investigations of vertical fluctuations of the outboard system. However, fluctuations in the system center of gravity (suspension points) in reality frequently perform complex oscillations along an elliptical trajectory. The slope of the trajectory corresponds to the period of transverse pitching of the vessel, as shown in Figure 1.

In addition to the geocentric and mobile coordinate systems (Figure 1), we will implement a third coordinate system "linked". The suspension point (point B) moves on the wave relatively to the mass center of the vessel. For these reasons, it is necessary to add the calculated values in the form of coordinate systems to the vertical coordinates in the form of increments to the mobile coordinate system associated with the center of mass gravity of the ship.

Modeling is feasible for these extreme cases with recalculation of the experiment data from reference data.

Hydrodynamic characteristics are calculated in a related coordinate system, the origin of which is taken at the point of the center of gravity of the joint garage-digger and ROV system.

The axis OY' coincides with the axis of the supporting frame and is directed upward along the cable-rope of the system's attachment, the axis OX' forms the right three vectors with the axis OY' at a perpendicular vector OZ'.

In practice, it is assumed that the attached mass of structural elements is approximately 20% of the total mass of the drum, plates and base. The center of the garage-digger is located in the center of the drum with the cable rope.

3.2. Computational models of the outboard system dynamics

It is quite problematic to solve the equation describing the dynamics of the outboard system. Due to errors in the calculated values of the attached water masses, flow, and the influence of waves, this problem can only be evaluated from the perspective of probability theory.

The study of workflow models allows us to obtain good results for the right part of expression (2). As follows from the results of the experiment, the dynamics of the outboard equipment is described by harmonic functions.

Spectral sets of registered realizations within the observed dynamics $[-\varphi, +\varphi]$ are defined as [8]:

$$\int_{-\varphi}^{+\varphi} \frac{d\sin(\varphi,t)}{dt} dt = 0. \quad (3)$$

Let us define a wavelet as:

$$\psi\left(\frac{\varphi-t}{a}\right) exp\left(jsin\left(\frac{\varphi-t}{a}\right)\right), \quad (4)$$

The transformation of the $\sin(\varphi, t)$ function takes place over the phase parameters so that:

$$\vartheta(\varphi,t) = arctan\left\{\frac{A_1\psi_i\left(\frac{\varphi_1-t_1}{a}\right)sin\left(\varphi_i\left(\frac{\varphi_1-t_1}{a}\right)\right)+A_2\psi_i\left(\frac{\varphi_2-t_2}{a}\right)sin\left(\varphi_i\left(\frac{\varphi_2-t_2}{a}\right)\right)}{A_1\psi_i\left(\frac{\varphi_1-t_1}{a}\right)cos\left(\varphi_i\left(\frac{\varphi_1-t_1}{a}\right)\right)+A_2\psi_i\left(\frac{\varphi_2-t_2}{a}\right)cos\left(\varphi_i\left(\frac{\varphi_2-t_2}{a}\right)\right)}\right\}, \quad (5)$$

for each set $\{\varphi-, \varphi+\}$: $\vartheta(\varphi-, t_1) = \vartheta(\varphi+, t_1)$ и $\vartheta(\varphi-, t_2) = \vartheta(\varphi+, t_2)$.

The initial values of the shift and scale transformation functions are equated [9]:

$$\psi_i\left(\frac{t_2-t_1}{\varphi-}\right) sin\left(\varphi_i\left(\frac{t_2-t_1}{\varphi-}\right)\right) = \psi_i\left(\frac{t_2-t_1}{\varphi+}\right) sin\left(\varphi_i\left(\frac{t_2-t_1}{\varphi+}\right)\right) \quad (6)$$

$$\psi_i\left(\frac{t_1-t_2}{\varphi-}\right) sin\left(\varphi_i\left(\frac{t_1-t_2}{\varphi-}\right)\right) = \psi_i\left(\frac{t_1-t_2}{\varphi+}\right) sin\left(\varphi_i\left(\frac{t_1-t_2}{\varphi+}\right)\right). \quad (7)$$

Taking the initial value of the phase equal to zero:

$$sin\left(\varphi_i\left(\frac{t_2-t_1}{\varphi-}\right)\right) = 0; \; sin\left(\varphi_i\left(\frac{t_2-t_1}{\varphi+}\right)\right) = 0; \quad (8)$$

which requires the condition $\varphi+ = \varphi-$, we get:

$$\varphi_i\left(\frac{t_2-t_1}{\varphi_k}\right) = k\pi. \quad (9)$$

If the base indicator is the Morlet wavelet [2]:

$$\varphi_k = \frac{\omega_0(t_1-t_2)}{k\pi}, \quad (10)$$

where the lowest iteration order is the value:

$$\varphi_0 = \frac{\omega_0(t_1-t_2)}{\pi}. \quad (11)$$

Thus, having the construction of the phase transformation of the observed dynamics and the scale coefficients, we can say that the Morlet wavelet decomposition provides the frequency set shown in Figures 3-4, which is a compression and mapping of the Morlet basis [10, 11].

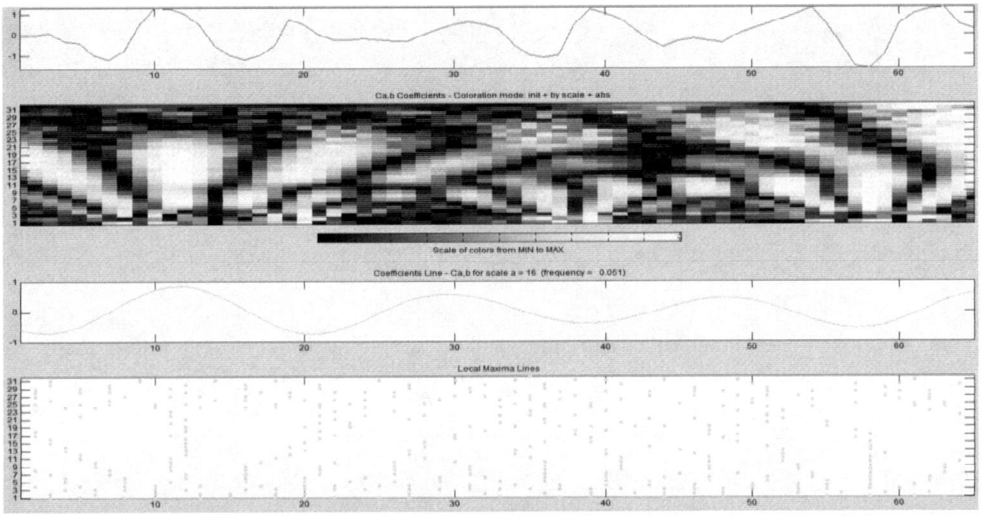

Figure 3. Decomposition of functions in the basis of Morlet wavelets, dynamics of the suspension point on the ship of outboard equipment

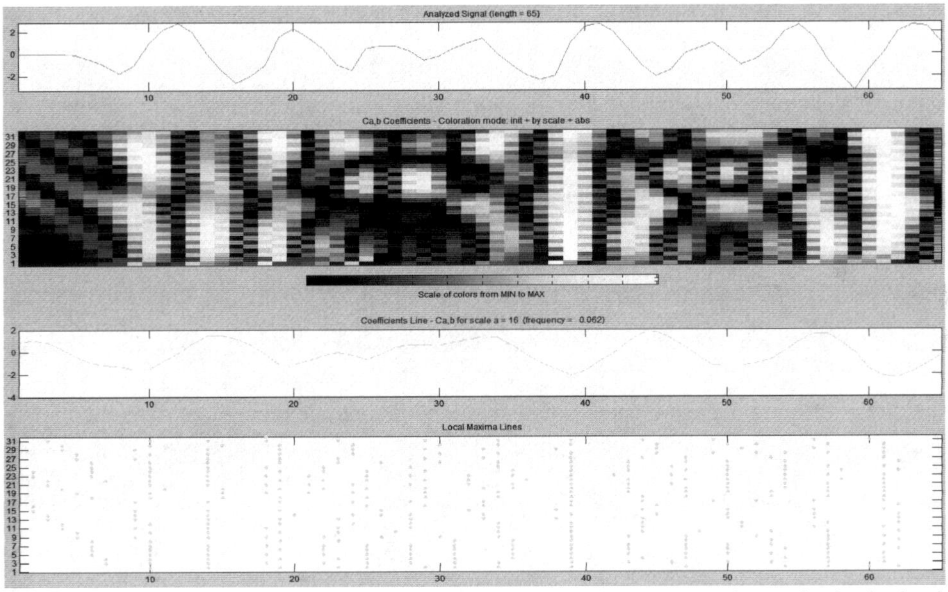

Figure 4. Expansion of the garage-digger dynamics driven by a suspension point in the basis of Morlet wavelets

For any system described by the algorithm (7-11), it is necessary to determine a set of coefficients of the Morlet basis expansion, as shown in table 1. We estimate the statistical averaged characteristics of the harmonics of the Morlaix decomposition and calculate the contribution to the overall dynamics of the process, which allows us to achieve compensation in descent-lifting devices (FHD).

IOP Conf. Series: Materials Science and Engineering **873** (2020) 012035 doi:10.1088/1757-899X/873/1/012035

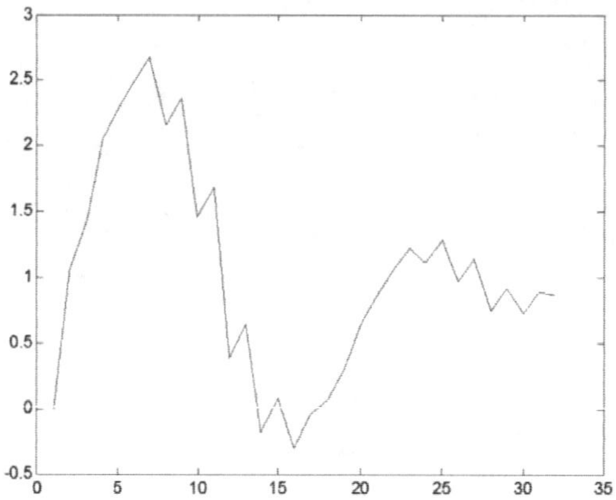

Figure 5. Morlet Wavelet "32-order", fragment, " Dynamics of decomposition coefficients in Morlet bursts"

Table 1. Calculated values of the Morlet wavelet fragment of the array (32x65)

The order of decomposition max "32»	Coefficients of polynomial expansion in the Morlet basis max "65»						
	1	2	3	4	5	…	64	65
	0	0	0	1,04083408	4,16333634	…	1,56125112	1,04083408
	0	0	0	558608e-17	234434e-17		837913e-16	558608e-16
	0	0	1,80277751	1,47196168	0,07071067		0,07071067	1,06066017
	0	0	848459e-17	001604e-17	81186548		81186548	177982
	0	2,32737576	2,08166817	0,05784779	0,28906980		0,92539550	1,44365758
	0	866932e-17	117217e-17	06434138	7630392		4707943	228519
	…	…	…	…	…		…	…
	…	…	…	…	…		…	…
	…	…	…	…	…		…	…
	2,753788	2,54951241	0,04476503	0,05000000	0,30000000		0,55000000	2,05000000
	14654010e-	658222e-17	27526815	00000001	0000000		0000000	000000
	17				…		…	0,71995562
					…		…	734
					…		…	…
					1595		4693	…
							0,44422494	…
							9651574	657
							0,07071067	0,88464381
							81186558	0383246
								0,86620580
								6953519

4. Discussion

The obtained results allow us to draw several important conclusions. Decomposition of the observed dynamics of hydrodynamic forces and moments in Morlet wavelets allows us to store information about the phase of the initial sequences of signals from the dynamics sensors.

Applying the tresholding technology to the spectral power of invariants calculated in the Morlet basis allows limiting the amount of calculations in the filters of the "fast" Morlet transform. Limiting the sets of decomposition coefficients that most affect the dynamics of the descent-ascent-ROV system is a crucial rule for identifying the control system.

5. Conclusion

Decomposition of the observed sequence in the Morlet wavelet basis gives a set of harmonic components of the packet wave spectrum. The estimation of the spectral power of the decomposition in the Morlet wavelet basis defines the main vector of motion as an average characteristic of the system vibrations caused by a packet sea wave. The stability of the system polynomials Morlet is determined by the harmonic component of the fundamental harmonic decomposition.

References

[1] Pérez T and Fossen T I 2008 Time-vs. frequency-domain Identification of parametric radiation force models for marine structures at zero speed *Model. Identif. Control* **29** 1–19

[2] Fossen T I and Pettersen K Y 2018 Modeling of Underwater Vehicles *Encyclopedia of Robotics* (Springer Berlin Heidelberg) pp 1–9

[3] Antonelli G 2006 1. Introduction *Underwater Robots – 2nd Edition* (Springer-Verlag) pp 1–13

[4] Antonelli G 2014 Modelling of underwater robots *Springer Tracts in Advanced Robotics* vol 96 (Springer Verlag) pp 23–63

[5] Andersen Bachelor K 2016 *Development of a Time-Domain Modeling Platform for Hybrid Marine Propulsion Systems*

[6] Bertram V 2012 *Practical Ship Hydrodynamics* (Elsevier Ltd)

[7] Postnikov E 2009 Wavelet transform with Morlet wavelet: calculation methods based on the solution of diffusive differential equations *Comput. Res. Model.* **1** 5–12

[8] Heuberger P, Van Den Hof P and Wahlberg B 2005 *Modelling and identification with rational orthogonal basis functions*

[9] Radchenko D 2011 Compensation of the impact of the ship's transverse pitching on the depth of loading of a non-Autonomous underwater object using a ship's descent device

[10] Slavič J, Mihalec M, Javh J and Boltežar M 2017 Morlet-wave damping identification: Application to high-speed video *Conference Proceedings of the Society for Experimental Mechanics Series* vol 10B (Springer New York LLC) pp 27–30

[11] Liner C 2010 An overview of wavelet transform concepts and applications *Univ. Houst.*

Mathematical model of network flow control

I V Zaitseva,[1,2,3] O A Malafeyev,[2] V V Zakharov,[1] T E Smirnova,[2] L M Novozhilova[2]

[1]Stavropol State Agrarian University, 12 Zootechnicheskiy Ln, Stavropol, 355017, Russia

[2]Saint Petersburg State University, 7/9 Universitetskaya nab. St. Petersburg, 199034, Russia

[3]Stavropol branch of the Moscow Pedagogical State University, 66G, Dovatortsev Street, Stavropol, 355042, Russia

E-mail irina.zaitseva.stv@yandex.ru

Abstract. In this paper, we consider a network whose nodes are regions between which it is necessary to distribute electricity coming from its source. Any power plant is a complex economic system that always combines simple concepts - the need to supply the power plant with the main resource for the production of electricity and the need to sell the generated electricity. In this model, due to the need to sell electricity to two regions, the prices for the sale of electricity do not matter. Profit maximization is achieved by minimizing production costs. The optimal policy for generating electricity in the i-th time period is found.

Keywords: mathematical model, control, flows, network, system, region, optimal policy.

1. Introduction

As you know, any power plant produces electricity through the conversion of natural energy. The main task of such power plants is to supply electricity to the regions of the country. The location of the power plant is selected taking into account economic and geographical convenience. Consumers are selected based on the power generated by the power plant and the region's electricity needs. To increase the efficiency and, therefore, maximize the profit of such power plants, the only way is to minimize the cost of generating electricity. To maximize the profit of such power plants, both the cost of generating electricity and the cost of selling are important. They also have to deal with the phenomenon of corruption, since they are competitors for monopolists. Depending on the remoteness of the power plant to the consumer, another problem arises – a loss in the transmission of electricity. These problems are associated with physical phenomena that occur when an electric current flows in a conductor.

2. Statement of the problem

We describe the general model of production and distribution of electricity. We introduce the basic terms. The primary resource is the material that is needed to generate electricity. Characteristics of the primary resource are: quantity and cost is the purchase or production price of the primary resource by the power plant. Power plant is a set of devices that produce electricity. Characteristics of the power plant are: power is the amount of electricity produced, storage capacity is the amount of the stored primary resource. A drive is a part of a power plant in which the primary resource necessary for the production of electricity is accumulated or stored. Characteristics of the drive are: drive capacity is the amount of the stored primary resource. Consumer is a set of devices that consume electricity. Characteristics of the consumer are: demand is the amount of electricity needed, cost is the price at which the power plant sells electricity to the consumer. Corruption is a cost necessary to bribe officials in the process of organizing the delivery of electricity to the consumer and / or arising from the sale of electricity to the consumer. Characteristic of corruption is the coefficient of corruption. Losses are costs incurred in electric current conductors during transmission of electricity to a consumer. Characteristic of losses is the loss coefficient [1-3].

We describe the model of production, distribution and transmission of electricity in a general way. There are 1, …, K power plants, 1, …, L primary resources, 1, …, M consumers. Power plants need to spend money on the production of electricity from primary resources. Primary resources are either bought or accumulated in the drive of the power plant, depending on the type of power plant. After production, electricity, through power lines, is transmitted to the consumer. During power transmission, losses occur. Each electricity production-sale transaction corresponds to the *i*-th loss coefficient. All processes occur in time. The prices of purchase, production, sale depend on time, as well as on the season. Typically, during cold seasons, energy needs increase, and in warm seasons, they decrease. Corruption rates are available that depend on the time period, season, and region in which the consumer is located. For each electricity production and sale transaction, the *i*-th corruption coefficient corresponds.

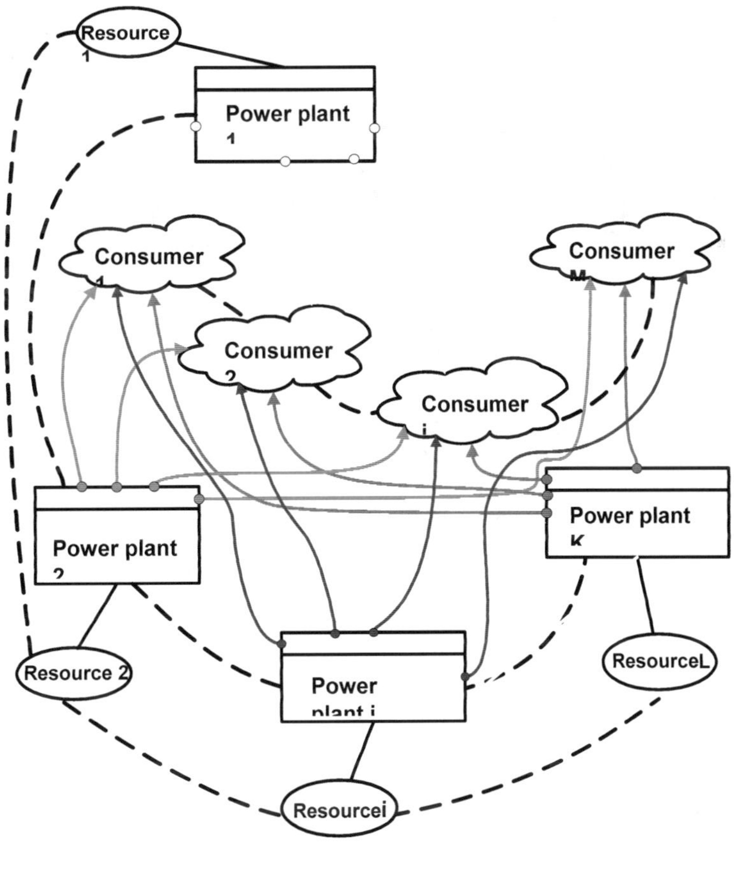

Figure 1. Model of electricity production and distribution

The profit of the power plant will be: the cost of selling electricity minus the cost of generating electricity multiplied by the coefficient of corruption multiplied by the coefficient of losses. Based on the source data, it is necessary to decide how to distribute the generated electricity between the regions in order to maximize profits [4-5]. Figure 1 shows a diagram of the resulting model.

The described model includes many subtasks. In this paper, we consider a model for the distribution of electricity between two regions with the following conditions: there is one power station, the capacity of the primary resource storage is limited, consumers are two regions, it is necessary to supply electricity to both regions, 6 periods of production and sale of electricity are considered, the cost of production is known electricity in the *i*-th period of time, the needs for electricity of the regions in the *i*-th period of time are known.

The purpose of studying the model is to maximize profits taking into account the source data. In this model, due to the need to sell electricity to two regions, the prices for the sale of electricity do not matter. Profit maximization is achieved by minimizing production costs. The optimal policy for generating electricity in the *i*-th time period is found [6-9].

3. A mathematical model of the distribution of electricity between regions with minimizing costs

We formulate the initial conditions: there is one power plant; the capacity of the primary resource drive is limited, we denote it by *S*; two regions are consumers of electricity; it is imperative to supply electricity to both regions; 6 periods of time of production and sale of electricity are considered; the cost of electricity production in the *i*-th time period is known, we denote it by pi; electricity needs of the regions in the *i*-th time period are known, we denote it by *d1i* and *d2i*.

The purpose of studying the model is to maximize profits taking into account the source data [10-12].

3.1. The formulation of the dynamic programming problem for the distribution of electricity between two regions with minimizing costs

In order for the power plant to be able to fulfill its plan for the supply of electricity, the procurement department must supply the primary resource at the beginning of each time period. Electricity production price p_i, where $i = 1, 2, 3, 4, 5, 6$, and the need of the regions $d_i = d1_i + d2_i$, $(i = 1, 2, 3, 4, 5, 6)$ which is equal to the sum of the needs of each region, are shown in table 1. The capacity of the primary resource drive is limited and must not exceed *S*.

Table 1. Electricity production price and the need of the regions

Period of time	i	1	2	3	4	5	6
1 Region's electricity demand	$d1_i$	15	10	12	17	14	20
2 Region's electricity demand	$d2_i$	20	13	17	22	10	15
Total electricity demand	d_i	35	23	29	39	24	35
Primary resource purchase price	p_i	8	12	13	11	17	11

The initial amount of primary resources in the drive allows you to produce power plants 5 units of electricity. After the 6th period, the number of primary resources in the drive should be 0.

It is necessary to decide the volume of purchases of the primary resource at the beginning of each period, so that the total cost of purchases is minimal.

We introduce the following notation: si is the number of primary resources in the drive in the period of time i, before purchase ai; ai is the volume of procurement of primary resources in time period i; xi is the balance of primary resources in the drive in the period of time i after purchase ai.

Figure 2 shows some solution that is not optimal.

Now our task will be to build the optimal solution.

$$x_i = s_i + a_i, i = 1,2,...,6 . \quad (1)$$

The state of the system at each time interval i can be described as the value si or xi; ai can be calculated from (1); it does not depend on the choice of a variable si or xi. We will make an arbitrary choice xi as a variable.

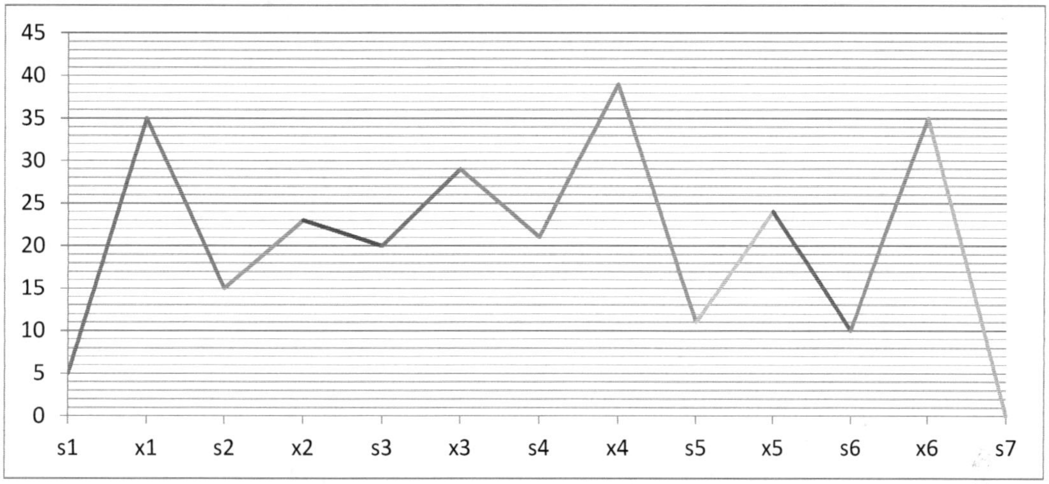

Figure 2. Some solution that is not optimal

The nature of the problem imposes the following restrictions, with numerical values limited between *0* and *S*:

$$0 \le s_i \le S, i = 1,2,...,6\,, \quad (2)$$

$$0 \le x_i \le S, i = 1,2,...,6\,, \quad (3)$$

$$s_i \le x_i, i = 1,2,...,6\,. \quad (4)$$

Moreover,

$$s_{i+1} = x_i - d_i, i = 1,2,...,6\,. \quad (5)$$

If we take $s_1=5$, then we get a set of restrictions related to x_i, which can be written as:

$$\max[d_i, x_{i-1} - d_{i-1}] \le x_i \le \min[S, x_{i+1} + d_i], i = 1,2,...,6, x_0 = 5, d_0 = 0, x_7 = 0\,. \quad (6)$$

The existing problem contains a linear program, which can be stated in the following form:

$S=50,\ x_0=s_1=0,\ d_0=0,\ x_6=35.$

$$[MIN]F = \sum_{i=1}^{6}(p_i \cdot a_i) = \sum_{i=1}^{6}(p_i(x_i - x_{i-1} + d_i)) = 2540 + 6x_5 + 6x_4 + 2x_3 + x_2 - 4x_1\,.$$

Notice, that x_i usually depends on x_{i-1} and x_{i+1}. Therefore, the calculation of the period variable of the next one below the considered one will be introduced as a parameter. It should be noted that in certain cases this fact can significantly simplify the optimization process [13-16].

3.2. The procedure for constructing an optimal policy for the distribution of electricity between two regions with minimizing costs

Now suppose that $S=50$. To find the best purchase policy

$$x* = [x_1^*, x_2^*, x_3^*, x_4^*, x_5^*, x_6^*], \quad (7)$$

which can be expressed through a_i:

$$a* = [a_1^*, a_2^*, a_3^*, a_4^*, a_5^*, a_6^*], \quad (8)$$

Arbitrarily start the optimization process with a period 6.

Imagine:

$$F(x_1,x_2,x_3,x_4,x_5,x_6)=v_1(x_1)+v_2(x_2)+v_3(x_3)+v_4(x_4)+v_5(x_5)+v_6(x_6), \quad (9)$$

where the quantity v_i represents the purchase price for the period i. We take $x_0=5$, as indicated in (6). After calculating, we get $x_6*=35$, the amount spent for period 6 has the value: $v_6(x_5, x_6) = 649 - 11x_5$. Calculate the boundaries x_5 to optimize periods 6 and 5. We get $24 \leq x_5 \leq 50$. Consider the case when periods 5 and 6 are combined.

3.3. Example

For example: $f_{6,5}(x_4) = min$ [$v_5(x_4, x_5) + v_6(x_5, x_6)$] при $24 \leq x_5 \leq 50$. We have $v_5 = 7x_5 - 17x_4 + 663$, $v_5(x_4, x_5) + v_6(x_5, x_6) = 1312 + 6x_5 - 17x_4$ и $f_{6,5}(x_4) = min$ [$1312 + 6x_5 - 17x_4$] with $24 \leq x5 \leq 50$. Since the right side (29) $1312 + 6x_5 - 17x_4$ this is a monotonically increasing function of x_5 we need to take lower bounds x_5: $x_5* = 24$, where $s_6* = 0$ and $a_6* = 35$ and $f_{6,5}(x_4) = 1071 - 17x_4$.

We calculate the boundaries for x_4 max [$d_4, x_3 - d_3$] $\leq x_4 \leq min$ [$S, x_5 + d_4$] gives max [$24, x_3 - 29$] $\leq x_4 \leq min$ [$40, 48$], so we get max [$24, x_3 - 29$] $\leq x4 \leq 40$. Therefore, after calculating, we obtain that the optimal policy is given from $x* =$ [$50, 50, 50, 40, 25, 35$] or from $a*=$ [$45, 35, 23, 20, 25, 35$] and minimum costs for 6 periods are $F* = 1709$.

4. Results

Thus, as a result, of studying the model, the following results were established:

1. The optimal income in N steps is a linear function, the coefficients of which depend on the cost of production and the amount of electricity produced.
2. In order to find the optimal policy, it is necessary to decide the volume of purchases of the primary resource at the beginning of each period, so that the total cost of purchases is minimal.

5. Conclusions

As a result, of the work carried out to study the methods of distributing electricity between regions, we came to the conclusion that in certain problems a quick optimal solution can be obtained through intuitive reasoning. But in the case of consideration of more voluminous problems, formalization of mathematical models and finding optimal policies by dynamic programming can significantly simplify calculations and lead to minimization of costs and maximization of profits

6. Acknowledgment

The work is partially supported by the RFBR grant # 18-01-00796.

References

[1] Malafeev O 1977 *Stationary strategies in differential games* (USSR Computational Mathematics and Mathematical Physics) **17(1)** 37-46

[2] Malafeev O 1974 *Equilibrium situations in dynamic games* (Cybernetics and System Analisys) **10(3)** 504-513

[3] Mirchandani P and Francis R 1990 Discrete Location Theory *Wiley New York NY* pp 100-125

[4] Drezner Zand Drezner T 2004 Finding the optimal solution to the Huff competitive location model *Computational Management Science* **1(2)** 193–208

[5] Vetta A 2002 Nash equilibria in competitive societies with applications to facility location, traffic routing and auctions *In Proceedings of the 43rd Annual IEEE Symposium on Foundations of Computer Science (FOCS)* pp 416–428

[6] Pichugin Yu 2015 Geometrical aspects of testing complex statistical hypotheses in mathematical simulation (*St. Petersburg Polytechnic University Journal Physics and Mathematics* **2(218)** 123-137

[7] Zubov A and Dikusar V and Zubov N 2010 Controllability criterion for stationary systems *Doklady Mathematics* vol **81** no **1** pp 6-7

[8] Malafeyev O and Redinskikh N and Zaitseva I and Smirnova T and Kolesov D 2019 Model of information multi-stage process in structured system *AIP Conference Proceedings*

[9] Murashko A and Orlov V and Zubov A and Bondarenko L and Petrova V 2019 Qualitative analysis of the behavior of one mechanical system *International Journal of Innovative Technology and Exploring Engineering* **8(7)** pp 653-658

[10] Malafeyev O and Lakhina J and Zaitseva I and Redinskikh N and Smirnova T and Smirnov N 2019 A mathematical model of production facilities location *Journal of Physics Conference Series* vol **1172** no 012090

[11] ReVelle C 1986 The maximum capture or sphere of influence location problem: Hotelling revisited on a network *Journal of Regional Science* **26** pp 343–357

[12] ReVelle C and Eiselt H 2005 Location analysis: A synthesis and survey. *European Journal of Operational Research* **165** pp 1–19

[13] Malashenko Y and Nazarova I and Novikova N 2020 Analysis of Cluster Damages in Network Systems *Computational Mathematics and Mathematical Physics*

[14] Malashenko Y and Nazarova I and Novikova N and Pospelova I 2019 A Network Flow Model for Power and Energy System with Changing Capabilities *International Journal of Public Administration*

[15] Plastria F 2001 Static competitive facility location: An overview of optimisation approaches *European Journal of Operational Research* **129** pp 461–470

[16] Ghiasi M and Esmaeilnamazi S and Ghiasi R and Fathi M 2020 *Role of Renewable Energy Sources in Evaluating Technical and Economic Efficiency of Power Quality* 5. 10.1007/s40866-019-0073-1

IOP Conf. Series: Materials Science and Engineering **873** (2020) 012037 doi:10.1088/1757-899X/873/1/012037

About one transportation technology of construction cargo

O A Malafeyev[1], N D Redinskikh[1], I V Zaitseva[2,3], Yu V Orel[2], D V Shlaev[2]

[1]Saint-Petersburg State University, Faculty of Applied Mathematics and Control Processes, 7/9 Universitetskaya nab., St. Petersburg, , 199034, Russia

[2]Stavropol State Agrarian University, 12 Zootechnicheskiy Ln, Stavropol, 355017, Russia

[3]Stavropol branch of the Moscow Pedagogical State University, 66G, Dovatortsev Street, Stavropol, 355042, Russia

E-mail: irina.zaitseva.stv@yandex.ru

Abstract. The model of shipping construction cargo from initial point to the destination point is formalized and studied in this paper. The possible shipping routes are represented as network. Each edge of such network is possible mode of cargo shipping (by rail transport, air transport). These edges compose the full route from the initial point to the final point. In the network there are several possible routes of cargo transportation from the initial point to the destination point. The choice of route by transportation organizer depends on the cost of cargo transportation. The transportation organizer chooses the optimal route according to the optimality principle, compromise solution.

Keywords: transport problem; construction cargo transportation; network; optimality principle

1. Introduction

The problem of transporting cargo by various modes of transport is often occurred because of the lack the direct connection between origin and destination point. Also the existing direct connection may not be suitable for consignee due to the high cost of transportation or the long cargo delivery time to the final point. As a rule, such transportation is carried out sequentially. The several methods of transportation in such case usually are used: by rail transport, car transport or air transport. Courier services may also be involved in such cargo transportation.

The problems of this types are previously considered in the papers [1-4]. A multi-agent interaction between different goods transporters is formalized in [5]. The problem where compromise solution using as the optimality principle is considered in [6]. The different approaches for finding the equilibrium situations in the games are proposed in [7]. The dynamic assignment problem according to the problem of finding the path of maximum cost is studied in [8]. Another methods and apparatus are proposed in the articles [5-9] can also be used in this paper.

2. Informal statement of the problem

IOP Conf. Series: Materials Science and Engineering **873** (2020) 012037 doi:10.1088/1757-899X/873/1/012037

In this paper, the problem of shipping construction cargo from origin to destination point by different modes of transport is studied. The set of possible transport routes is a network, each edge of which is a possible mode of transport (air transport, rail transport), which is assumed to be set for this edge of the network.

The choice of network edge by the transportation organizer for cargo transportation depends on the several factors. First it is the cost of cargo transportation on the specified network edge. Second it is the duration of transportation along the selected network edge. And the third it is the distance from the start point to the end point of cargo transportation. The general route of cargo transportation from the initial point to the final point is determined by the transportation organizer. Each transportation organizer provides such route for cargo transportation from the initial to the final point, at which he receives the maximum profit from transportation.

3. Formalization and analysis of the problem

The problem of construction cargo transportation from the initial point s to the final point s' is considered.

Let $I = \{1, 2, ..., n\}$ be the set of participants (organizers of transportation) that provide transportation of cargo from the initial point s to the final point s'.

Let $R_j, j = 1, 2, ..., m$ are the possible routes of construction cargo transportation from the starting point s to the final point s'. It is assumed that the cargo is transported from the initial point s to the final point s' by various types of transport: by rail transport, car transport, etc. In this case, the route $R_j, j = 1, 2, ..., m$ can be represented as composite routes $(s, x_1), (x_1, x_2), ..., (x_k, s')$ (i.e. Figure 1).

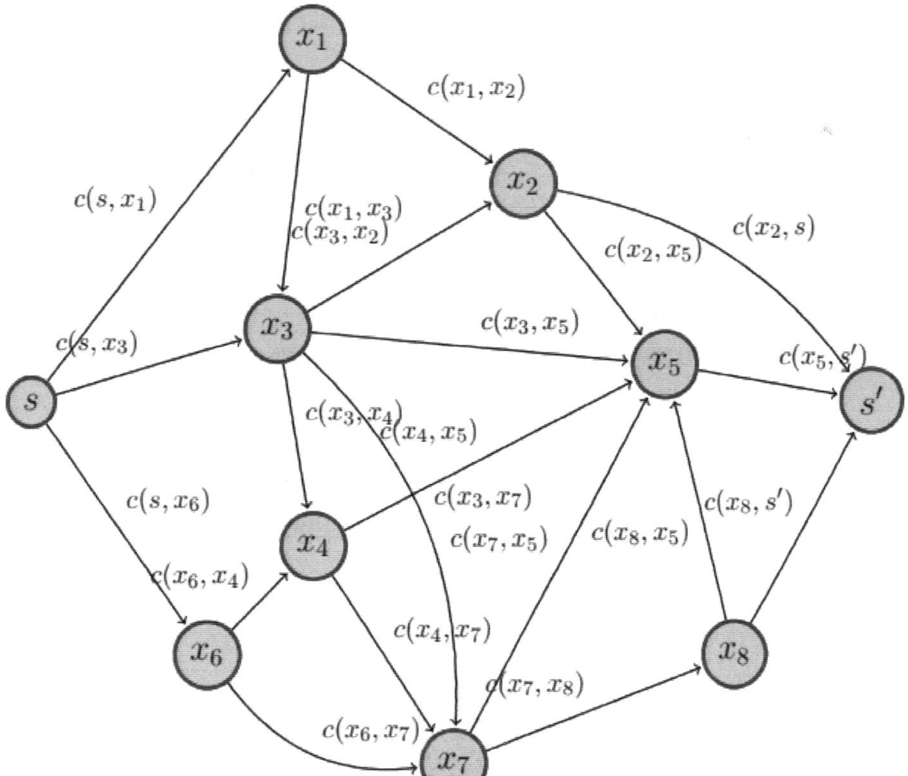

Figure 1. The network of routes.

Each composite route has a start and end point. Each endpoint of a composite route is the starting point of the next composite route.

So the starting point s is the initial departure point of the composite route (s, x_1), the point x_1 is the initial point of the composite route (x_1, x_2) etc.

For the k-th composite route (x_k, s'), the starting point is the point x_k, and the end point is the final destination point s' [10-11].

In the carriage of cargo for each composite part of the route the participant takes the transportation costs $c_i(x_{k-1}, x_k) > 0, i \in I$. By carrying out transportation on a composite part of the route $R_j, j = 1,2,...,m$, the transportation participant receives a certain income. Each transport participant maximizes the total net income $G_i(x_{k-1}, x_k), i \in I$ [12-13].

This income is received by transportation of cargo from the initial point s to the final point S'. To determine the overall route that allows each participant to maximize the total net income, the optimality principle a compromise solution is used.

3.1 Algorithm for finding a compromise solution

Algorithm for finding a compromise solution [14-15]:

Step 1. The ideal vector $M = [M_1,...,M_n]$ is constructed.

The value $M_i = \max_i(G_i(R_j))$ is the maximum net income of transport participant i, choosing the possible transportation route $R_j, j = 1,2,...,m$ from the initial point s to the final point s'.

Step 2. For each route $R_j, j = 1,2,...,m$ the deviation $M_i - G_i(R_j)$ of net income values from the maximum value M_i are determined.

Step 3. From the obtained values of net income deviations from the maximum value ($M_i - G_i(R_j)$) the maximum value $A_j = \max_i(M_i - G_i(R_j)), j = 1,2,...,m, k = 1,2,...,l$ for each route $R_j, j = 1,2,...,m$ is determined.

Step 4. The minimum value is determined from the obtained maximum deviations of the net income $\min_j A_j$.

4. Numerical example

Let us suppose in the transportation of construction cargo are four participants $i = 1,2,3,4$.

It is necessary to transport the construction cargo from the initial point s to the final point s' (Figure 1). Let us suppose there are 12 possible routes for cargo transportation $R_j, j = \overline{1,12}$.

On the each route $R_j, j = \overline{1,12}$ the transportation can be carried out by any participant of the carriage.

The set of possible rotes of cargo transportation from the initial point s to the final point s' is the next:

$R_1 : (s, x_1), (x_1, x_3), (x_3, x_2), (x_2, s')$;

$R_2 : (s, x_1), (x_1, x_3), (x_3, x_4), (x_4, x_7), (x_7, x_8), (x_8, s')$;

$R_3 : (s, x_1), (x_1, x_2), (x_2, x_5)(x_5, s')$;

$R_4 : (s, x_1), (x_1, x_2), (x_2, s')$;

$R_5 : (s, x_3), (x_3, x_2), (x_2, s')$;

$R_6 : (s, x_3), (x_3, x_2), (x_2, x_5), (x_5, s')$;

$R_7 : (s, x_3), (x_3, x_5), (x_5, s')$;

$R_8 : (s, x_3), (x_3, x_4), (x_4, x_5), (x_5, s')$;
$R_9 : (s, x_3), (x_3, x_4), (x_4, x_7), (x_7, x_8), (x_8, s')$;
$R_{10} : (s, x_6), (x_6, x_4), (x_4, x_5), (x_5, s')$;
$R_{11} : (s, x_6), (x_6, x_7), (x_7, x_8), (x_8, s')$;
$R_{12} : (s, x_6), (x_6, x_4), (x_4, x_7), (x_7, x_8), (x_8, s')$.

By the construction cargo transportation each participant of the transportation receives the income from cargo transportation from the initial point s to the final point s'.

The net income of each participant of the transportation is determined in the Table 1.

Table 1. The net income of each participant of the transportation

The route\the participant of the route	1 participant	2 participant	3 participant	4 participant
R_1	10	9	14	8
R_2	8	7	3	6
R_3	9	6	7	5
R_4	6	5	6	7
R_5	12	4	7	9
R_6	14	8	10	12
R_7	13	7	4	10
R_8	11	9	5	12
R_9	10	12	6	14
R_{10}	5	4	8	6
R_{11}	7	6	9	12
R_{12}	4	7	11	10

According to the algorithm for finding compromise solution, let us found compromise solution for the problem.

1. Let us make the ideal vector $M = [M_1, ..., M_n]$, where $M_i = \max_i (G_i(R_j))$ — maximal net income of the transportation participant i, choosing the possible route of cargo transportation $R_j, j = 1, 2, ..., m$ from the initial point s to the final point s'.

$$M = (14,8,9,7,12,14,13,12,14,8,12,11)^T.$$

2. For each route $R_j, j = 1, 2, ..., m$ let us determine the deviations of the values of the net income from the maximum value M_i, i. e. $M_i - G_i(R_j)$.

The net income of each participant of the transportation is determined in the Table 2.

3. From the obtained values of net income deviations from the maximum value ($M_i - G_i(R_j)$) let us choose the maximum value

$$A_j = \max_i (M_i - G_i(R_j)), j = 1, 2, ..., m, k = 1, 2, ..., l \text{ for each route } R_j, j = 1, 2, ..., m;$$

$$A_j = (6, 5, 4, 2, 8, 6, 9, 7, 8, 4, 5)^T.$$

4. The minimum value of the deviations of the net income from the maximum value equals 2. This value corresponds to the route R_4.

Table 2. The net income of each participant of the trasnportation

The route\the participant of the route	1 participant	2 participant	3 participant	4 participant
R_1	4	5	0	6
R_2	0	1	5	2
R_3	0	3	2	4
R_4	1	2	1	0
R_5	0	8	5	3
R_6	0	6	4	4
R_7	0	6	9	3
R_8	1	3	7	0
R_9	4	2	8	0
R_{10}	3	4	0	2
R_{11}	4	5	3	1
R_{12}	7	4	2	1

So each of the transport participants receives the maximum possible net income (a compromise solution) by transporting the construction cargo from the starting point s to the final point s' by choosing the route R_4.

5. Conclusion

In this paper, the problem of construction cargo transportation by different transport modes is studied. Such type of construction cargo transportation is very popular because of the lack the direct route between origin and destination point. Such approach allow constructing nonstandard routes that help transportation organizer to minimize the time of delivery or maximize the net profit from transportation.

6. Acknowledgment

The work is partially supported by the RFBR grant # 18-01-00796.

References

[1] Beiglböck Mathias and Griessler Claus (2014) *An optimality principle with applications in optimal transport* arXiv: 1404.7054

[2] Hari Krishna Gaddam and K. Ramachandra Rao 2019 *Modelling vehicular behaviour using trajectory data under non-lane based heterogeneous traffic conditions* Archives of transport vol 52 is 4 pp 95-108 DOI: 10.5604/01.3001.0014.0211

[3] Ambarwati L and Adam J P and Robert V and Bart Van A 2014. *Empirical analysis of heterogeneous traffic flow and calibration of porous flow model* Transportation Research Part C: Emerging Technologies 48 pp 418–436 doi:https://doi.org/10.1016/j.trc.2014.09.017

[4] Dey P and Chandra S and Gangopadhyay S 2008 *Simulation of mixed traffic flow on two-lane roads* Journal of Transportation Engineering 134(9) pp 361-369.

[5] Chanut S and Buisson C 2003 *Macroscopic Model and Its Numerical Solution for Two-Flow Mixed Traffic with Different Speeds and Lengths* Transportation Research Record 1852(1) pp 209–219

[6] Gupta A K and Dhiman I 2014 *Analyses of a continuum traffic flow model for a nonlanebased system* International Journal of Modern Physics C 25(10) 1450045. doi: http://dx.doi.org/10.1142/S0129183114500454

[7] Jiang R and Wu Q-S and Zhu Z-J 2002 *A new continuum model for traffic flow and numerical tests* Transportation Research Part B: Methodological 36(5) pp 405–419. doi: https://doi.org/10.1016/S0191-2615(01)00010-8

[8] Khan S and Maini P 1999 *Modeling Heterogeneous Traffic Flow. Transportation Research Record* Journal of the Transportation Research Board 1678(1) pp 234–241 doi: http://dx.doi.org/10.3141/1678-28.
[9] Logghe S and Immers L 2003 Heterogeneous traffic flow modelling with the lwr-model using passenger-car equivalents *Proceedings of the 10th World congress on ITS* pp 1–15 Available at: http://www.kuleuven.be/traffic/dwn/P2003E.pdf
[10] Malafeyev O and Saifullina D and Ivaniukovich G and Marakhov V and Zaytseva I 2017 *The model of multi-agent interaction in a transportation problem with a corruption component* (AIP Conference Proceedings) http://dx.doi.org/10.1063/1.4992360.
[11] Malafeyev O and Lakhina J and Redinskikh N and Smirnova T and Smirnov N and Zaitseva I 2019 A mathematical model of production facilities locatio *Journal of Physics: Conference Series*, https://doi.org/10.1088/1742-6596/1172/1/012090
[12] Semenov I and Filina-Dawidowicz L 2017 Topology-based approach to the modernization of transport and logistics systems with hybrid architecture. Part 1. Proof-of-concept study Archives of transport vol 43 is 3 pp 105-124
[13] Malafeyev O and Onishenko V and Zubov A and Bondarenko L and Orlov V and Petrova V and Kirjanen A and Zaitseva I 2019 Optimal location problem in the transportation network as an investment project: A numerical method *AIP Conference Proceedings* https://doi.org/10.1063/1.5114525.
[14] Chen B and Cheng H H, 2010 A Review of the Applications of Agent Technology in Traffic and Transportation Systems *IEEE Transactions on Intelligent Transportation Systems* 11(2) pp 485-497
[15] Chen Y M and Wang B-Y 2009 Towards Participatory Design of Multi-agent Approach to Transport Demands *IJCSI International Journal of Computer Science Issues* 4(1) pp 10-15

Simulation of multi-agent interaction between mobile operators and users

I V Zaitseva[1,2], O A Malafeyev[1], I F Kefeli[3], N V Poddubnaya[2], A V Shuvaev[2]

[1]Saint-Petersburg State University, 7/9 Universitetskaya nab., St. Petersburg, 199034, Russia

[2]Stavropol State Agrarian University, Zootekhnichesky Lane, 12, Stavropol, 355017, Russia

[3] Center for geopolitical expertise and publishing projects of the North-Western Institute of management Russian, Academy of national economy and state service under the President of the Russian Federation, Saint-Petersburg, Russia, 4 Pesochnaya Embankment, Saint-Petersburg, 197376, Russia

E-mail irina.zaitseva.stv@yandex.ru

Abstract. Game-theoretic models for two classes of players are considered in this paper. The first class will consist of mobile operators, who, based on information about competitors' actions, will plan changes in prices for services. The second class will consist of dedicated user cohorts who, depending on the current situation, will plan their actions, namely, change or not change the operator. The presented models solve the problem of maximizing the income of mobile operators and minimizing the cost of communication by users. A turn-based game is considered where a possible strategy for a first-class player will be considered a non-zero price vector for the provided communication services. A possible strategy for a second-class player will be considered the sequence of choosing a mobile operator for each of the services. The payoff function of both classes players will be built at each stage. According to the winnings, there is a compromise solution for different classes of players.

Keywords: Game-theoretic model, multi-agent interaction, operator, mobile communication, user, game tree

1. Introduction

There are companies on the market that provide various communication services for users, let's call them mobile operators. Each mobile operator is characterized by a price vector for various communication services.

The following types of communication are understood as communication services in the model: voice communication in the "home region", long-distance communication, international communication, SMS, mobile Internet. The "home" region is understood as the subscriber connection region.

If a user of mobile communications services uses different mobile operators for different services, he incurs costs compared with the user who has one operator for all mobile communication services.

If the user changes the mobile operator to one of the services that the other mobile operator previously provided to him, he incurs slightly different costs. These costs can vary from small values, for those users who, for example, come for a short time to another city and want to save on roaming, so they temporarily change the operator. To large amounts, when, for example, the company incurred large costs for advertising a particular subscriber number and cannot stop using it, so as not to lose customers. Also, these costs may not be material, and their monetary equivalent will be taken into account in the model. Such costs, for example, may arise if the user of communication services has not changed the operator for a long time and, changing the number, his friends who were not informed, may lose contact with him.

2. Informal statement of the problem

Since the number of users of communication services has already exceeded the number of residents in Russia, and in order not to consider these users, they are divided into cohorts (groups) by the number and type of consumed services, by the value of the operator change cost parameter. Thus, the dimension of the problem can be significantly reduced.

In this model, we will consider two classes of players. The first class will consist of mobile operators, who based on information about competitors' actions, will plan changes in prices for services. The second class will consist of dedicated user groups, which depending on the current situation will plan their actions (change or not change the operator). In this model, in order for both classes of players to have a maximization problem, suppose that second-class players have a constant income that exceeds communication costs, and they maximize the difference between income and communication costs [1-3].

Incomes of mobile operators are calculated by multiplying the total number of users who have chosen this particular operator for this communication service by the price per unit of service. The goal of players of both classes is to maximize their income function. We will consider a turn-based game, i.e. first-class players alternately change prices for the provided communication services, after each step of the first, second-class players either change the operator to certain types of communications, or decide to keep it. A possible strategy for a first-class player will be considered a non-zero price vector for the provided communication services. A possible strategy for a second-class player will be considered the sequence of choosing a mobile operator for each of the services.

After each of the mobile operators makes one move, the final wins of the players of both classes are calculated. At each stage, the payoff function of players of both classes is built, by which we find a compromise solution for different classes of players.

3. Problem formalization of choosing a vector of prices by mobile operators

The following task is considered: there are n mobile operators T_1, T_2, T_3, ..., T_i, ... T_n, $i=\overline{1,n}$, which provide k different services U_1, U_2, U_3, ..., U_j, ... U_k, $j=\overline{1,k}$. Each mobile operator T_i, $i=\overline{1,n}$, is defined by a vector of dimension (k+2), $T_i=(t_{i1},t_{i2},...,t_{ij},...,t_{ik},p_i,q_i)$ in which t_{ij}, $i=\overline{1,n}$, $j=\overline{1,k}$, are the prices of the mobile operator T_i, $i=\overline{1,n}$, for the service U_j, $j=\overline{1,k}$. p_i, $i=\overline{1,n}$, is the cost of the operator T_i, $i=\overline{1,n}$ for advertising, and q_i, $i=\overline{1,n}$, is the cost of the operator T_i, $i=\overline{1,n}$ for the development of mobile communications network [4-5].

Each user of communication services is set by the vector of their needs $I=(i_1,i_2,i_3,...i_k)$ and also by the parameter of the cost of changing the operator $s=s[t_i,t_j,f(I)]$, $i\neq j$ where t_i, whose operator is changed, t_j, the operator to which they change, the function f(I) takes different values, depending on the type of communication services user [6].

Since the number of users of communication services significantly exceeds the number of mobile operators and the services they provide, we group them. Denote by S the set of values of the parameter s. We divide this set exactly on R disjoint subsets and enumerate them.

$$\bigcup_{i=1}^{R} s_i = S; s_i \bigcap s_j = 0, \forall i, j \in R; \forall i, j \in R.$$

Next, we will group communication service users into cohorts by the parameter of operator change cost, so that if for some group of users this parameter is in the subset s_i, we will say that this communication service user is in the i-th cohort. Moreover, the entire cohort can be considered as one user of communication services, whose needs are equal to the sum of the needs of the cohort participants, and the cost of changing an operator is equal to the average for the cohort. Thus, each cohort is denoted by Q_l, $l = \overline{1,r}$ and is a vector $Q_1 = (q_{l1}, q_{l2}, ..., q_{lk}, s_l)$ which contains the need for communication services of this cohort as well as the cost of changing the operator for this cohort.

At the initial moment of time, a matrix is defined, $W = W^{k \times n}$, in which the element $w_{i,j}$ denotes the price of the i-th mobile operator I_i, $i = \overline{1,n}$ for the j-th service U_j, $j = \overline{1,k}$.

By the payoff function H_i $i = \overline{1,n}$ of each mobile operator, we consider the income received from users of communication services who have chosen for their service U_j $j = \overline{1,k}$, operator T_i $i = \overline{1,n}$.

For uniformity, we assume that each cohort of users receives a certain stable income B_l=B, $l = \overline{1,r}$, with $B >> \max(q_l) \times \max(t_{ij})$, $l = \overline{1,r}$, $i = \overline{1,n}$, $j = \overline{1,k}$. The cohort gain function will be the difference between *B* and the cost of communication services and a possible change of operators.

By the top of the tree we mean the value of the matrix $W = W^{(k+2) \times n}$, an indication of which of the operators will go next, as well as the entire history of the change in the matrix W [7-9].

Consider the actions of mobile operators. For the many possible strategies of the operator I_i, $i = \overline{1,n}$ we take the set of price vectors for communication services that this operator can offer.

4. Dimension analysis of the practically solved multi-agent interaction problem

Let us consider an example of the interaction of 4 mobile communication operators providing 4 different communication services: voice communication, long-distance communication, SMS and mobile Internet. As an example, take a multi-step game with full information on a tree-like finite graph. By the top of the graph in this model we mean the price matrix of mobile operators for communication services, $W = W^{4 \times 4}$, in which the element $w_{i,j}$ denotes the price of the i-th mobile operator, I_i, $i = \overline{1,n}$ for the j-th service, U_j, $j = \overline{1,k}$. In the initial period of time, the matrix W_0 is determined. Further, the first operator, Tele2, offers 2 different strategies, in which it indicates 2 different tariffs for communication services and in each cases its costs for advertising and network development. After that, Megafon makes a move, after which MTS move and Beeline completes.

As a result, we get a game tree in which for each vertex, as well as for the set of final vertices, we write down the profit of each operator, taken from a quarterly report published by mobile operators. The winning function for each player will be equal to the profit earned for the reporting period.

In the practical implementation of this example, cohorts of users can choose their own operator for each of the services in an arbitrary way. In the interaction of 4 mobile operators providing 4 services, in the first step we get $4^4 = 64$ the choice for each of the cohorts, in the second step the cohort of communication service users can change the operator at random, which will give an option $64 \times 64 \times 2 = 8192$. For all the terminal vertices of the game tree, we get $64^5 = 1073741824$ the options for getting to this terminal vertex. It is also necessary to take into account that there are only 16 such vertices, which increases the dimension of the problem by one more order.

To reduce the dimension of the problem, in the examples considered, a model with two mobile operators providing two services and a model with three mobile operators providing three services will be used. In the latter case, we get 27 options for choosing a mobile operator to meet demand at one vertex. And 531441 choices for each of the 8 end vertices.

5. Examples of user choices for a mobile operator

Let us consider the number of different choices for a mobile operator by the user of communication services. With the interaction of 2 operators providing 2 services, we get 4 options: ((1,1), (1,2), (2,1), (2,2)). The first option means that for voice communication in the home region the user of communication services chooses the first mobile operator for himself, for using mobile Internet the user of communication services chooses the first mobile operator for himself. The second option means that for voice communication in the home region the user of communication services chooses the first mobile operator for himself, for using mobile Internet the user of communication services chooses the second mobile operator for himself. The third option means that for voice communication in the home region the user of communication services chooses a second mobile operator for himself, for using mobile Internet the user of communication services chooses his first mobile operator. The fourth option means that for voice communication in the home region the user of communications services chooses a second mobile operator, for using mobile Internet the user of communications services chooses a second mobile operator.

5.1. An example of a positional game for 2 mobile operators

In this example, we will only consider price changes by the mobile operator and we will not consider the costs of advertising and network development.

The initial price matrix for communication services W_0 of mobile operators looks like $W_0 = \begin{pmatrix} 10 & 15 \\ 7 & 8 \end{pmatrix}$. The first operator offers two different price vectors for communication services. Thus, we obtain the matrix W_1 and the matrix W_2: $W_1 = \begin{pmatrix} 6{,}9 & 17{,}5 \\ 7 & 18 \end{pmatrix}$, $W_2 = \begin{pmatrix} 6 & 16 \\ 7 & 18 \end{pmatrix}$.

After that, the user of communication services has 4 choice strategies ((1,1), (1,2), (2,1), (2,2)), i.e. he can leave everything as it is, can change one operator for one service, can exchange one operator for another service and can exchange both operators for both services.

There is a second operator and offers two price vectors for communication services. After that, the user of communication services is again redistributed between operators:

$$W_3 = \begin{pmatrix} 6{,}9 & 17{,}5 \\ 6 & 17 \end{pmatrix},\ W_4 = \begin{pmatrix} 6{,}9 & 17{,}5 \\ 6{,}5 & 15 \end{pmatrix},\ W_5 = \begin{pmatrix} 6 & 16 \\ 5 & 15{,}5 \end{pmatrix},\ W_6 = \begin{pmatrix} 6 & 16 \\ 5{,}9 & 15{,}9 \end{pmatrix}.$$

For a mobile operator, the game tree will look like this (Figure 1).

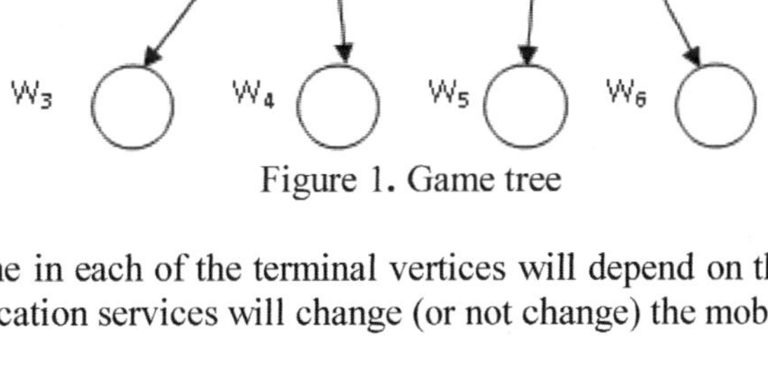

Figure 1. Game tree

In this case, the income in each of the terminal vertices will depend on the sequence in which and how the user of communication services will change (or not change) the mobile operator.

IOP Conf. Series: Materials Science and Engineering **873** (2020) 012038 doi:10.1088/1757-899X/873/1/012038

Consider the following user of communication services: voice communication, the first type of service he needs 3 units, mobile Internet, the second type of service he needs 2 units. The cost of changing the operator for him will be 1 unit. We will write the winnings of all the players in each of the terminal vertices, subject to a different choice of the user of communication services, the mobile operator. We denote by H_1 the gain of the communication services user, H_2 is the gain of the first mobile operator, H_3 is the gain of the second mobile operator.

Vertex W_3

H_1	H_2	H_3	User way
164,4	141,4	21	(2,1) → (1,1) → (1,1)
164,4	106,4	55	(2,2) → (1,1) → (1,2)
161	100	60	(2,2) → (2,1) → (2,1)
161	65	94	(2,1) → (2,1) → (2,2)
166	0	166	(2,2) → (2,2) → (2,2)
171,4	171,4	0	(1,1) → (1,1) → (1,1)

Vertex W_4

H_1	H_2	H_3	User way
164,4	141,4	21	(2,1) → (1,1) → (1,1)
160,4	106,4	51	(2,1) → (1,1) → (1,2)
162,5	100	61,5	(1,2) → (1,2) → (1,2)
158,5	65	91,5	(2,1) → (2,1) → (2,2)
163,5	0	163,5	(2,2) → (2,2) → (2,2)
171,4	171,4	0	(1,1) → (1,1) → (1,1)

Vertex W_5

H_1	H_2	H_3	User way
153	130	21	(2,1) → (1,1) → (1,1)
153	98	52	(2,1) → (1,1) → (1,2)
151	112	36	(2,1) → (1,1) → (2,1)
151	80	67	(2,1) → (1,1) → (2,2)
160	0	160	(2,2) → (2,2) → (2,2)
160	160	0	(1,1) → (1,1) → (1,1)

Vertex W_6

H_1	H_2	H_3	User way
153	130	21	(2,1) → (1,1) → (1,1)
153,8	98	52,8	(2,1) → (1,1) → (1,2)
153,7	112	38,7	(2,1) → (1,1) → (2,1)
154,5	80	70,5	(2,1) → (1,1) → (2,2)
163,5	0	163,5	(2,2) → (2,2) → (2,2)
160	160	0	(1,1) → (1,1) → (1,1)

Thus, we obtain a solution for the cohort minimum cost of communication services users (Figure 2). This will be the vertex of W_5, and here for the cohort there are two strategies with the same gain: (2,1) → (1,1) → (2,1); (2.1) → (1.1) → (2.2).

It is in this option that you can see how even insignificant costs for advertising or network development will help the first mobile operator to attract this particular cohort, because her costs will increase by only 2 units, while the profit of the first operator will increase by 50 units. The second mobile operator at the vertex of W_6 has a similar opportunity, the costs of the cohort will grow by only 1.5 units, and the operator's profit will grow by 49.9 units.

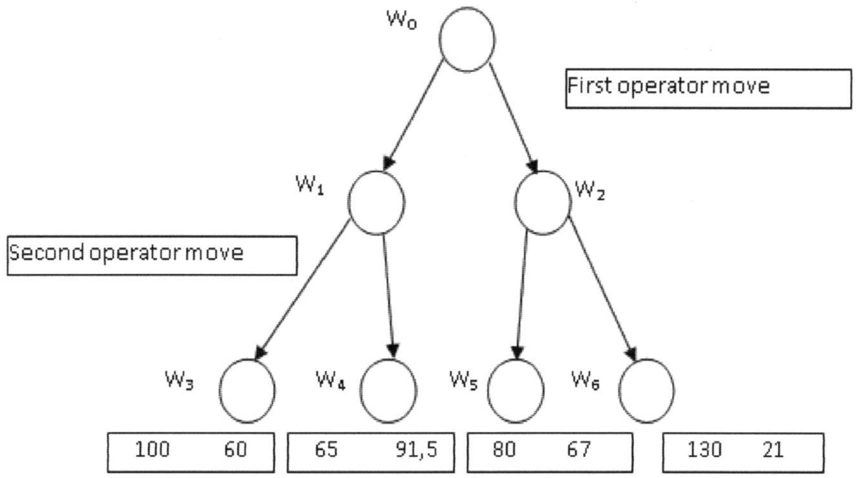

Figure 2. Solution for the cohort minimum cost of communication services users

In this example, the compromise solution for mobile operators coincides with the solution that would have been learned if the first operator chooses a cautious strategy. So at the first step, evaluating your minimum profit, the first operator chooses the W_2 strategy, where this minimum profit is greater. In the second step, the second operator chooses a strategy that maximizes his profit, and this is the W_5 strategy, which coincides with the compromise solution of this game tree.

5.2. An example of a positional game for 2 mobile operators providing 2 services and 6 different user cohorts

In the second example, we consider the interaction of 2 mobile operators and 6 user cohorts of communications services with different needs for voice and Internet traffic. We write down each of the cohorts needs in the following table 1.

Table 1. Cohort Needs.

Cohort number	Voice Requirements	Mobile Internet Needs	Operator Change Cost
1	4	5	2
2	5	2	2
3	6	3	5
4	7	4	5
5	5	10	30
6	10	5	30

Now we describe the payoff functions each of the cohorts and mobile operators at each of the end vertices W_3, W_4, W_5, W_6. H_1 - H_6 payoffs of the cohorts from the first to the sixth, H_7 and H_8 are the payoffs of mobile operators. And also for each of the cohorts we will write the path by which she came to the current vertex.

For the vertex W_3

		H_7	H_8	Way
H_1	331,5	162,5	165	$(2,1) \to (2,1) \to (2,2)$
H_2	202	100	100	$(2,1) \to (2,1) \to (2,1)$
H_3	275	150	120	$(2,1) \to (2,1) \to (2,1)$
H_4	345	200	140	$(2,1) \to (2,1) \to (2,1)$
H_5	619	619	0	$(1,1) \to (1,1) \to (1,1)$
H_6	465	0	465	$(2,2) \to (2,2) \to (2,2)$
	H_7 and H_8	1231,5	990	

For the vertex W_4

		H_7	H_8	Way
H_1	323,5	162,5	157	(2,1) → (2,1) → (2,2)
H_2	201,5	65	132,5	(2,1) → (2,1) → (2,2)
H_3	275,5	97,5	168	(2,1) → (2,1) → (2,2)
H_4	343,5	130	203,5	(2,1) → (2,1) → (2,2)
H_5	612,5	0	612,5	(2,2) → (2,2) → (2,2)
H_6	460	0	460	(2,2) → (2,2) → (2,2)
	H_7 and H_8	455	1733,5	

For the vertex W_5

		H_7	H_8	Way
H_1	312,5	155	153,5	(2,1) → (2,1) → (2,2)
H_2	190	124	60	(2,1) → (1,1) → (2,1)
H_3	260	141	114	(2,1) → (2,1) → (2,1)
H_4	326	188	133	(2,1) → (2,1) →2,1)
H_5	580	580	0	(1,1) → (1,1) →1,1)
H_6	447,5	0	447,5	(2,2) → (2,2) → (2,2)
	H_7 and H_8	1188	908	

For the vertex W_6

		H_7	H_8	Way
H_1	315	283	28	311
H_2	191	154	35	311
H_3	265	213	42	311
H_4	331	272	49	311
H_5	580	580	0	(1,1) → (1,1) → (1,1)
H_6	458,5	0	458,5	(2,2) → (2,2) → (2,2)
	H_7 and H_8	1502	612,5	

Based on the payoffs of the mobile operators, we will find a compromise solution. To do this, we define at each vertex of the operator that made the least profit, and find the vertex at which this smaller value will be the largest. Such a vertex W_3, in which the first mobile operator makes a profit of 1231.5 units, and the second mobile operator receives 990 units. Moreover, for each of the cohorts, the maximum profit is brought by the vertex W_5. Unfortunately, user cohorts of communication services can only indirectly influence mobile operators, therefore, when choosing their move, mobile operators are more focused on the moves of competitors than on the actions of user cohorts of communication services.

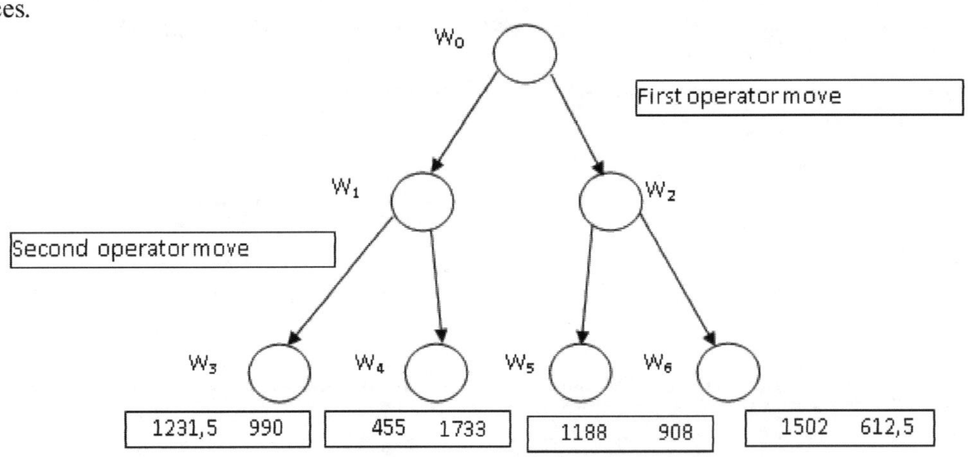

Figure 3. Compromise solution

With such a game tree (Figure 3), a compromise solution will not be reached, because predicting its profit in the first move, the first mobile operator will choose the strategy W2. During the course of the second mobile operator, he will also strive to maximize his profit and, when the first operator chooses the strategy W1, the second operator will choose W4 to maximize his profit. The first mobile operator in this case will receive its minimum profit. Fearing just this, the first operator will choose the W2 strategy in its first move, and the second mobile operator will choose the W5 strategy to maximize its profit. At the same time, the user cohorts will receive the maximum of their profit.

6. Conclusion

Dividing into cohorts (groups) (by the number and type of consumed services, by the value of the operator change cost parameter) has significantly reduced the dimension of the multi-agent interaction model between mobile operators and users. Two classes of players are distinguished in the work: mobile operators and dedicated user cohorts. In this model, the goal of both class players was to maximize the profit function. The presented turn-based game contains strategies for players of the first and second classes. The payoff function of both class players built at every stage of the game made it possible to find a compromise solution for different classes of players.

7. Acknowledgment

The work is partially supported by the RFBR grant # 18-01-00796.

References

[1] Johnson L and Yannakakis G and Togelius J 2010 Cellular Automata for Real-time *Generation of Infinite Cave Levels Proceedings of the Workshop on Procedural Content Generation in Games* PCGames '10 pp 10.1–10.4.

[2] Antonelli G 2013 Interconnected dynamic systems: An overview on distributed control *IEEE Control Systems* Vol 33 no 1 pp 76–88

[3] Ma Hang and Kumar Satish and and Koenig S 2017 Multi-Agent Path Finding with Delay Probabilities *Proceedings of the Thirty-First AAAI Conference on Artificial Intelligence* (AAAI-17) USA pp 3605–3612

[4] Amir O and Sharon G and Stern R 2015 Multi-agent Pathfinding As a Combinatorial Auction *Proceedings of the Twenty-Ninth AAAI Conference on Artificial Intelligence* AAAI'15 AAAI Press pp 2003–2009

[5] Malafeev O 1977 Stationary strategies in differential games, *USSR Computational Mathematics and Mathematical Physics* 17(1) pp37-46

[6] Malafeev O A 1974 *Equilibrium situations in dynamic games* (Cybernetics and System Analisys) 10(3) pp 504-513

[7] Batool K and Niazi M 2017Modeling the internet of things: a hybrid modeling approach using complex networks and agent-based models *Complex Adaptive Systems Modeling* Vol 5 no 1 p 4

[8] Malafeyev O and Lakhina J and Zaitseva I and Redinskikh N and Smirnova T and Smirnov N 2019 A mathematical model of production facilities location *Journal of Physics Conference Series* vol 1172 no 012090

[9] Zubov A and Dikusar V and Zubov N 2010 Controllability criterion for stationary systems *Doklady Mathematics* vol 81 no 1 pp 6-7

Non-linear Mathematical Model of a Quad-Tiltrotor UAV Spatial Motion

D B Litvin[1], I P Shepet[2], L N Korolkova[2], D L Vinokursky[3], L A Lyutikova[4]

[1]Stavropol State Agrarian University, Zootekhnichesky Lane, 12, Stavropol, 355017, Russia

[2]Technological Institute of service (branch of don state technical University), Kulakov Avenue, 41/1, Stavropol, 355000, Russia

[3]North Caucasus Federal University, Pushkin street, 1, Stavropol, 355017, Russia

[4]Institute of Applied Mathematics and Automation of Kabardin-Balkar Scientific Centre of RAS (IAMA KBSC RAS), Shortanov street, 89 A, Nalchik, 360000, Russia

E-mail litvin-372@yandex.ru

Abstract: The article presents a design scheme of a quad-tiltrotor unmanned aerial vehicle (UAV), which combines a safe vertical take-off and landing procedures and a relatively large flight range inherent in the aircraft. In order to reduce the cost and weight of the quad-tiltrotor UAV, to increase its reliability, a circuit with minimization of movable executive drives is considered. Namely, in the proposed structural scheme of the quad-tiltrotor UAV there are no deflectable airfoils. In addition, the rotors of all four engines are rotated by the same setting angle in the longitudinal plane using a single actuator. The research results showed that the quad-tiltrotor UAV of the proposed scheme as a whole is statically controlled. All the necessary balancing forces and moments are formed independently of each other only by a coordinated change in engine speed and control of the overall setting angle of their thrust vectors. It is also shown that there is a so-called degenerate flight mode in which static controllability is not provided. Theoretically, this is reflected in the degeneracy of the linear equation system matrix for determining the balance values of engine thrusts. Physically, the specified mode corresponds to the case when the resulting thrust vector of the engines lies on a straight line passing through the centre of gravity of the aircraft, and therefore it cannot create a pitch moment. The paper proposes two ways to overcome this limitation. The constructive method offers a special arrangement of rotor bushings. The algorithmic method carries out a quick transition through a degenerate balancing mode using a control system.

Keywords: Quadcopter, Quad-Tiltrotor, Unmanned Aerial Vehicle (UAV), Model of Motion, Static Controllability.

1. Introduction

For research we chose a model of a quad-tiltrotor UAV. The design of the drone is a flying wing with 4 rotary engines [9]. Such constructive scheme is chosen in order to provide a greater radius of action compared to the quadrocopter [1, 7,8,11,14,17] while maintaining a safe vertical take-off and landing modes [4,14,18]. It is assumed that in the steady cruising mode, the wing will create most of the lift. The use of four engines provides the quad-tiltrotor UAV with a wide range of alignments [2, 12].

IOP Conf. Series: Materials Science and Engineering **873** (2020) 012039 doi:10.1088/1757-899X/873/1/012039

An important feature of the proposed design, in comparison with the known ones [5, 9,12-16], is the absence of airfoils control. The second feature of the quad-tiltrotor UAV is that the rotors of all four engines are rotated by the same installation angle in the longitudinal plane using a single actuator [14, 18]. These features are designed to facilitate, simplify and cheapen the quad-tiltrotor UAV design [13].

At the same time, the formation of both the lifting force and all the necessary control forces and moments should be provided only by coordinated changes in engine revolutions and a joint turn of their thrust vectors [6, 10].

The purpose of the study is to develop a mathematical model of the quad-tiltrotor UAV movement, to show the fundamental possibility of its controllability over all channels and to identify the existing restrictions on the control in transient flight modes [14, 18].

2. The validation of motion quad-tiltrotor UAV model structure and assumptions

The first step in the study of the stability and controllability characteristics of the "aircraft - autopilot" system is to build a model of the aircraft's movement. To obtain the model, the following assumptions were used:

- the design of the quad-tiltrotor UAV was considered as a solid [3], aeroelastic deformations were not taken into account;
- the design of the aircraft was assumed symmetrical, in which the centre of mass is located in the plane of symmetry and coincides with the origin of the coordinate system associated with the object (according to Russian State Standard 20058-80 Aircraft dynamics in atmosphere. Terms, definitions and symbols);
- the axis of the associated coordinate system is directed along the main axes of aircraft inertness, the centrifugal moments of inertness are zero;
- the atmosphere was supposed to be calm with a known dependence of pressure on altitude.

A non-linear mathematical model of the spatial motion of the quad-tiltrotor UAV includes dynamic and kinematic equations of motion, equations of motion parameters relations in various coordinate systems, as well as supporting (exterior) models [2].

Dynamic equations of spatial motion of the quad-tiltrotor UAV as a solid in a connected coordinate system under the above assumptions:

$$m\left(\begin{bmatrix}\dot{V}_{Kx}\\ \dot{V}_{Ky}\\ \dot{V}_{Kz}\end{bmatrix}+\begin{bmatrix}0 & -\omega_z & \omega_y\\ \omega_z & 0 & -\omega_x\\ -\omega_y & \omega_x & 0\end{bmatrix}\begin{bmatrix}V_{Kx}\\ V_{Ky}\\ V_{Kz}\end{bmatrix}\right)=M_{ca}(\alpha,\beta)\begin{bmatrix}-X_a\\ Y_a\\ Z_a\end{bmatrix}+\begin{bmatrix}P_x\\ P_y\\ P_z\end{bmatrix}+M_{cg}(\psi,\vartheta,\gamma)\begin{bmatrix}0\\ -G\\ 0\end{bmatrix}, \quad (1)$$

$$\begin{aligned}I_x\dot{\omega}_x+\omega_y\omega_z\left(I_z-I_y\right)&=M_x+M_{Px};\\ I_y\dot{\omega}_y+\omega_x\omega_z\left(I_x-I_z\right)&=M_y+M_{Py};\\ I_z\dot{\omega}_z+\omega_x\omega_y\left(I_y-I_x\right)&=M_z+M_{Pz}-4I_{Dz}\cdot\ddot{\delta},\end{aligned} \quad (2)$$

where M_{ca} is the transfer matrix from the wind-axis system $OX_aY_aZ_a$ to body axis system $OXYZ$ of the following form:

$$M_{ca}=M_\alpha M_\beta;\quad M_\alpha=\begin{bmatrix}\cos\alpha & \sin\alpha & 0\\ -\sin\alpha & \cos\alpha & 0\\ 0 & 0 & 1\end{bmatrix};\quad M_\beta=\begin{bmatrix}\cos\beta & 0 & -\sin\beta\\ 0 & 1 & 0\\ \sin\beta & 0 & \cos\beta\end{bmatrix}, \quad (3)$$

α,β are incidence and gliding angles;

M_{cg} is the transfer matrix from the normal axis system $OX_gY_gZ_g$ to body axis system $OXYZ$ of the form:

$$M_{cg}=M_{\gamma}M_{\vartheta}M_{\psi};\; M_{\gamma}=\begin{bmatrix}1 & 0 & 0\\ 0 & \cos\gamma & \sin\gamma\\ 0 & -\sin\gamma & \cos\gamma\end{bmatrix};\; M_{\vartheta}=\begin{bmatrix}\cos\vartheta & \sin\vartheta & 0\\ -\sin\vartheta & \cos\vartheta & 0\\ 0 & 0 & 1\end{bmatrix};\; M_{\psi}=\begin{bmatrix}\cos\psi & 0 & -\sin\psi\\ 0 & 1 & 0\\ \sin\psi & 0 & \cos\psi\end{bmatrix},$$

ϑ,γ,ψ are pitch, bank and yaw angles;

$\vec{P}=\begin{pmatrix}P_x & P_y & P_z\end{pmatrix}^T$ is the resultant engine thrust vector;

$\vec{M}=\begin{pmatrix}M_x & M_y & M_z\end{pmatrix}^T$ и $\vec{M_P}=\begin{pmatrix}M_{Px} & M_{Py} & M_{Pz}\end{pmatrix}^T$ are vectors of aerodynamic and engine thrust moments;

I_{Dz} is the engine inertness rotary moment.

Kinematic equations of translational and rotational movements are:

$$\begin{bmatrix}\dot{X}_g\\ \dot{Y}_g\\ \dot{Z}_g\end{bmatrix}=\begin{bmatrix}\dot{L}\\ \dot{H}\\ \dot{Z}\end{bmatrix}=\begin{bmatrix}V_{K_{Xg}}\\ V_{K_{Yg}}\\ V_{K_{Zg}}\end{bmatrix}=M_{gc}\begin{bmatrix}V_{Kx}\\ V_{Ky}\\ V_{Kz}\end{bmatrix}=\left(M_{\gamma}M_{\vartheta}M_{\psi}\right)^T\begin{bmatrix}V_{Kx}\\ V_{Ky}\\ V_{Kz}\end{bmatrix}, \tag{4}$$

$$\begin{aligned}&\dot{\gamma}=\omega_x+tg\vartheta\left(\omega_z\sin\gamma-\omega_y\cos\gamma\right);\\ &\dot{\vartheta}=\omega_y\sin\gamma+\omega_z\cos\gamma;\\ &\dot{\psi}=\left(\omega_y\cos\gamma-\omega_z\sin\gamma\right)\frac{1}{\cos\vartheta}.\end{aligned} \tag{5}$$

Equations of constraints are:

$$\begin{bmatrix}V_{Kxg}\\ V_{Kyg}\\ V_{Kzg}\end{bmatrix}_g=M_{gc}\left(\psi,\vartheta,\gamma\right)\begin{bmatrix}V_{Kx}\\ V_{Ky}\\ V_{Kz}\end{bmatrix}_c=M_{gk}\left(\Psi,\theta\right)\begin{bmatrix}V_K\\ 0\\ 0\end{bmatrix}_k=V_K\begin{bmatrix}\cos\theta\cos\Psi\\ \sin\theta\\ -\cos\theta\sin\Psi\end{bmatrix}_g, \tag{6}$$

M_{gk} is the transfer matrix from flight path axis system $OX_kY_kZ_k$ to normal axis system $OX_gY_gZ_g$ of form:

$$M_{gk}=M_{\Psi}M_{\theta};\;\; M_{\Psi}=\begin{bmatrix}\cos\Psi & 0 & \sin\Psi\\ 0 & 1 & 0\\ -\sin\Psi & 0 & \cos\Psi\end{bmatrix};\;\; M_{\theta}=\begin{bmatrix}\cos\theta & -\sin\theta & 0\\ \sin\theta & \cos\theta & 0\\ 0 & 0 & 1\end{bmatrix},$$

θ,Ψ are climb and path angles;

$$V_K=\sqrt{V_{Kx}^2+V_{Ky}^2+V_{Kz}^2};\;\; \theta=arctg\frac{V_{Kyg}}{\sqrt{V_{Kxg}^2+V_{Kzg}^2}};\;\; \Psi=-arctg\frac{V_{Kzg}}{V_{Kxg}}. \tag{7}$$

$$\vec{V}=\begin{bmatrix}V_x\\ V_y\\ V_z\end{bmatrix}_c=M_{ca}\begin{bmatrix}V\\ 0\\ 0\end{bmatrix}_a=\begin{bmatrix}V\cos\alpha\cos\beta\\ -V\sin\alpha\cos\beta\\ V\sin\beta\end{bmatrix}_c=\begin{bmatrix}V_{Kx}\\ V_{Ky}\\ V_{Kz}\end{bmatrix}_c-M_{ca}\begin{bmatrix}W_{xa}\\ W_{ya}\\ W_{za}\end{bmatrix}_a. \tag{8}$$

$$V=\sqrt{V_x^2+V_y^2+V_z^2};\;\; \alpha=-arctg\frac{V_y}{V_x};\;\; \beta=arctg\frac{V_z}{\sqrt{V_x^2+V_y^2}}. \tag{9}$$

Supporting models are:

- aerodynamic:

$$\vec{R}_a=[-X_a\;\; Y_a\;\; Z_a]^T=\rho\frac{V^2}{2}S[-c_{Xa}\;\; c_{Ya}\;\; c_{Za}]^T;\quad c_{Xa,Ya,Za}=c_{Xa,Ya,Za}\left(x,u,w\right);$$

$$\vec{M} = \begin{bmatrix} M_x & M_y & M_z \end{bmatrix}^T = \rho\frac{V^2}{2}S\begin{bmatrix} m_x l & m_y l & m_z b_a \end{bmatrix}^T; \quad m_{x,y,z} = m_{x,y,z}(x,u,w); \tag{10}$$

- traction electric engines:

$$\vec{P} = \begin{bmatrix} P_x & P_y & P_z \end{bmatrix}^T = \sum_{i=1}^{4} \vec{P}_i(x,u_i,\delta,w); \tag{11}$$

$$\vec{M}_P(x,u_i,\delta,w) = \begin{bmatrix} M_{Px} & M_{Py} & M_{Pz} \end{bmatrix}^T = \sum_{i=1}^{4}\left(\vec{r}_i \times \vec{P}_i + \overrightarrow{M}_{Ri}\right); \tag{12}$$

- atmosphere:

$$\rho = \rho(H);$$

- aircraft control system:

$$\delta = \delta(u_\delta); \quad P_i = P_i(u_i), \ \text{i=1,4}; \tag{13}$$

where u_δ, u_i are signals of control.

3. Model of a system for generating control forces and moments of the quad-tiltrotor UAV

In order for the motion model presented above to take on a complete form, it must be supplemented by the equations of forces and moments inherent in the particular quad-tiltrotor UAV.

The vector of the resulting engine thrust in projections on the axis of the associated coordinate system will take the form:

$$\overrightarrow{P} = \begin{pmatrix} P_x \\ P_y \\ P_z \end{pmatrix} = \vec{p} \cdot P_\Sigma = \begin{pmatrix} \cos\delta \\ \sin\delta \\ 0 \end{pmatrix} P_\Sigma; \qquad \vec{p} = \begin{pmatrix} \cos\delta \\ \sin\delta \\ 0 \end{pmatrix}, \tag{14}$$

where $\overrightarrow{P}, \vec{p}$ are full and normalized vectors of total engine thrust;

δ is engine installation angle in the body axis system;

$P_\Sigma = P_1 + P_2 + P_3 + P_4$ is the module of the total engine thrust vector.

The resulting moment of engine thrust force can be represented as:

$$\overrightarrow{M}_P = \sum_{i=1}^{4} \overrightarrow{M}_{Pi} = \sum_{i=1}^{4}\left(\vec{r}_i \times \overrightarrow{P}_i + \overrightarrow{M}_{Ri}\right), \tag{15}$$

where $\vec{r}_i = \begin{pmatrix} r_x & r_y & r_z \end{pmatrix}^T$ is the radius vector of the mass center position of the i-th engine in the body axis system;

$\overrightarrow{P}_i = \begin{pmatrix} \cos\delta & \sin\delta & 0 \end{pmatrix}^T \cdot P_i$ is thrust vector;

$\overrightarrow{M}_{Ri} = \begin{pmatrix} \cos\delta & \sin\delta & 0 \end{pmatrix}^T \cdot M_{Ri}$ is reactive moment vector of i-th engine in the body axis system.

If the vector product is presented in matrix form, then the moment created by the i-th engine will take the form:

$$\overrightarrow{M}_{Pi} = \vec{r}_i \times \overrightarrow{P}_i + \overrightarrow{M}_{Ri} = \begin{pmatrix} 0 & -r_z & r_y \\ r_z & 0 & -r_x \\ -r_y & r_x & 0 \end{pmatrix}\begin{pmatrix} \cos\delta \\ \sin\delta \\ 0 \end{pmatrix} P_i \pm a_m \begin{pmatrix} \cos\delta \\ \sin\delta \\ 0 \end{pmatrix} P_i = \begin{pmatrix} \pm a_m & -r_z & r_y \\ r_z & \pm a_m & -r_x \\ -r_y & r_x & \pm a_m \end{pmatrix}\begin{pmatrix} \cos\delta \\ \sin\delta \\ 0 \end{pmatrix} P_i. \tag{16}$$

Here $a_m = \dfrac{M_{Ri}}{P_i}$ is the coefficient of proportionality between the engine thrust and the reactive moment created by it. The sign of the moment is determined by the direction of rotor rotation.

For the considered quad-tiltrotor UAV scheme, the coordinates of the radius vectors of the position of the engines are as follows:

$$r=\left(\overrightarrow{r_1},\overrightarrow{r_2},\overrightarrow{r_3},\overrightarrow{r_4}\right)=\begin{pmatrix}-l_x & -l_x & l_x & l_x\\ -l_y & -l_y & l_y & l_y\\ -l_z & l_z & l_z & -l_z\end{pmatrix}. \qquad (17)$$

Given the accepted direction of motor matrix configuration (16) the defining moments are expressed as follows:

$$R_1=\begin{pmatrix}a_m & l_z & -l_y\\ -l_z & a_m & l_x\\ l_y & -l_x & a_m\end{pmatrix};\ R_2=\begin{pmatrix}-a_m & -l_z & -l_y\\ l_z & -a_m & l_x\\ l_y & -l_x & -a_m\end{pmatrix};\ R_3=\begin{pmatrix}a_m & -l_z & l_y\\ l_z & a_m & -l_x\\ -l_y & l_x & a_m\end{pmatrix};\ R_4=\begin{pmatrix}-a_m & l_z & l_y\\ -l_z & -a_m & -l_x\\ -l_y & l_x & -a_m\end{pmatrix}. \qquad (18)$$

Then expression (15) can be represented as:

$$\overrightarrow{M}_P=\sum_{i=1}^{4}\left(R_i\begin{pmatrix}\cos\delta\\ \sin\delta\\ 0\end{pmatrix}P_i\right)=\begin{pmatrix}l_z\sin\delta\left(P_1-P_2-P_3+P_4\right)\\ l_z\cos\delta\left(-P_1+P_2+P_3-P_4\right)\\ \left(l_x\sin\delta-l_y\cos\delta\right)\left(-P_1-P_2+P_3+P_4\right)\end{pmatrix}+a_m\begin{pmatrix}\cos\delta\\ \sin\delta\\ 0\end{pmatrix}\left(P_1-P_2+P_3-P_4\right). \qquad (19)$$

The nonlinear model presented above is closed with respect to its phase coordinates, but due to its complexity, it is unsuitable for practical use. Let to consider a special case of spatial motion in the form of a straight, uniform flight at a constant height in a calm atmosphere, in which part of the lifting force is created by the wing and part by the screws.

We have the following reference values for motion parameters:

$$\theta_o=\Psi_o=\dot{\theta}_o=\dot{\Psi}_o=\gamma_o=\psi_o=\dot{\gamma}_o=\dot{\vartheta}_o=\dot{\psi}_o=\beta_o=\omega_{x_o}=\omega_{y_o}=\omega_{z_o}=0;$$
$$\vartheta_o=\alpha_o=const\neq 0;\qquad W_{xa_o}=W_{ya_o}=W_{za_o}=0, \qquad (20)$$

In this case, the reference phase trajectory degenerates to a point in the phase space, and the model of motion becomes stationary. In this mode, the dynamic equations of translational motion (1) are usually represented in projections on the axis of the wind-axes system $OX_aY_aZ_a$, which for the reference motion become non-differential balance equations:

$$\begin{pmatrix}0\\0\\0\end{pmatrix}=\begin{pmatrix}-X_{a0}\\ Y_{a0}\\ 0\end{pmatrix}+\begin{pmatrix}\cos A_0\\ \sin A_0\\ 0\end{pmatrix}P_{\Sigma 0}-\begin{pmatrix}0\\ G_0\\ 0\end{pmatrix}, \qquad (21)$$

where $P_{\Sigma 0}=\left(P_{10}+P_{20}+P_{30}+P_{40}\right)$, $A_0=\vartheta_0-\theta_0+\delta_0=\alpha_0+\delta_0$.

From these equations we find the module and direction of the resulting thrust, as well as the angle of the engines:

$$P_{\Sigma 0}=P_{10}+P_{20}+P_{30}+P_{40}=\sqrt{X_{a0}^2+\left(G_0-Y_{a0}\right)^2};\quad tgA_0=\frac{G_0-Y_{a0}}{X_{a0}},\quad \delta_0=A_0-\alpha_0. \qquad (22)$$

The balance equations of rotational motion (2), (19), considering that the resultant aerodynamic force is applied to the center of mass of the aircraft, take the form:

$$\overrightarrow{M}_{P0}=\begin{pmatrix}l_z\sin\delta_0\left(P_{10}-P_{20}-P_{30}+P_{40}\right)\\ l_z\cos\delta_0\left(-P_{10}+P_{20}+P_{30}-P_{40}\right)\\ \left(l_x\sin\delta_0-l_y\cos\delta_0\right)\left(-P_{10}-P_{20}+P_{30}+P_{40}\right)\end{pmatrix}+a_m\begin{pmatrix}\cos\delta_0\\ \sin\delta_0\\ 0\end{pmatrix}\left(P_{10}-P_{20}+P_{30}-P_{40}\right)=\begin{pmatrix}0\\0\\0\end{pmatrix}. \qquad (23)$$

Three equations (23) and first equation (22) are generated the system of linear equations as regard to $P_{10},P_{20},P_{30},P_{40}$. Dividing these equations by $P_{\Sigma 0}$, we get:

$$\begin{pmatrix} l_z\sin\delta_0+a_m\cos\delta_0 & -l_z\sin\delta_0-a_m\cos\delta_0 & -l_z\sin\delta_0+a_m\cos\delta_0 & l_z\sin\delta_0-a_m\cos\delta_0 \\ -l_z\cos\delta_0+a_m\sin\delta_0 & l_z\cos\delta_0-a_m\sin\delta_0 & l_z\cos\delta_0+a_m\sin\delta_0 & -l_z\cos\delta_0-a_m\sin\delta_0 \\ -l_x\sin\delta_0+l_y\cos\delta_0 & -l_x\sin\delta_0+l_y\cos\delta_0 & l_x\sin\delta_0-l_y\cos\delta_0 & l_x\sin\delta_0-l_y\cos\delta_0 \\ 1 & 1 & 1 & 1 \end{pmatrix}\begin{pmatrix} p_{10} \\ p_{20} \\ p_{30} \\ p_{40} \end{pmatrix}=\begin{pmatrix} 0 \\ 0 \\ 0 \\ 1 \end{pmatrix}, \quad (24)$$

where $p_{i0}=\dfrac{P_{i0}}{P_{\Sigma 0}}$ is the thrust fraction of the i-th engine in the total thrust of all engines.

It is known that this system has a unique solution if its determinant is not equal to zero. It can be shown that the determinant of the original system of equations is determined by the expression:

$$\Delta=16a_m l_z\left(l_x\sin\delta_0-l_y\cos\delta_0\right),$$

which becomes zero at:

$$tg\delta_0=\frac{l_y}{l_x}; \qquad \delta_0=arctg\frac{l_y}{l_x}. \quad (25)$$

With this value of the engine setting angle, the system does not have a unique solution. Physically, this means that the resulting thrust vector passes through the aircraft mass centre. Engines do not create a longitudinal moment, and the possible distribution of the components of the total thrust becomes ambiguous.

To exclude the possibility of such a "degenerate" steady-state flight mode is possible by constructive improvements. It is proposed that the front pair of engines be positioned higher, and the rear pair, below the OXZ plane of the body axis system, symmetrically with respect to its origin, as shown in the figure. The motor bushings here are indicated by bold dots.

Then, with a positive incidence angle α, the thrust vector in the steady-state should deviation below OX, i.e. at a negative angle δ.

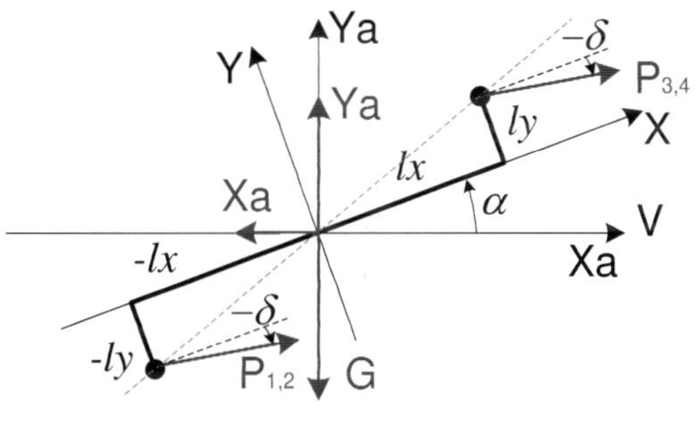

Figure 1. Recommended positioning of engines in the OXY plane of the body axis system

It should be noted that, unlike an airplane, the quad-tiltrotor UAV does not have a strict relationship between the steady-state flight mode (H, V) and the balanced value of the incidence angle. For the quad-tiltrotor UAV, the incidence angle can be set within certain limits in the interests of solving tasks external to the control.

The calculation of the steady-state flight mode is based on task H and V. Then it is determined from the condition of ensuring the maximum aerodynamic quality of the wing, the aerodynamic forces of the fuselage *Xa* and *Ya*, then, according to formulas (22) and (24), the balancing values of the engine thrusts (total thrust) are calculated. Moreover, if the balancing value of the angle of inclination of the engines is close to the critical value (25), then it is necessary to change the incidence angle (pitch) in one direction or another. If the settings angle of the engines δ_0 (22) satisfies the controllability condition (25), then the balanced values of the engine thrusts are determined from the system (24).

4. Discussion

The design scheme of the quad-tiltrotor UAV proposed in the work compares with the known ones in that it has only one movable control drive, which provides a change in the overall settings angle of all four engines. Aerodynamic control surfaces are absent. Minimization of the number of movable controls should provide cheaper and easier construction, increase its reliability. The paper studies the static controllability of the quad-tiltrotor UAV. It has been established that the proposed structural scheme of the quad-tiltrotor UAV is controllable since both the lifting force and all the necessary control forces and moments can be formed independently. However, there is a flight mode in which the resulting thrust vector of the engines lies on a straight line passing through the center of gravity of the aircraft, and therefore it cannot create a control moment. It is possible to reduce the risk of such a "degenerate" flight mode by the special placement of the propellers, as suggested above, but it cannot be completely eliminated in this way. The solution seems to be the creation of a control system that, when the flight mode approaches critical, should ensure the fastest possible transition through the critical angle of engine installation to another balancing mode.

5. Conclusions

Thus, the paper shows the fundamental possibility of static controllability of the quad-tiltrotor UAV of the proposed design scheme and defines condition (25) under which longitudinal controllability at the time of pitch is not provided. Methods are proposed for minimizing the influence of the indicated degenerate flight regime. In the future, to analyse the dynamic properties of the proposed aircraft, it is necessary to develop its linear dynamic model of spatial motion in deviations from the reference trajectory.

References

[1] Campo L V, Ledezma A and Corrales J C 2020 Optimization of coverage mission for lightweight unmanned aerial vehicles applied in crop data acquisition *Expert Syst. Appl.* **149** 113227

[2] Azouz N, Benzemrane K, Damm G and Pradel G 2007 Modeling and development of a 4 rotors helicopter UAV *IFAC Proceedings Volumes (IFAC-PapersOnline)* vol 6 (IFAC Secretariat) pp 215–20

[3] Us K Y, Cevher A, Sever M and Kirli A 2019 On the Effect of Slung Load on Quadrotor Performance *Procedia Computer Science* vol 158 (Elsevier B.V.) pp 346–54

[4] Anweiler S and Piwowarski D 2017 Multicopter platform prototype for environmental monitoring *J. Clean. Prod.* **155** 204–11

[5] Wandrie L J, Klug P E and Clark M E 2019 Evaluation of two unmanned aircraft systems as tools for protecting crops from blackbird damage *Crop Prot.* **117** 15–9

[6] Ferdaus M M, Pratama M, Anavatti S G and Garratt M A 2019 Online identification of a rotary wing Unmanned Aerial Vehicle from data streams *Appl. Soft Comput. J.* **76** 313–25

[7] Meshcheryakov R V., Trefilov P M, Chekhov A V., Diane S A K, Rusakov K D, Lesiv E A, Kolodochka M A, Shchukin K O, Novoselskiy A K and Goncharova E 2019 An application of swarm of quadcopters for searching operations *IFAC-PapersOnLine* vol 52 (Elsevier B.V.) pp 14–8

[8] Ribeiro T T, Conceição A G S, Sa I and Corke P 2015 Nonlinear Model Predictive Formation Control for Quadcopters *IFAC-PapersOnLine* **48** 39–44

[9] Bonyan Khamseh H, Janabi-Sharifi F and Abdessameud A 2018 Aerial manipulation—A literature survey *Rob. Auton. Syst.* **107** 221–35

[10] Thien R T Y and Kim Y 2018 Decentralized Formation Flight via PID and Integral Sliding Mode Control *IFAC-PapersOnLine* **51** 13–5

[11] Pitchiah H K and Moshi A A M 2019 Design and analysis of roto – Cylindrical wing for a drone aircraft *Mater. Today Proc.* **22** 393–9

[12] Sanchez A, Escareño J, Garcia O and Lozano R 2008 Autonomous Hovering of a Noncyclic

Tiltrotor UAV: Modeling, Control and Implementation *IFAC Proc. Vol.* **41** 803–8

[13] De Almeida M M and Raffo G V. 2015 Nonlinear Control of a TiltRotor UAV for Load Transportation *IFAC-PapersOnLine* **48** 232–7

[14] LIU N, CAI Z, ZHAO J and WANG Y 2020 Predictor-based model reference adaptive roll and yaw control of a quad-tiltrotor UAV *Chinese J. Aeronaut.* **33** 282–95

[15] Zhao A, He D and Wen D 2020 Structural design and experimental verification of a novel split aileron wing *Aerosp. Sci. Technol.* **98** 105635

[16] Zhu Z, Guo H and Ma J 2019 Aerodynamic layout optimization design of a barrel-launched UAV wing considering control capability of multiple control surfaces *Aerosp. Sci. Technol.* **93** 105297

[17] Wu Q, Sun P and Boukerche A 2019 Unmanned aerial vehicle-assisted energy-efficient data collection scheme for sustainable wireless sensor networks *Comput. Networks* **165** 106927

[18] Liu Z, He Y, Yang L and Han J 2017 Control techniques of tilt rotor unmanned aerial vehicle systems: A review *Chinese J. Aeronaut.* **30** 135–48

Cognitive modeling of the development of complex managed systems

Baidakov A N, Nazarenko A V, Babkina O N

Stavropol State Agrarian University, Zootekhnichesky Lane, 12, Stavropol, 355017, Russia

E-mail Nazarenkoa.v@yandex.ru

Abstract. The purpose of this research is to form a set of methodological provisions related to the use of cognitive modeling in the management of complex systems in conditions of high dynamism of studied processes. The rapid development of information and analytical tools provides new opportunities for applying cognitive modeling to a wider range of systems. The methodological basis of the study is the system methodology and its direct implementation is based on the system sequence "mission – goal – functions - structure – behavior – result". Cognitive modeling in the management of complex systems takes into account their hierarchical structure, which entails the hierarchy of the corresponding models, which is directly reflected in the generated models or considered when building local models for systems of a certain hierarchy level. Complex systems also determine the complexity of the models that represent them. Therefore, the paper proposes stratification of the modeling process and its results on five levels: the representation model, the understanding model, the interaction model, the management model and the behavior model. The dynamics of simulated systems and their external environment assumes an iterative (cyclic) nature of cognitive modeling. The paper presents a structural cognitive model of managing the functioning and development of complex systems as its result.

Keywords: modeling in management, system modeling, cognitive modeling, cognitive model of management of complex systems.

1. Introduction.

Our ideas about any objects and processes are always some of their models formed using a particular modeling language. Therefore, research on theoretical and methodological problems of modeling is always relevant. At present, the importance of solving such problems and their scale, as well as the volume of relevant information sets and knowledge, have increased enormously. But on the other hand, new modeling tools are appearing. All this together determines the necessity and productivity of using a cognitive approach to modeling in relation to the management of complex systems [1].

A large number of papers have been devoted to modeling [2, 3, 4], however, there is a lack in research of cognitive implementation in the context of the relevance of solving management problems, especially in the knowledge economy. These circumstances determine the significance and scientific novelty of the presented results.

The purpose of the research is to form a set of methodological provisions of cognitive modeling aimed at creating conditions for its wide application in the practice of managing complex systems.

2. Materials and methods.

The methodological apparatus of the research is based on a system methodology and a cognitive

IOP Conf. Series: Materials Science and Engineering **873** (2020) 012040 doi:10.1088/1757-899X/873/1/012040

approach to the problems of management and modeling. The structural cognitive model of managing the functioning and development of complex systems presented in this paper allows us to develop algorithms for managing such systems using a variety of modeling tools.

3. Stratification of modeling.

People in their activities operate with models, carry out their activities based on their models of the universe, society, state, enterprise, family, etc. Its success depends on the adequacy of such modeling.

Let us clarify that by model we mean the subject's representation of an image of some object or process through the use of some verbal, visual, mathematical language.

We can distinguish the following levels (strata) of the purpose of modeling and accordingly its course, goals and results in terms of their perception and use by the subject of management [5]:

- representation model as a formalized or often nonformalized appearance of a certain phenomenon in the subject of management, preliminary from the point of view of purposeful modeling;
- understanding model as semantic and causal (if possible) picture of the state and dynamics of changes in the studied phenomenon, common to the customer and performer of modeling;
- interaction model as a structural model represented by the main elements (subsystems) of the system under study and its connections, including connections with the external environment;
- control model as the actual model of the phenomenon under study as a control tool with inputs (control actions) and outputs (expected results of the functioning and development of the controlled system);
- behavior model as representations of the modeling customer about the place, role, order of use and effectiveness of the management model in the processes of functioning and development of the managed system-operational and evaluation.

Depending on the tasks to be solved, such differentiation can be considered both as a set of submodels of the same object of research and as independent results of the formation stages and use of models in management. Additionally, each subsequent submodel necessarily includes the previous one, but already expanded with new knowledge.

These levels, in fact, represent the phases of the cycle of formation and development of the model in management from the unconscious use of the representation model to understanding the need for purposeful modeling, setting its goals and objectives, and then to understanding the need to build a new model, based on the previous or fundamentally different one.

Thus, modeling in management is a continuous cyclic (iterative) process. The transition to the next level of modeling is carried out as a result of purposeful increment of knowledge about the phenomenon under study, that is, it is inherently cognitive. This means that this is also the process of modeling.

The first three model levels (modeling phases) form a model image of the perception of controlled objects and processes, the last two form a tool and mechanism for management actions using modeling.

Let us pay attention to inevitable and unremovable subjective component of any model-based approach to managing complex systems [6, 7].

The world is constantly changing and its perception by customers and modeling performers does the same. This means that the ideas about objects and modeling processes and their understanding are also changing. Therefore, either the model of interaction between subjects and objects of management or models used in management must constantly change.

4. The consistency of the simulation

Continuous changes in the characteristics of the managed system and its environment, their high dynamism, the need to reduce the negative consequences of subjectivity in management determine the conditionality. Moreover, the vital need to develop new knowledge about these processes in order to ensure timely adaptation to them. This means that the coordinated formation and competent use of the complex of the above five model levels (modeling phases) for complex systems is possible only on the basis of a cognitive approach to the processes of their functioning and development.

The specified modeling cycle should reflect the characteristic features of the managed system and its management mechanism, take into account the influence of significant supersystems, rationally use the available diverse resource opportunities for the functioning and development of the system, the structure and content of the implemented rules and management procedures, including their instrumental implementation.

In general, the models used in managing a complex system can be represented as a tuple

$$M_i = \langle R_i, Re_i, Y_i, X_i, I_i, D_i, F_i, R_{ti} \rangle , \quad (1)$$

where i is hierarchical level of the modeled object or process;

R_i is model reduction level;

Re_i is modeling resource support;

$Y_i = (y_{1i},..., y_{ni})$ is output variables;

$X_i = (x_{1i},..., x_{mi})$ is factor variables;

I_i is modeling tools;

D_i is information support;

$F_i = (F_{1i}, F_{2i}, F_{3i}, F_{4i}, F_{5i})$ is the form of providing results that corresponds to each of the five phases of modeling;

R_{ti} is rules for implementing procedures for using, updating, and developing the model.

System regularities of isomorphism and isofunctionalism allow us to use the above provisions in the formation of models and mechanisms for managing the functioning and development of systems at any level.

The structure of complex systems and their management mechanisms and, therefore, the corresponding models used in management, has a hierarchical structure, which is reflected in the corresponding hierarchy of the management system. This position also means that a hierarchy of management models is required. We are also talking about cyclic stratification of models with five modeling phases, presented above.

As a basis for forming a vertical hierarchy for studied systems, we can consider, for example, the construction "country-region-enterprise". It would seem that it would be possible to build a corresponding model with a tree structure. However, in practice, modeling is usually performed separately at each level. And the problem is not the reluctance to build such a large model that covers all levels and subsystems. Several "simple" questions arise at once: What is the purpose of such a model? Who are the customer and the contractor? How much time and resources are needed to build, operate, and update it? Is it possible to force businesses to participate in this simulation? The list of such questions can be continued.

It is clear, that such bulky structures often do not justify themselves. At least for the time being. And for each level of the hierarchy, such models are needed. And even if they are built seemingly independently of each other, the system approach to modeling requires that they take into account the relationships of the system under study with adjacent (and not only) levels of the hierarchy. This means that there is at least a latent hierarchy in the simulation, even if it is localized. In other words, each systemically constructed model in management necessarily assumes its "embeddedness" in the existing or possible hierarchy of the corresponding management models.

Violation of this condition should be perceived as a disregard for the system approach, and therefore, potentially dangerous use of modeling methods in the practice of managing fairly complex systems.

Modeling in the management of complex hierarchical systems should be considered from the perspective of a dynamic correlation of their various components and properties: general, specific, certain; functional; structural; behavioral and effective. The integrating position in the definition and systematic use of these positions and categories in modeling is their inclusion in the system sequence "goal – functions – structure – behavior – result", based on the well – known approach of D. Bain "structure – behavior – result" [1].

The fractality property, which we consider in two aspects as the fractality of systems and the fractality of processes, deserves special attention within the framework of the system and, as a result, the cognitive approach to management.

The fractal approach, as we know, proceeds from the self-similarity of the studied objects and processes: in our case not geometric self-similarity, but structural and semantic. Among the main fractal features of complex controlled systems, we include:

- system properties and patterns;
- purposeful functioning and development;
- hierarchical certainty;
- availability of management;
- openness.

The number of process fractal features of controlled systems should also include their inherent multi-level cyclical functioning and development, which determines the cyclical nature of cognitive modeling.

5. The cognitive control model

The practical implementation of the system approach should be based on the methodology of cognitive modeling, since it is not enough to use the accumulated knowledge and methods. It is necessary to constantly generate and effectively use in practice new knowledge and methods that meet emerging problems and challenges.

The concept of "cognition" in relation to modeling is understood in a broad sense. This means using in management not only the existing knowledge about the objects and processes under study, but also necessarily creating prerequisites, defining subjects and ensuring the processes of their formation [8].

In the context of globalization and high dynamism of social and economic processes, the identification and determination of ways to solve emerging problems based on a cognitive approach in the course of functioning and development of a managed system should be based on modern methods of generating, practically oriented representation and targeted interpretation and projecting existing and new knowledge to solve problems. And the rapidly increasing power of modern information and analytical tools allows us to increasingly translate the cognitive approach into a broad practical plane.

The proposed structural cognitive model for managing the development of complex systems (Figure 1) is based on a systematic approach to research. It defines the place and role of cognitive research in the structure of the "mission-goal-function-structure-behavior-result" system sequence.

The factors of influence on the studied system in the cognitive model are grouped in the main directions:

- natural global and local environment;
- social environment by level: personal, group, enterprise, municipal, regional, state, international;
- the institutional environment that determines the conditions for the functioning and development of the studied systems: social, economic, political and legal.

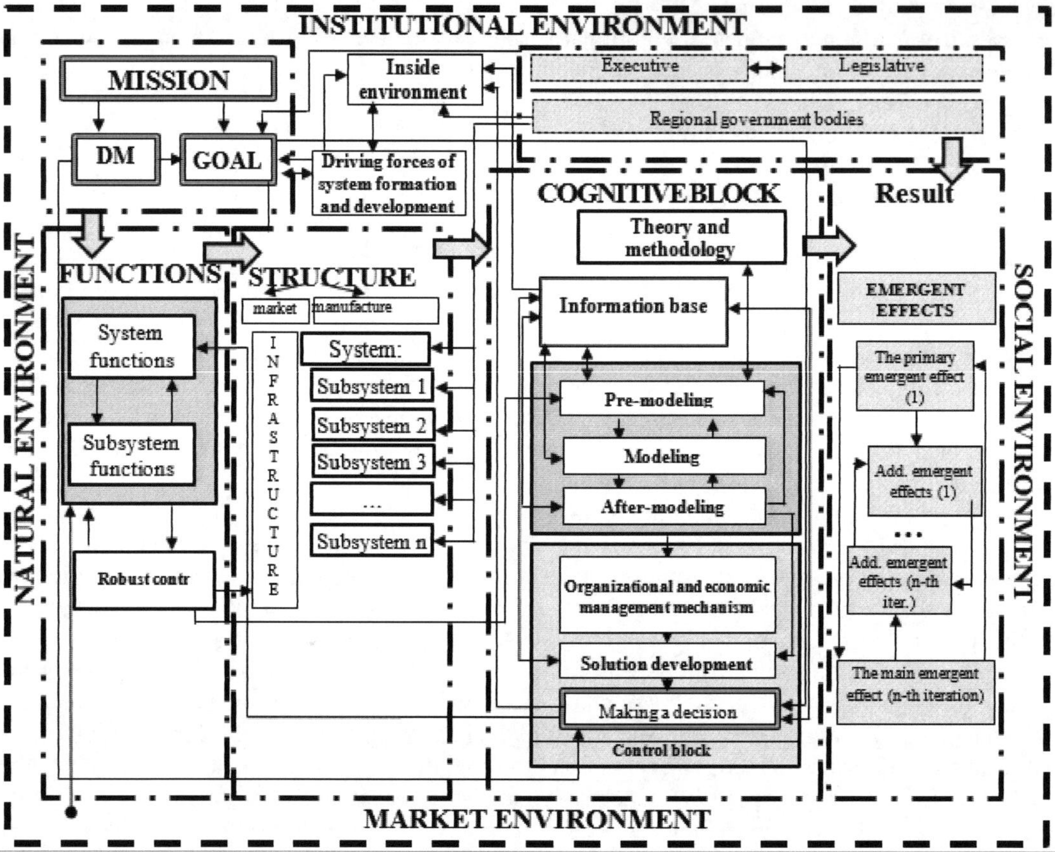

Figure 1. Structural cognitive model for managing the functioning and development of complex systems

The cognitive model of controlled systems development is structured according to semantic blocks: target, cognitive, functional, structural, effective and regulatory. This approach allows us to combine the analytical and synthetic components of system research, its theoretical and methodological tools, as well as the methodology of its practical implementation in the model.

The target block is a key component of targeted development of the managed system. It combines the mission, goals of the system and the subject of its management – the DM (decision-maker). A complex of internal and external conditions of functioning and development influence the target settings of the functioning and development of the studied systems [9].

The adequacy and success of goal formation processes is directly dependent on the level of awareness of the DM about cause-and-effect relationships both within the managed system and with its external environment. However, modern cognitive capabilities are not always used in goal formation processes, even if they are available to the DM.

A separate factor in the success of management is the level of analytical skills of people who develop and make decisions, their awareness, readiness and practical desire to regularly use modern information and analytical tools.

The formation of a hierarchical structure of the goals of the system and its subsystems should proceed from their existing or being formed organizational and functional hierarchy. This means that it is necessary to ensure organic connections between two hierarchical structures-goals and functions.

The formation and updating of these structures is a non-trivial and contradictory task. The set of contradictions between the functions and structure of the system combines both the contradictions of various functions of the managed system, its subsystems and its inter-level contradictions.

The cognitive block of the model is an active theoretical and methodological component of the formation and use of knowledge. It is directly involved in cognitive activity during the development, decision-making and implementation processes. The decision block is included to the cognitive module of the model.

The possibility of a different composition of the cognitive block is also allowed.

In this study, the cognitive block of the model is represented by the following semantic components:
- theoretical and methodological module combines theoretical and methodological provisions and knowledge used to ensure the functioning and development of the managed system, both existing and new, generated in the course of relevant activities within the system under study and/or outside it;
- the information base, although organically included in the block under consideration, is intended for information support of all processes of functioning and development of the managed system;
- pre-modeling mainly refers to the phases of representation and understanding:
  • in the case of the initial modeling cycle the formation of an information base, identification and structuring of the problems to be solved, their preliminary qualitative and quantitative analysis, the formation and structuring of information aggregates, time series analysis, multidimensional statistical and trend analysis, etc.; setting the modeling problem, establishing its resource capabilities and reduction level; determining the list of effective and factor indicators, the order of use, verification and updating of models, the formation of contours of forecast scenarios;
  • in the case of subsequent cycles (iterations) of modeling-updating the information base, clarifying the problem to be solved and setting the modeling task, increasing the resource capabilities of modeling, updating the list of performance indicators and factors taken into account;
- modeling refers to the understanding, interaction, and management phases:
  • in the case of the initial modeling cycle the construction of a verbal conceptual model, its subsequent formalized representation, the formation of a synthetic mathematical model based on models for each level of the studied systems, determining the instrumental capabilities of the numerical implementation of the model, its verification and validation, the numerical implementation of the model;
  • in the case of subsequent cycles (iterations) of modeling-updating and development, numerical implementation of the management model using new knowledge obtained during the previous stage of modeling and new information; analysis and evaluation of changes in the modeling conditions, including its resource capabilities; assessment of the adequacy of the model to the changed conditions of functioning and development of the system;
- postmodeling refers to the behavior phase for all modeling cycles, means the use of the built model in management practice: conducting calculation operations, forecasting, including scenario, virtual economic experiments, stress testing, determining ways to improve the model and its implementation tools, identifying trends that are practically significant for the development of the system under study, conclusions and conclusions developed during modeling;
- solution block – development of solutions using the results of modeling and forecasting; decision making and implementation with the involvement of the DM in cognitive processes.

The cognitive functional-structural approach to managing a complex system is focused primarily on the purposeful formation of its required emergent effect, or rather, a set of emergent main and additional effects, as a result of "behavior", that is, the goal-oriented scientifically based activity of the controlled system.

The practical implementation of cognitive activity in the framework of the presented cognitive block is carried out through the formation and implementation of appropriate algorithms for system information and analytical activities that meet the structural cognitive model (Figure 1).

6. Discussion
Despite the fact that the obtained results fully meet the goal of the study, in the future, in our opinion, when forming cognitive models used in management, more attention should be paid to its cyclical nature. It is also necessary to consider the issues of personnel support for cognitive modeling, since this part, in our opinion, is currently its most problematic area.

7. Conclusion
The study of the problems of cognitive modeling in the management of complex systems was devoted to the consideration of its two main aspects: organizational and chronological. The first one allowed us to determine the role and place of cognitive modeling in the mechanism of managing complex systems. The second aspect of the research is devoted to the consideration of the internal structure of such modeling processes, represented by five levels, the implementation of which quite definitely sets the chronological sequence of the corresponding phases of modeling. The same aspect is related to the requirement of continuity of modeling processes, with its necessary iterative character as a reaction to constantly changing conditions for the functioning and development of complex systems.

Thus, cognitive modeling of functioning and development of complex managed systems is an effective and practically significant tool for information and analytical support of improving their efficiency at the present stage of socio-economic development of society.

References

[1] Bain J 1951 Relation of Profit Rate to Industry Concentration American Manufacturing, 1936 - 1940 *Quarterly Journal of Economics* Vol 65 (**3**) 293 – 324
[2] Hyun J-Y 2019 Using a coupled agent-based modeling approach to analyze the role of risk perception in water management decisions *Hydrology and Earth System Sciences* 23 2261-2278
[3] Kucherova H 1951 Modelling tax consciousness evaluation in the context of economic development uncertainty *Periodicals of Engineering and Natural Sciences* 7 567-579
[4] Baydakov A, Chernikova L, Nazarenko A, Zaporozhets D, Sidorova D 2017 Scenario forecasting of the reproduction process in the agriculture in view of inflation *Journal of Environmental Management and Tourism* 8 (**3**) 620-628
[5] Lutsenko E, Troshin L, Zviagin A, Zviagin A, Milovanov A 2018 Application of the automated system-cognitive analysis for solving problems of genetics *Journal of Mechanical Engineering Research and Developments* 41(**2**) 1-8
[6] Raymond D 2006 Cognitive Analysis of Problems of Comprehension in a Learning of Mathematics *Educational Studies in Mathematics* 61(**1**) 103-131
[7] Gorelova G Pankratova N 2018 Strategy of Complex Systems Development Based on the Synthesis of Foresight and Cognitive Modelling Methodologies *IEEE First International Conference on System Analysis & Intelligent Computing (SAIC)*
[8] Schwartz M Cook C Pressey R Pullin A Runge, etc 2017 Decision Support Frameworks and Tools for Conservation *Conservation Letters* 11(**2**) 1-12
[9] Raymond D 2006 A Cognitive Analysis of Problems of Comprehension in a Learning of Mathematics *Educational Studies in Mathematics* 61(**1**) 103-131

IOP Conf. Series: Materials Science and Engineering **873** (2020) 012041 doi:10.1088/1757-899X/873/1/012041

A Modeling of Battery Discharge Within Different Temperature Operating Conditions

V V Samoylenko

North-Caucasus Federal University, Department of Oil and Gas Wells Construction, 1 Pushkin str., Stavropol, 355000, Russian Federation

E-mail vvsamoylenko@ncfu.ru

Abstract. The article presents the results of investigations including the battery characteristics in conditions of changing temperature conditions. The possibility of applying logistic functions for approximation depending on the capacity of the battery from the load current is substantiated. An analytical overview of approaches to power modeling depending on the type of battery is presented. The simulation of the discharge process of the CR2 battery in the MATLAB & Simulink™ environment was carried out.

Keywords: Battery, Electrical Capacity, Discharge Current, Operating Conditions, Logistic Function, MATLAB & Simulink™

1. Introduction

The development of devices using power sources in the form of rechargeable batteries is increasing annually. The global consumer battery market is projected to reach $ 50 billion by 2025 [1]. Electric batteries are actively used in portable electronics and industry. In that connection, power supplies for the Internet of Things (IoT) peripherals, smart homes have an interest for researches.

Manufacturers maximize capacity, service life while reducing weight and dimensions. The most promising in the near future are lithium-ion batteries (Figure 1). Compared to other rechargeable batteries, such as Ni-Cd, Ni-MH and lead-acid battery, the lithium-ion battery has a high energy density and power, long life [2].

When operating rechargeable power supplies, it is necessary to consider the discharge processes, including the evaluate energy reserves. Various analytical and empirical characteristics are developed and used to describe these processes. These models describe the change in parameters (State of Charge - SoC, State of Health - SoH, etc.) of autonomous power sources depending on the load and operating conditions.

For NiCd batteries, the following methods [3] for approximating discharge functions are typically used.

The Peukert law in general view [4,5] is:

$$C = \frac{A}{i^n},$$

where: C is discharge capacity;

i is discharge current;

A, n are empirical coefficients.

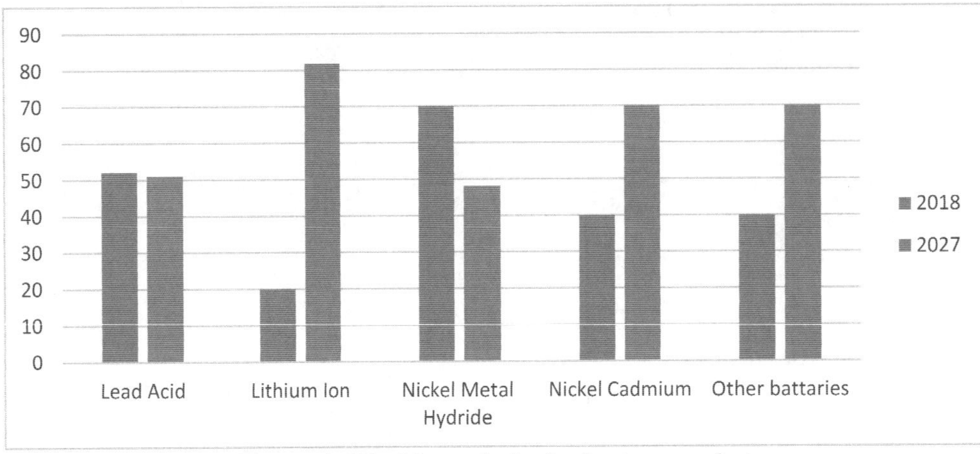

Figure 1. World trends in the battery market
(https://www.inkwoodresearch.com/reports/global-battery-market/)

This expression loses its significance when currents tend to zero. In this connection, in practice, the Peukert ratio is more often used:

$$C = \frac{A}{1+B \cdot i^n}, \quad [1]$$

where: C is discharge capacity;
i is discharge current;
A, B, n are empirical coefficients.

Libenov`s Equation [6] is:

$$C = \frac{A}{1+B \cdot i}, \quad [2]$$

where: C is discharge capacity (A*h);
i is discharge current (A);
A, B are empirical coefficients.

Aguf`s Equation [6] is:

$$C = a0 + \frac{a1}{i} + \frac{a2}{i^2} + \frac{a3}{i^3} + \cdots, \quad [3]$$

where: C is discharge capacity (A*h);
i is discharge current (A);
$a1, a2, a3$ are empirical coefficients.

Korovin-Skundin`s Equation [7] is:

$$C = \frac{A}{i^n} th\left(\frac{i^n}{B}\right), \quad [4]$$

where: C is discharge capacity (A*h);
i is discharge current (A);
A, B are empirical coefficients.

IOP Conf. Series: Materials Science and Engineering **873** (2020) 012041 doi:10.1088/1757-899X/873/1/012041

To approximate the discharge functions of alkaline batteries, the dependences of Shepherd, Haskina-Danilenko, Romanova [6] are used.

The disadvantages of the above approximating functions of the lithium batteries discharge, which are currently the "standard" due to their high power density, include insufficient accuracy [8]. In this connection, it is of interest to study the logical function for approximating these dependencies in order to increase the accuracy of modeling this process.

2. Approximation of the battery discharge using the logistic function

2.1. Experimental data

For research, the claimed battery characteristics [9] Tekcell (Manufacture Vitzrocell Co. Ltd, г. Saint-Peterburg, Russian Federation) were selected.

Table 1. Technical specification of the CR2 battery

Characteristic	Value
Nominal Voltage	3,0 V
Nominal Capacity (at the 10 mA, 20 °C ; cutoff 2,0 B)	850 mA*h
Current battery	800 mA
Weight	11,5 g
Operational temperature	-30 °C … +60 °C

According to the claimed technical characteristics [9], this battery has the following parametric dependence of capacity on current consumption:

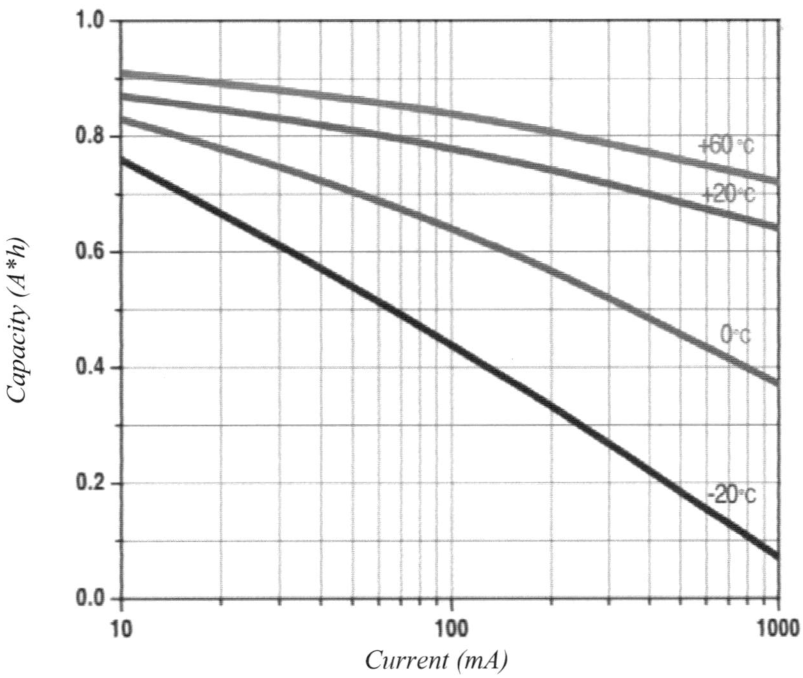

Figure 2. The diagram of the CR2 battery capacity and current [9]

2.2. Linearization of the logistic function

The logistic function (sigmoid) has the following form:

$$y = 1 - \frac{K}{1+b \cdot e^{-cx}}, \qquad [5]$$

To convert the function to the linear form, it is necessary to perform several transformations:

$$\frac{1-y}{k} - 1 = be^{-cx}, \qquad [6]$$

Logarithm of the expression [6], we obtain:

$$\ln\left(\frac{1-y}{k} - 1\right) = \ln(b) - \mathrm{c} \cdot \ln(x), \qquad [7]$$

Introducing new variables $z = \ln\left(\frac{1-y}{k} - 1\right)$; $A = \ln(b)$; $B = \mathrm{c}$; $\mathrm{t} = \ln(x)$ into expression [7], we obtain the classical linear regression equation:

$$Z = A - B \cdot t. \qquad [8]$$

The result of the linear function parameterization [8] by the least-squares method are the formulas:

$$B = \frac{\overline{X \cdot Y} - \bar{X} \cdot \bar{Y}}{\overline{X^2} - (\bar{X})^2}; \; A = \bar{Y} - B \cdot \bar{X} \qquad [9]$$

where $\bar{X} = \frac{1}{n}\sum_{i=1}^{n} X_i$; $\bar{Y} = \frac{1}{n}\sum_{i=1}^{n} Y_i$; $\overline{X \cdot Y} = \frac{1}{n}\sum_{i=1}^{n} X_i Y_i$.

After calculating the parameters A and B, the inverse transformation of the variables is performed to obtain the coefficients b and c of the function [5].

2.3. Verification of modeling results

To verify the obtained analytical dependencies MATHLAB & Simulink ™ was used. Curve Fitting Tool was involved. This package has the following features:

- primary processing of data, such as division into regions and smoothing;
- parametric and nonparametric approximation with the possibility of using function libraries or creating your own function;
- standard linear least squares method, nonlinear least squares method, weighted least squares method, constrained least squares method and stable approximation methods;
- the ability to assess the quality of the approximation.

3. Comparative analysis of the obtained empirical dependence

Based on the results of the study, the following results were obtained:

Table 2. Experimental data for approximating the dependence of the relative capacitance on the load current at temperature -20 °C

Approximate Graph	Goodness of fit
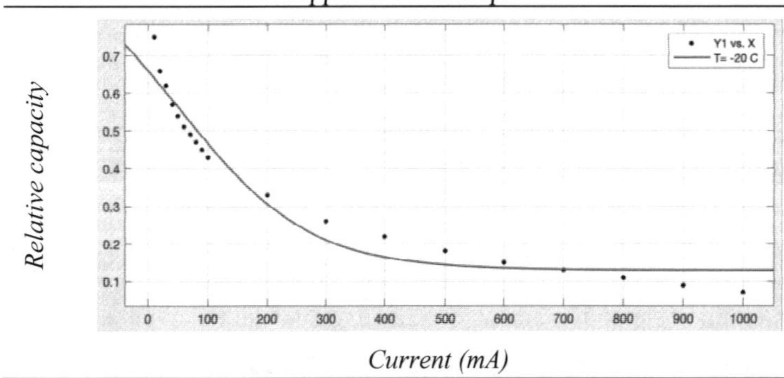	SSE: 0.03345 R-square: 0.9596 Adjusted R-square: 0.9546 RMSE: 0.04573

IOP Conf. Series: Materials Science and Engineering **873** (2020) 012041 doi:10.1088/1757-899X/873/1/012041

Table 3. Experimental data for approximating the dependence of the relative capacitance on the load current at temperature 0 °C

Approximate Graph	Goodness of fit
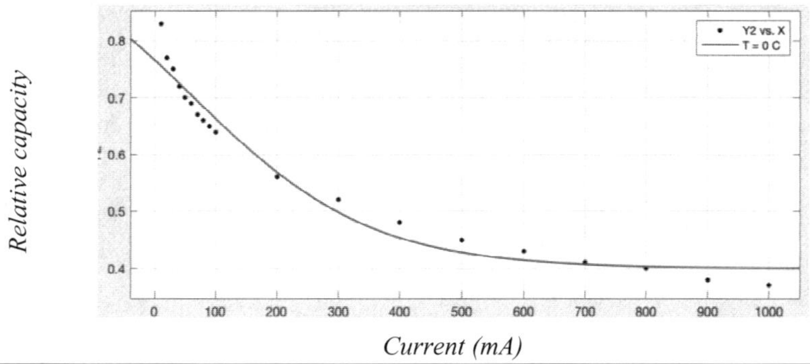	SSE: 0.01244 R-square: 0.9685 Adjusted R-square: 0.9646 RMSE: 0.02788

Table 4. Experimental data for approximating the dependence of the relative capacitance on the load current at temperature +20 °C

Approximate Graph	Goodness of fit
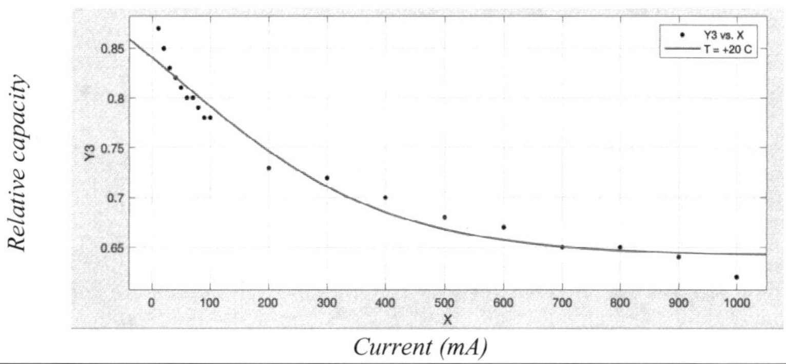	SSE: 0.003763 R-square: 0.9663 Adjusted R-square: 0.9621 RMSE: 0.01534

Table 5. Experimental data for approximating the dependence of the relative capacitance on the load current at temperature +60 °C

Approximate Graph	Goodness of fit
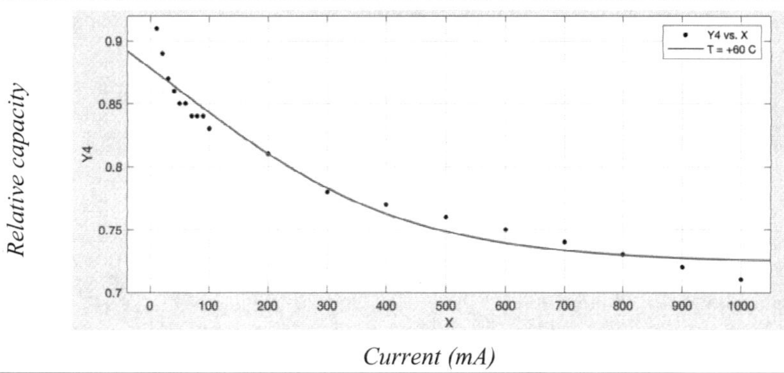	SSE: 0.002947 R-square: 0.9558 Adjusted R-square: 0.9503 RMSE: 0.01357

Tables 2-5 present the simulation results of the dependence of the relative capacity on the load current value of the CR2 battery. To approximate the s-shaped lines (Figure 2), a logistic function was adopted.

By the methodology (Section 2.2), linearization was carried out, the values of the coefficients were determined. The data obtained were investigated in the Curve Fitting Tool MATLAB & Simulink ™.

To assess the accuracy of the approximating model, standard SSE (Sum Squared Error) criteria were used; R-square; Adjusted R-square, RMSE (Root Mean Square Error). The presented data show the preference for using the logistic function in comparison with other ratios (1-4). The standard deviation in assessing the relative capacitance is 0.013-0.046. This indicator is lower than in the case of applying the equations of Peikert (0.02-0.03), Libenov (0.03-0.05), Aguf (0.01-0.02) [10]. In this connection, we can conclude that the use of a logistic function for the approximation of discharge processes of rechargeable batteries.

4. Conclusion

The objective of the research presented in the article was a comparative analysis of the dependences of the parameters of electric batteries obtained during the simulation. In particular, modeling of the process of discharging the battery when changing the temperature conditions of operation was considered. Based on the analysis of open sources, the methods used for approximating these dependencies were identified, their advantages and disadvantages were determined. We continued to use the s-shaped logistic function, which generally describes these graphical dependencies, to model this process. The simulation of the discharge process of the CR2 battery was carried out in MATLAB & Simulink™.

5. Acknowledgment

The work is financed by the Fung of President of Russian Federation, Grant № SP-5142.2018.

Reference:

[1] Anon Consumer Batteries: Global Market Insights and Projections to 2025
[2] Ahmadi L, Young S B, Fowler M, Fraser R A and Achachlouei M A 2017 A cascaded life cycle: reuse of electric vehicle lithium-ion battery packs in energy storage systems *Int. J. Life Cycle Assess.* **22** 111–24
[3] Galushkin N E, Yazvinskaya N N and Galushkina I A 2012 Analysis of the use of empirical relations for assessing the capacity of a CFT NiCd battery of a long discharge mode *Basic Res.* **11** 1180–4
[4] Xie J, Ma J and Chen J 2018 Peukert-equation-based state-of-charge estimation for LiFePO4 batteries considering the battery thermal evolution effect *Energies* **11** 1–14
[5] Hausmann A and Depcik C 2013 Expanding the Peukert equation for battery capacity modeling through inclusion of a temperature dependency *J. Power Sources* **235** 148–58
[6] Galushkin N E and Galushkina I A 2005 An analysis of empirical relationships describing a discharge of alkaline batteries *Electrochem. Energy* **5** 43–50
[7] Korovin N V and Skudin A M 2003 *Chemical Power Supply Sources: Handbook* (Moscow: Moscow Power Energy Institure Publ.)
[8] Traub L 2016 Calculation of Constant Power Lithium Battery Discharge Curves *Batteries* **2** 17
[9] Anon Datasheet CR2 Li Battery
[10] Galushkin N E, Yazvinskaya N N and Galushkina I A 2013 Generalized Model of the Dependence of the Capacity of NiCd Batteries on the Discharge Current *Electrochem. Energy* **13** 96–102

AUTHOR INDEX